Single Cell Analysis: Technologies and Future Prospects

Single Cell Analysis: Technologies and Future Prospects

Edited by Forest Pearson

Syrawood Publishing House
New York

Published by Syrawood Publishing House,
750 Third Avenue, 9th Floor,
New York, NY 10017, USA
www.syrawoodpublishinghouse.com

Single Cell Analysis: Technologies and Future Prospects
Edited by Forest Pearson

International Standard Book Number: 978-1-64740-364-5 (Hardback)

Cataloging-in-publication Data

Single cell analysis : technologies and future prospects / edited by Forest Pearson.
p. cm.
Includes bibliographical references and index.
ISBN 978-1-64740-364-5
1. Cytology--Technique. 2. Cells--Analysis. 3. Cell--Physiology. I. Pearson, Forest.
QH585 .S56 2023
574.072 4--dc23

TABLE OF CONTENTS

PREFACE

Single cell analysis (SCA) is a widely recognized method for elucidating cellular functions that are not accessible through bulk measurements on the population-level. The understanding of genomic sequence at the single cell level can help in developing a better understanding of how various parts of a cell are combined in space and time in order to create dynamic cellular processes. SCA has affected and influenced a variety of scientific fields including immunology, cancer biology, and neuroscience. It involves the miniaturization of conventional engineering concepts in order to fit within the confines of a single cell. The major technologies used for single cell analysis include single cell sequencing, single cell imagers, microfluidic droplet technology platforms and flow cytometry. This book explores all the important aspects of single cell analysis in the present day scenario. Those in search of information to further their knowledge will be greatly assisted by it.

Various studies have approached the subject by analyzing it with a single perspective, but the present book provides diverse methodologies and techniques to address this field. This book contains theories and applications needed for understanding the subject from different perspectives. The aim is to keep the readers informed about the progresses in the field; therefore, the contributions were carefully examined to compile novel researches by specialists from across the globe.

Indeed, the job of the editor is the most crucial and challenging in compiling all chapters into a single book. In the end, I would extend my sincere thanks to the chapter authors for their profound work. I am also thankful for the support provided by my family and colleagues during the compilation of this book.

Editor

1

Evaluating the Effect of Cell Culture on Gene Expression in Primary Tissue Samples using Microfluidic-Based Single Cell Transcriptional Analysis

Michael Januszyk, Robert C. Rennert, Michael Sorkin, Zeshaan N. Maan, Lisa K. Wong, Alexander J. Whittam, Arnetha Whitmore, Dominik Duscher and Geoffrey C. Gurtner *

Department of Surgery, Division of Plastic and Reconstructive Surgery, 257 Campus Drive West, Hagey Building GK-201, Stanford, CA 94305-5148, USA; E-Mails: januszyk@stanford.edu (M.J.); rrennert@stanford.edu (R.C.R.); msorkin@med.umich.edu (M.S.); zmaan@stanford.edu (Z.N.M.); Lisakanata@gmail.com (L.K.W.); alexander.whittam@gmail.com (A.J.W.); arnethaw@stanford.edu (A.W.); dominikduscher@me.com (D.D.)

* Author to whom correspondence should be addressed; E-Mail: ggurtner@stanford.edu

Academic Editor: Zvi Loewy

Abstract: Significant transcriptional heterogeneity is an inherent property of complex tissues such as tumors and healing wounds. Traditional methods of high-throughput analysis rely on pooling gene expression data from hundreds of thousands of cells and reporting a population-wide average that is unable to capture differences within distinct cell subsets. Recent advances in microfluidic technology have permitted the development of large-scale single cell analytic methods that overcome this limitation. The increased granularity afforded by such approaches allows us to answer the critical question of whether expansion in cell culture significantly alters the transcriptional characteristics of cells isolated from primary tissue. Here we examine an established population of human adipose-derived stem cells (ASCs) using a novel, microfluidic-based method for high-throughput transcriptional interrogation, coupled with advanced bioinformatic analysis, to evaluate the dynamics of single cell gene expression among primary, passage 0, and passage 1 stem cells. We find significant differences in the transcriptional profiles of cells from each group, as well as a considerable shift in subpopulation dynamics as those subgroups better able to adhere and proliferate under these culture conditions gradually emerge as dominant. Taken together, these findings reinforce the importance of using primary or very early passage cells in future studies.

Keywords: microfluidics; singe-cell; transcriptional analysis; gene expression; bioinformatics; subpopulations; cell culture

---

1. Introduction

Functional heterogeneity is an inherent characteristic of complex tissues such as tumors, healing wounds, and various stem cell niches [1,2]. These functional differences are subsumed by variation in both cellular transcriptomics and proteomics [3–5]. Although the extent to which single-observation transcriptional or, to a lesser extent, proteomic differences predict physiologic-level functional variations remains controversial [6], significant scientific insights have been obtained though studies evaluating and ultimately validating the functional significance of discrepancies observed at the transcriptional stage [7,8].

Traditional methods of transcriptional analysis, such as DNA microarray, rely on pooling the mRNA from hundreds of thousands of individual cells and reporting a population-wide average. Although this approach has been widely successful at advancing our knowledge of disease physiology over the preceding two decades, these methods lack the granularity to observe differential expression among cellular subsets within the target population [9]. Recent evidence strongly suggests that this limitation severely hampers the utility of such technologies to interrogate the cellular subgroup dynamics inherent to the complex populations described above [10].

Only in the last several years have high-throughput techniques, such as that embodied by the Fluidigm Biomark system, evolved to interrogate tissue with single cell resolution [11]. These approaches make use of microfluidic technology to achieve massively parallel qPCR, permitting the simultaneous interrogation of hundreds of individual cells against large numbers of genes [12]. Such techniques have been successfully employed by our laboratory and others to answer key questions about the fundamental mechanisms underlying stem cell biology, tumor evolution, and the pathobiology of chronic disease [7,13–16].

One extant question in designing these studies is whether analyzed cells need to be freshly isolated from primary tissue or can be subjected to a priori expansion in cell culture. This is a particularly important issue for scarce populations, such as rare or rarely-resected tumors, for which obtaining sufficient cells is often difficult, whereas heavily passaged commercially available cell lines are readily available. Similarly, rare stem cell populations, such as the putative long-term hematopoietic cell (LT-HSC), which represent only 0.0001% of whole bone marrow, would be more easily studied following culture expansion [17]. Early-gestation embryonic stem cells represent yet another area where the number of cells harvested from a single sample can be as low as four or eight [18]. For these cases, among others, an approach that permits passaging cells in culture before transcriptional interrogation would be considerably attractive. However, the impact of cell culture on single cell transcriptomics remains poorly understood.

In this study, we apply our novel approach to high-throughput, microfluidic-based single cell gene expression analysis to evaluate the effect of cell culture on the transcriptional profiles of stem cells.

In particular, we examine one established population of human adipose-derived stem cells, defined as CD45−/CD31−/CD34+, which has been shown to have considerable vasculogenic and angiogenic potential, making it a potentially attractive therapeutic target [14]. We apply partitional clustering, gene enrichment analysis, and network modeling to compare the transcriptional variations observed across culture passaging and evaluate the impact of this *in vitro* expansion.

2. Experimental Section

After acquiring informed consent, in accordance with the Stanford University Institutional Review Board guidelines, human abdominoplasty specimens were obtained from a single adult female patient without major medical conditions undergoing an elective abdominoplasty procedure. Adipose-derived stem cells (ASCs) were isolated as described previously by Zuk *et al.* [19]. Briefly, raw samples were manually minced, washed, and treated with 0.075% collagenase type I (Sigma-Aldrich, St. Louis, MO, USA) in Hank's balanced salt solution (Life Technologies, Grand Island, NY, USA) for 1 h at 37 °C with gentle agitation. The reaction was stopped with the addition of fetal bovine serum (FBS), and after centrifugation, the pelleted stromal vascular fraction was prepared for fluorescence activated cell sorting (FACS) as described below.

Fresh human ASCs were sorted on a FACS Aria II instrument (BD Biosciences, San Jose, CA, USA) with the use of a 100-μm nozzle. Briefly, cells were isolated as described above, and incubated for 20 min in FACS buffer (PBS supplemented with 2% FBS) containing anti-human ef-450-conjugated CD45 (eBioscience, San Diego, CA, USA), allophycocyanin (APC)- or phycoerythrin (PE)-conjugated CD34 (BD Biosciences), and fluorescein isothyocianate (FITC)-conjugated CD31 (BD Biosciences). Using a Becton Dickinson flow cytometric cell sorter, cells were either sorted as single cells into 6 μL of lysis buffer for single cell transcriptional analysis, or as populations for subsequent culture. ASCs were defined with the surface marker profile CD45−/CD31−/CD34+ (to exclude contaminating hematopoietic and endothelial cells found within the stromal vascular fraction (SVF). Cells sorted for culture were plated onto conventional tissue culture plates in the Dulbecco's modified Eagle's medium (DMEM; Life Technologies) supplemented with 10% FBS and 1% Penicillin/Streptomycin. Plated cells were cultured under standard conditions (37 °C and 5% CO_2) and used at passage zero or one.

Single cell reverse transcription and low cycle pre-amplification of individual cells were performed as described previously [20]. Briefly cell suspensions of freshly isolated human SVF were sorted as single cells into each well of a 96-well plate using a Becton Dickinson FACSAria flow cytometer into 6 μL of lysis buffer and SUPERase-In RNAse inhibitor (Applied Biosystems, Waltham, MA, USA). Live/dead gating was performed based on propidium iodide exclusion. Reverse transcription and low cycle pre-amplification was performed following addition of Superscript III reverse transcriptase enzyme (Invitrogen, Waltham, MA, USA), Cells Direct reaction mix (Invitrogen), and target gene-specific TaqMan assay (primer/probe) sets (Applied Biosystems) (Table S1). Resultant single cell cDNA was mixed with sample loading agent (Fluidigm, South San Francisco, CA, USA) and Universal PCR Master Mix (Applied Biosystems) and loaded into 48.48 Dynamic Array chips (Fluidigm) along with TaqMan assays and assay loading agent according to the manufacturer's instructions. Products were analyzed on the BioMark reader system (Fluidigm) using a hot start protocol to minimize primer-dimer formation, with 40 quantitative PCR cycles performed.

Analysis of single cell data was performed as previously described [8,20]. Briefly, expression data from all chips/passages were normalized relative to the median expression for each gene in the pooled sample and converted to base 2 logarithms. Absolute bounds of ±5 cycle thresholds from the median were set, and non-expressers were assigned to this floor (−5). Clustergrams were then generated using hierarchical clustering in order to facilitate data visualization using MATLAB (R2011b, MathWorks, Natick, MA, USA).

In order to detect subpopulations based on single cell transcriptional data, k-means clustering was employed, assigning each cell to membership in a single cluster as dictated by similarities in expression profiles (minimizing the within-cluster sum of square distances in the 48-dimensional gene hyperspace). Optimally partitioned clusters were then sub-grouped using hierarchical clustering in order to facilitate visualization of data patterning within and across these clusters [20].

Two-sample Kolmogorov-Smirnov (K-S) tests were used to identify genes whose expression patterns differed significantly between population clusters and/or passage groups, using a strict cutoff of $p < 0.05$ following Bonferroni correction for multiple samples. For comparisons among subgroups, the empirical distribution of cells from each cluster was evaluated against that of the remaining cells in the experiment.

Ingenuity Pathway Analysis (IPA, Ingenuity Systems, Redwood City, CA, USA) was used to construct transcriptome networks based on genes that were significantly increased following passage from fresh→P0 and P0→P1, as well as among the 3 individual subpopulations determined above. For this analysis, the 48 genes included in the corresponding single cell analysis (rather than the entire transcriptome) were used as the reference set in order to avoid biasing the associated enrichment calculations in IPA's internal network generation algorithm.

3. Results and Discussion

3.1. Cell Culture Alters the Transcriptional Profiles of Human Adipose-Derived Stem Cells

Utilizing a microfluidic-based single-cell gene expression platform previously described by our laboratory [20], the transcriptional profiles of 60–100 individual cells from fresh, P0, and P1 adipose-derived stem cells (ASCs) were simultaneously evaluated for 48 surface-marker coding gene (Table S1). In this analysis, ASCs isolated from all passages displayed significant heterogeneity at the single-cell level, both within and across groups (Figure 1).

Differences in the transcriptional profiles of genes related to cell stemness, proliferation, and tumor biology, such as the CD47, CD151, DPPA3, and IGF2R were also observed at the single cell level (Figure 2), suggesting a potential evolution of these properties throughout expansion in cell culture. These differences are embodied in both the fraction of cells expressing these genes, as well as the characteristics of their distributions across cell types.

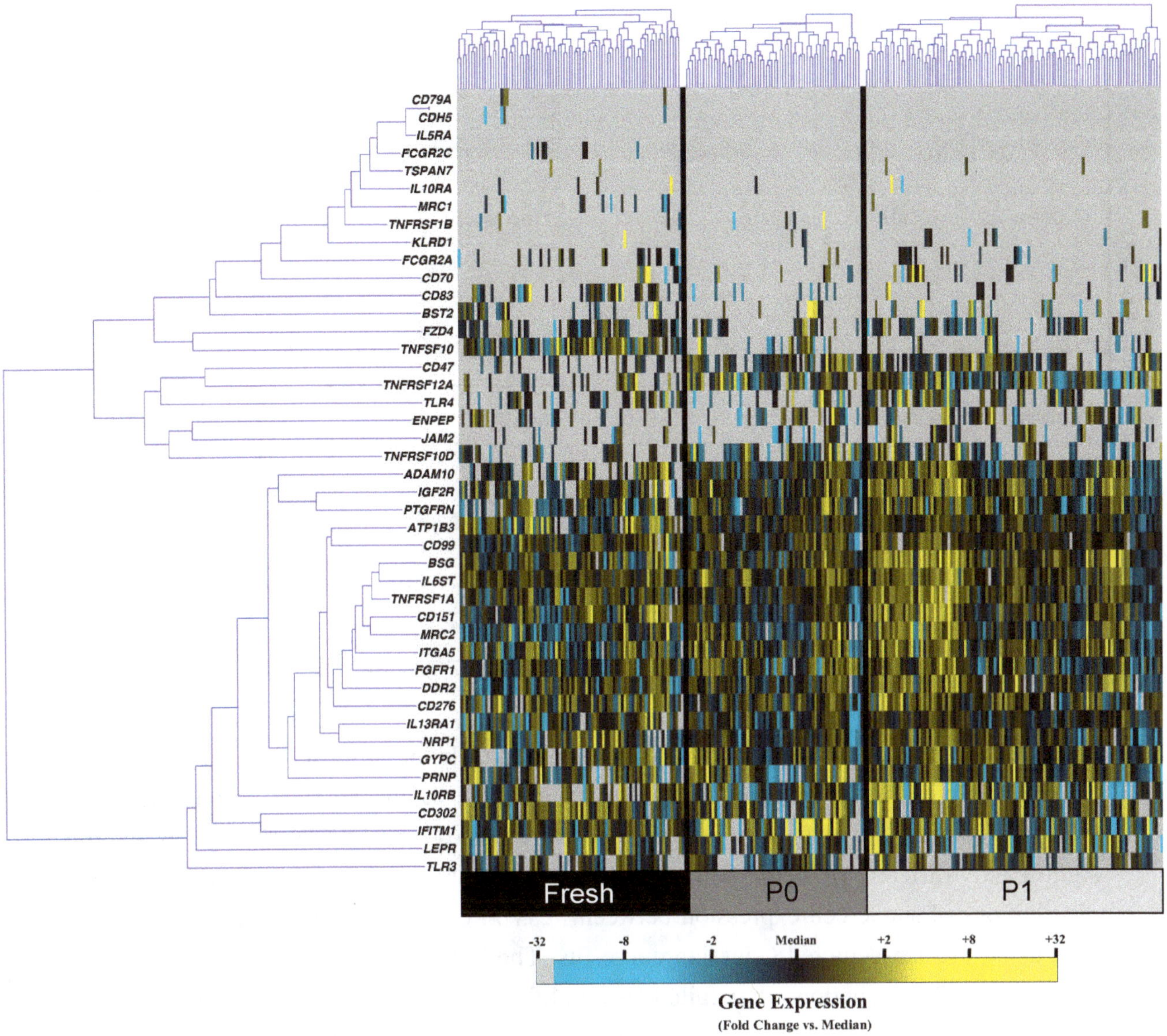

Figure 1. Single cell transcriptional analysis of fresh, passage 0, and passage 1 adipose-derived stem cells. Hierarchical clustering of simultaneous gene expression for single cells from fresh (left; black), passage 0 (middle; dark grey), and passage 1 (right; light grey) human adipose-derived stem cells (ASCs). Gene expression is presented as fold change *vs.* median on a color scale from yellow (high expression, 32-fold above median) to blue (low expression, 32-fold below median). Cell/gene qPCR reactions failing to amplify after 40 cycles are designated as non-expressers and represented in grey.

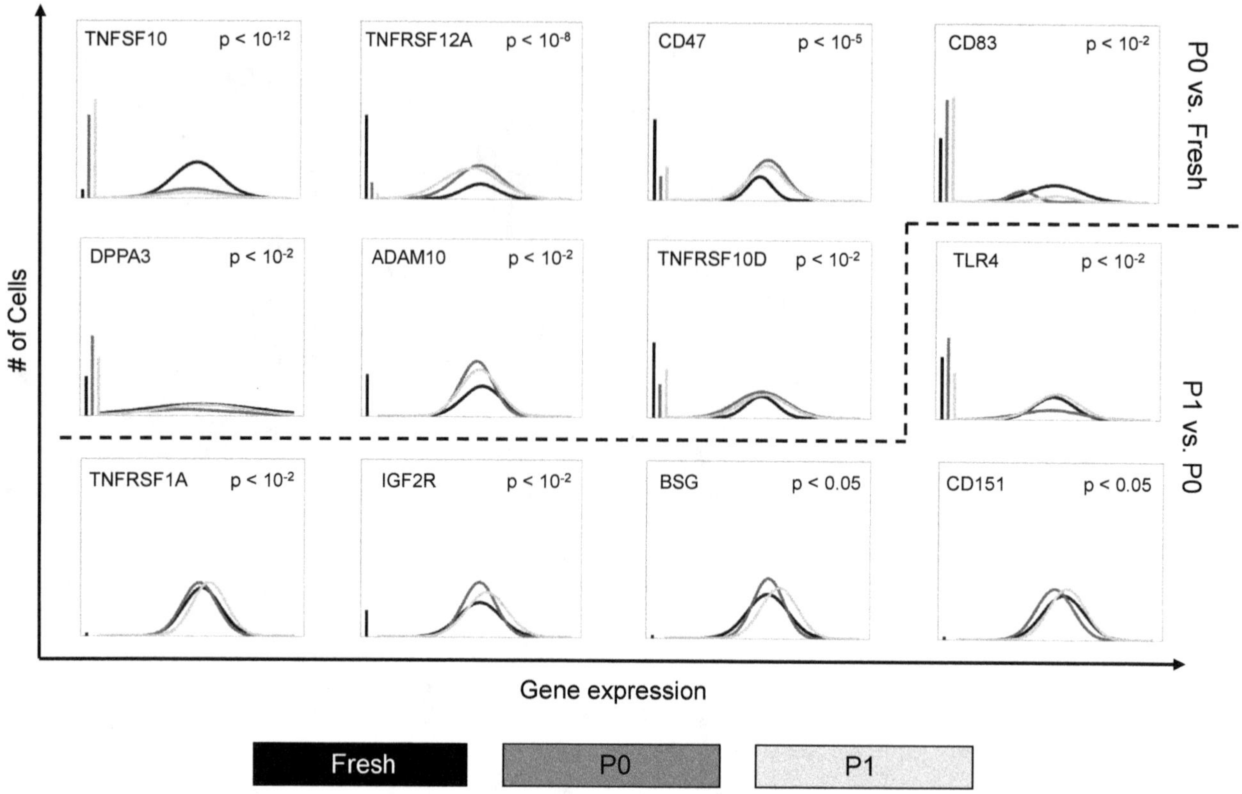

Figure 2. Cell culture alters the transcriptional profile of adipose-derived stem cells. Differentially-expressed genes among primary, passage 0, and passage 1 cells identified using non-parametric two sample Kolmogorov–Smirnov testing. Twelve genes exhibit significantly different ($p < 0.01$ following Bonferroni correction for multiple comparisons) distributions of single cell expression between fresh→P0 or P0→P1 populations, illustrated here using median-centered Gaussian curve fits. The left bar for each panel represents the fraction of qPCR reactions that failed to amplify in each group. The curves above the dashed line represent transcriptional comparisons between primarily isolated cells versus cells at passage 0, whereas the curves below the dashed line represent comparisons between cells at passage 1 versus passage 0.

Furthermore, the top molecular networks associated with each set of significantly-altered genes, generated using the Ingenuity Knowledge Base, appear to link key mediators of cell proliferation, embryonic development, and organogenesis (Figure 3).

Collectively, these data support the transcriptional evolution of cultured stem cells toward gene expression profiles characterized by more robust, proliferative, and stem-like properties.

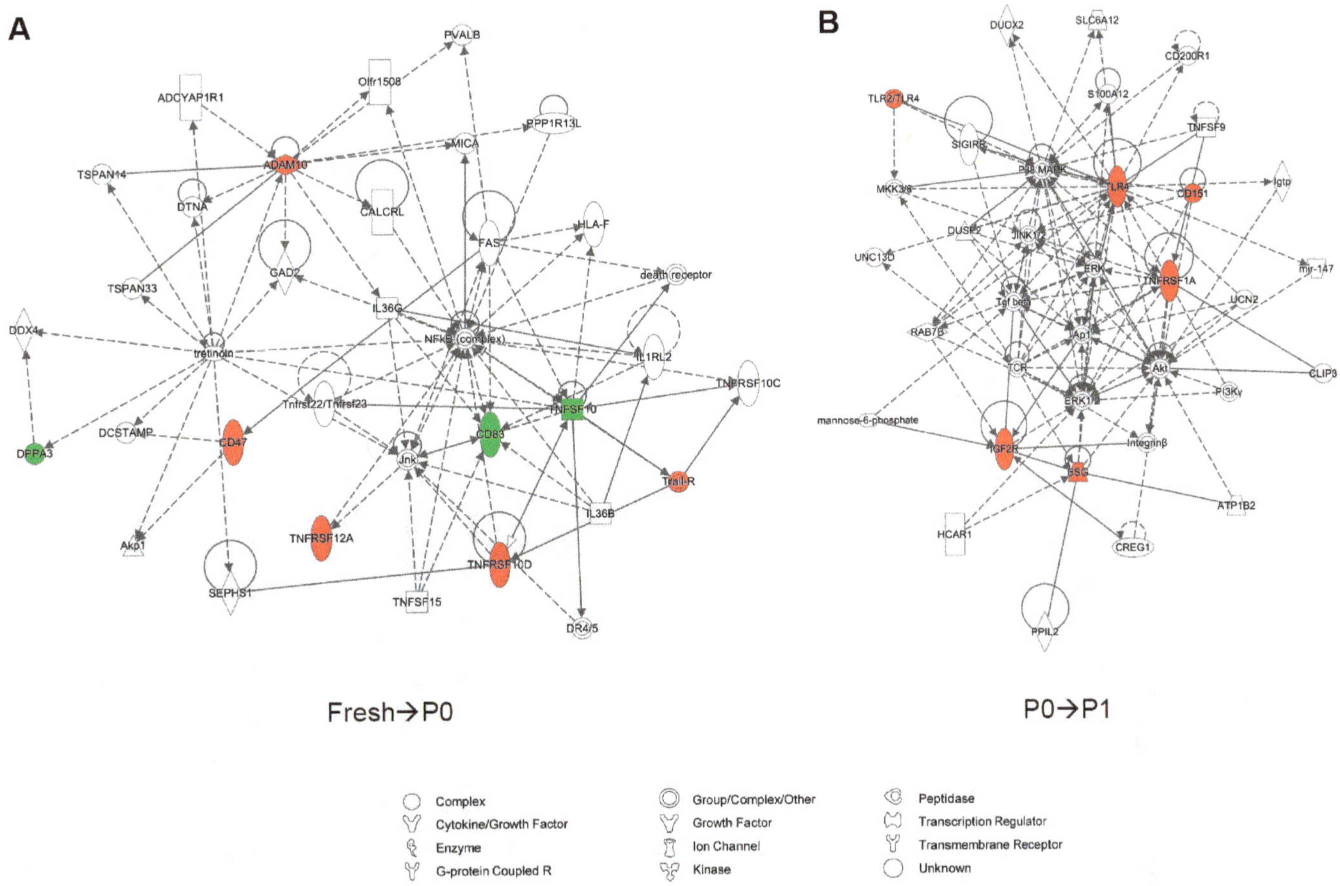

Figure 3. Network analysis of gene expression changes in of fresh, passage 0, and passage 1 adipose-derived stem cells. Top scoring Ingenuity Pathway Analysis (IPA)-constructed transcriptome networks based genes that were significantly up-regulated in fresh→P0 (**A**) and P0→P1 (**B**) human adipose-derived stem cells (ASCs). Direct relationships are indicated by solid lines, and dashed lines represent indirect relationships.

3.2. Partitional Cluster Analysis Reveals Subpopulations of Adipose-Derived Stem Cells that Evolve Throughout Cell Culture Passage

To further examine this niche, the super-set of transcriptional profiles for primary, passage 0, and passage 1 cells was evaluated using unbiased partitional clustering of gene expression data, as previously described [20]. This analysis identified three distinct, transcriptionally-defined ASC subpopulations or clusters within each phenotypic group (Figure 4). Interestingly, the first subgroup was considerably enriched for primary cells, the second subgroup dominated by a majority of passage 0 cells and a strong minority of passage 1 cells, and the third subgroup comprised primarily of passage 1 cells.

To predict the functional significance of these subpopulations, we evaluated the top molecular networks associated with genes up-regulated in each cellular subgroup using known transcriptional relationships derived from the Ingenuity Knowledge Base (IPA). The network generated from the first subpopulation (cluster 1), dominated by fresh ASCs, was characterized by non-specific cell-to-cell signaling and interactions (Figure 5A). The second network, derived from cluster 2, was associated with cellular proliferation and inflammatory pathways, which may represent the temporal shift of cells in this subgroup from primary to early passage (Figure 5B), and the third network (corresponding to cluster 3 and dominated by P1/P0 cells) was similarly characterized by cell development, proliferation, and movement (Figure 5C).

To further explore the significance of these signaling pathways, we used IPA to merge these three networks to generate a super-network including all genes associated with either pathway (Figure S1). Interestingly, the central elements linking these networks were extracellular-related kinase (ERK)1/2 and focal adhesion kinase (FAK), both closely implicated in the hyperproliferative response in cutaneous scarring [21], and tumor necrosis factor receptor superfamily member 12A (TNFRSF12A), which is a key regulatory element of cellular invasion in breast cancer [22].

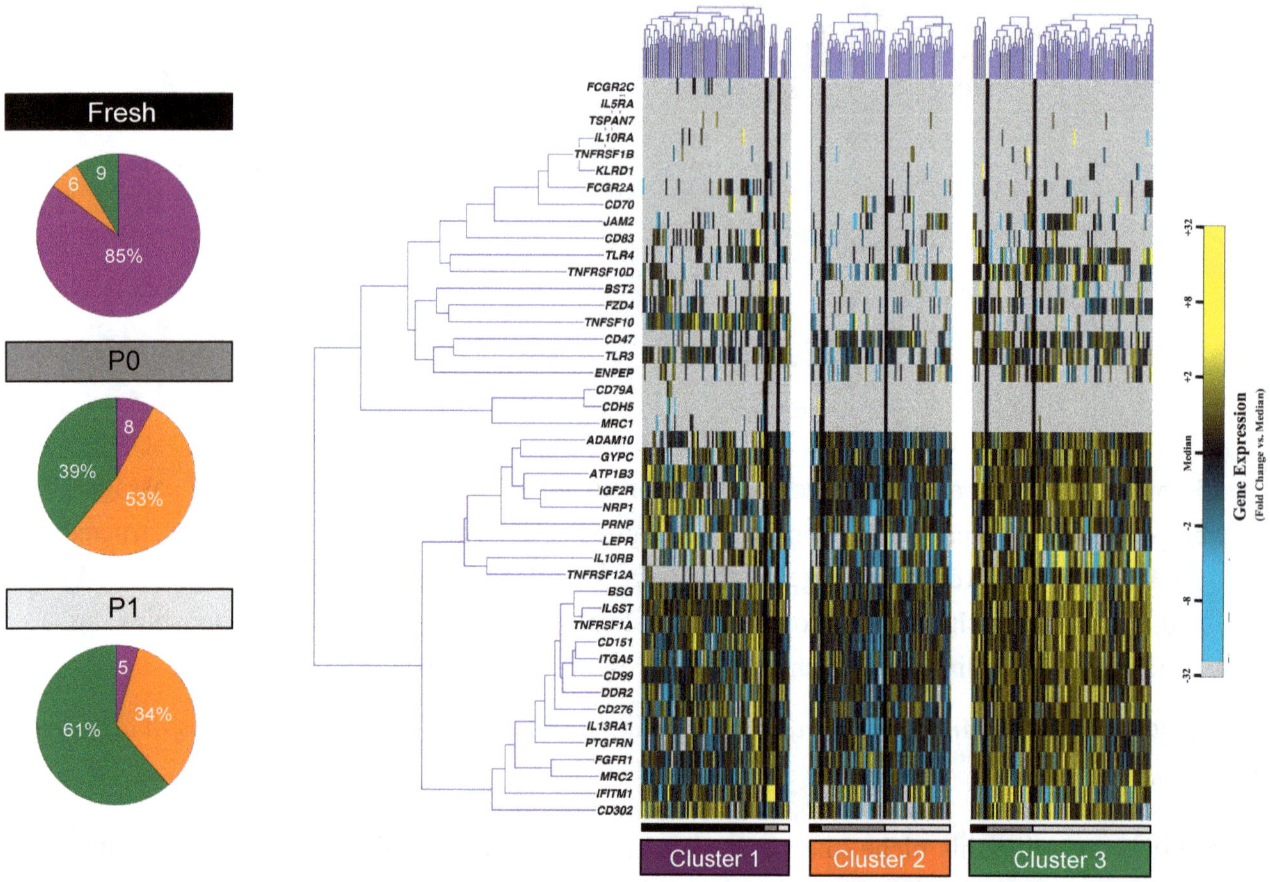

Figure 4. Partitional cluster analysis of single cell adipose-derived stem cell data across cell culture passage. K-means clustering of fresh (black), passage 0 (dark grey), and passage 1 (light grey) human adipose-derived stem cells (ASCs). Gene expression is presented as fold change from median on a color scale from yellow (high expression, 32-fold above median) to blue (low expression, 32-fold below median). Pie graphs represent the fraction of cells comprising each cluster from fresh, passage 0 (P0), and passage 1 (P1) cells, colored in purple, orange, and green in accordance with clusters 1, 2, and 3, respectively.

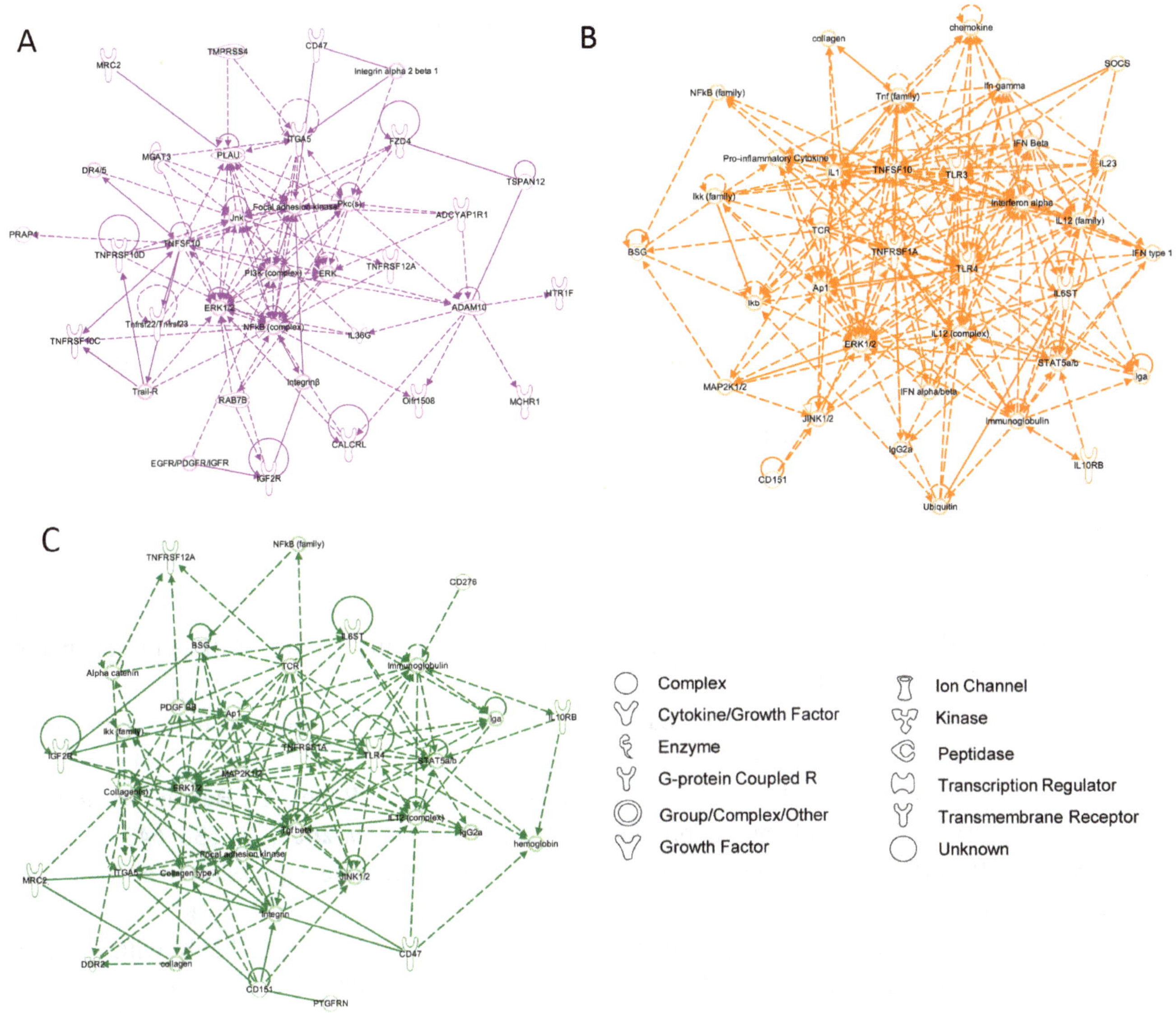

Figure 5. Network analysis of adipose-derived stem cell subpopulations. Top scoring Ingenuity Pathway Analysis (IPA)-constructed transcriptome networks based genes that were significantly up-regulated in cluster 1 (**A**; purple), cluster 2 (**B**; orange), and cluster 3 (**C**; green). Direct relationships are indicated by solid lines, and dashed lines represent indirect relationships.

4. Conclusions

Functional heterogeneity is a characteristic feature of complex cell populations that is frequently overlooked in studies employing traditional population-averaging measurement techniques. Single cell analytical methods provide the granularity to examine cellular subgroups within these complex samples, but harvesting and isolating sufficient quantities of rare or highly-purified populations is frequently challenging. In this study, we examined the transcriptional consequences of pre-passaging a well-studied human stem cell population prior to gene expression profiling. Our results demonstrate that these cells undergo significant transcriptional changes throughout cell culture as early as passage 0 and passage 1. Further, we observed a considerable shift in subpopulation dynamics, with comparative gene expression profiles suggestive of the gradual emergence of more active and proliferative subgroups. Taken together, these findings reinforce the importance of using primary or very early passage cells in future studies.

Author Contributions

Michael Januszyk and Geoffrey C. Gurtner designed the study. Michael Januszyk, Michael Sorkin, and Robert C. Rennert performed the experiments. Michael Januszyk and Geoffrey C. Gurtner developed the mathematical framework. Michael Januszyk, Lisa K. Wong, and Geoffrey C. Gurtner wrote the manuscript. Zeshaan N. Maan, Alexander J. Whittam, Arnetha Whitmore, and Dominik Duscher provided intellectual input and aided in experimental design.

References

1. Raaijmakers, M.H.; Scadden, D.T. Divided within: Heterogeneity within adult stem cell pools. *Cell* **2008**, *135*, 1006–1008.
2. Januszyk, M.; Gurtner, G.C. High-throughput single-cell analysis for wound healing applications. *Adv. Wound Care* **2013**, *2*, 457–469.
3. Bar-Even, A.; Paulsson, J.; Maheshri, N.; Carmi, M.; O'Shea, E.; Pilpel, Y.; Barkai, N. Noise in protein expression scales with natural protein abundance. *Nat. Genet.* **2006**, *38*, 636–643.
4. Elowitz, M.B.; Levine, A.J.; Siggia, E.D.; Swain, P.S. Stochastic gene expression in a single cell. *Science* **2002**, *297*, 1183–1186.
5. Maheshri, N.; O'Shea, E.K. Living with noisy genes: How cells function reliably with inherent variability in gene expression. *Ann. Rev. Biophys. Biomol. Struct.* **2007**, *36*, 413–434.
6. Vogel, C.; Marcotte, E.M. Insights into the regulation of protein abundance from proteomic and transcriptomic analyses. *Nat. Rev. Genet.* **2012**, *13*, 227–232.
7. Dalerba, P.; Kalisky, T.; Sahoo, D.; Rajendran, P.S.; Rothenberg, M.E.; Leyrat, A.A.; Sim, S.; Okamoto, J.; Johnston, D.M.; Qian, D.; *et al.* Single-cell dissection of transcriptional heterogeneity in human colon tumors. *Nat. Biotechnol.* **2011**, *29*, 1120–1127.
8. Levi, B.; Wan, D.C.; Glotzbach, J.P.; Hyun, J.; Januszyk, M.; Montoro, D.; Sorkin, M.; James, A.W.; Nelson, E.R.; Li, S.; *et al.* CD105 protein depletion enhances human adipose-derived stromal cell osteogenesis through reduction of transforming growth factor beta1 (TGF-β1) signaling. *J. Biol. Chem.* **2011**, *286*, 39497–39509.
9. Eisen, M.B.; Spellman, P.T.; Brown, P.O.; Botstein, D. Cluster analysis and display of genome-wide expression patterns. *Proc. Natl. Acad. Sci. USA* **1998**, *95*, 14863–14868.
10. Levsky, J.M.; Singer, R.H. Gene expression and the myth of the average cell. *Trends Cell Biol.* **2003**, *13*, 4–6.
11. Warren, L.; Bryder, D.; Weissman, I.L.; Quake, S.R. Transcription factor profiling in individual hematopoietic progenitors by digital RT-PCR. *Proc. Natl. Acad. Sci. USA* **2006**, *103*, 17807–17812.
12. Melin, J.; Quake, S.R. Microfluidic large-scale integration: The evolution of design rules for biological automation. *Ann. Rev. Biophys. Biomol. Struct.* **2007** *36*, 213–231.

13. Januszyk, M.; Sorkin, M.; Glotzbach, J.P.; Vial, I.N.; Maan, Z.N.; Rennert, R.C.; Duscher, D.; Thangarajah, H.; Longaker, M.T.; Butte, A.J.; *et al.* Diabetes irreversibly depletes bone marrow-derived mesenchymal progenitor cell subpopulations. *Diabetes* **2014**, *63*, 3047–3056.
14. Rennert, R.C.; Sorkin, M.; Januszyk, M.; Duscher, D.; Kosaraju, R.; Chung, M.T.; Lennon, J.; Radiya-Dixit, A.; Raghvendra, S.; Maan, Z.N.; *et al.* Diabetes impairs the angiogenic potential of adipose derived stem cells by selectively depleting cellular subpopulations. *Stem Cell Res. Ther.* **2014**, *5*, doi:10.1186/scrt468.
15. Suga, H.; Rennert, R.C.; Rodrigues, M.; Sorkin, M.; Glotzbach, J.P.; Januszyk, M.; Fujiwara, T.; Longaker, M.T.; Gurtner, G.C. Tracking the elusive fibrocyte: Identification and characterization of collagen-producing hematopoietic lineage cells during murine wound healing. *Stem Cells* **2014**, *32*, 1347–1360.
16. Sawada, H.; Saito,T.; Nickel, N.P.; Alastalo, T.-P.; Glotzbach, J.P.; Chan, R.; Haghighat, L.; Fuchs, G.; Januszyk, M.; Cao, A.; *et al.* Reduced BMPR2 expression induces GM-CSF translation and macrophage recruitment in humans and mice to exacerbate pulmonary hypertension. *J. Exp. Med.* **2014**, *211*, 263–280.
17. Wilson, A.; Laurenti, E.; Oser, G.; van der Wath, R.C.; Blanco-Bose, W.; Jaworski, M.; Offner, S.; Dunant, C.F.; Eshkind, L.; Bockamp, E.; Lió, P.; *et al.* Hematopoietic stem cells reversibly switch from dormancy to self-renewal during homeostasis and repair. *Cell* **2008**, *135*, 1118–1129.
18. Guo, H.; Zhu, P.; Wu, X.L.; Li, X.L.; Wen, L.; Tang, F. Single-cell methylome landscapes of mouse embryonic stem cells and early embryos analyzed using reduced representation bisulfite sequencing. *Genome Res.* **2013**, *23*, 2126–2135.
19. Zuk, P.A.; Zhu, M.; Ashjian, P.; de Ugarte, D.A.; Huang, J.I.; Mizuno, H.; Alfonso, Z.C.; Fraser, J.K.; Benhaim, P.; Hedrick, M.H. Human adipose tissue is a source of multipotent stem cells. *Mol. Biol. Cell* **2002**, *13*, 4279–4295.
20. Glotzbach, J.P.; Januszyk, M.; Vial, I.N.; Wong, V.W.; Gelbard, A.; Kalisky, T.; Thangarajah, H.; Longaker, M.T.; Quake, S.R.; Chu, G.; Gurtner, G.C.; *et al.* An information theoretic, microfluidic-based single cell analysis permits identification of subpopulations among putatively homogeneous stem cells. *PLoS ONE* **2011**, *6*, e21211.
21. Wong, V.W.; Rustad, K.C.; Akaishi, S.; Sorkin, M.; Glotzbach, J.P.; Januszyk, M.; Nelson, E.R.; Levi, K.; Paterno, J.; Vial, I.N.; *et al.* Focal adhesion kinase links mechanical force to skin fibrosis via inflammatory signaling. *Nat. Med.* **2012**, *18*, 148–152.
22. Gaudineau, B.; Fougère, M.; Guaddachi, F.; Lemoine, F.; de la Grange, P.; Jauliac, S. Lipocalin 2, the TNF-like receptor TWEAKR and its ligand TWEAK act downstream of NFAT1 to regulate breast cancer cell invasion. *J. Cell Sci.* **2012**, *125*, 4475–4486.

2

Single Cell Confocal Raman Spectroscopy of Human Osteoarthritic Chondrocytes

Rajesh Kumar [1,*], Gajendra P. Singh [2], Kirsten M. Grønhaug [3], Nils K. Afseth [4], Catharina de Lange Davies [1], Jon O. Drogset [5] and Magnus B. Lilledahl [1]

1 Department of Physics, Norwegian University of Science and Technology (NTNU), N-7491 Trondheim, Norway; E-Mails: catharina.davies@ntnu.no (C.L.D.); magnus.lilledahl@ntnu.no (M.B.L.)

2 Physical Optics and Electronics Group, Massachusetts Institute of Technology (MIT), Cambridge, MA 02139, USA; E-Mail: gpsingh@mit.edu

3 Orthopaedic Department, Levanger Hospital, Kirkegata 2, N-7600 Levanger, Norway; E-Mail: kirstengroenhaug@gmail.com

4 Nofima, Postbox 210, N-1431 Ås, Norway; E-Mail: nils.kristian.afseth@nofima.no

5 Department of Orthopaedic Surgery, Trondheim University Hospital, N-7491 Trondheim, Norway; E-Mail: jon.o.drogset@ntnu.no

* Author to whom correspondence should be addressed; E-Mail: 101rajesh@gmail.com

Academic Editor: Ritva Tikkanen

Abstract: A great deal of effort has been focused on exploring the underlying molecular mechanism of osteoarthritis (OA) especially at the cellular level. We report a confocal Raman spectroscopic investigation on human osteoarthritic chondrocytes. The objective of this investigation is to identify molecular features and the stage of OA based on the spectral signatures corresponding to bio-molecular changes at the cellular level in chondrocytes. In this study, we isolated chondrocytes from human osteoarthritic cartilage and acquired Raman spectra from single cells. Major spectral differences between the cells obtained from different International Cartilage Repair Society (ICRS) grades of osteoarthritic cartilage were identified. During progression of OA, a decrease in protein content and an increase in cell death were observed from the vibrational spectra. Principal component analysis and subsequent cross-validation was able to associate osteoarthritic chondrocytes to ICRS Grade I, II and III with specificity 100.0%, 98.1%, and 90.7%

respectively, while, sensitivity was 98.6%, 82.8%, and 97.5% respectively. The overall predictive efficiency was 92.2%. Our pilot study encourages further use of Raman spectroscopy as a noninvasive and label free technique for revealing molecular features associated with osteoarthritic chondrocytes.

Keywords: Raman spectroscopy; osteoarthritis; chondrocytes; biomedical optical diagnosis

---

1. Introduction

Osteoarthritis (OA) is a complex musculoskeletal disorder. It is a leading cause of disability among elderly people and is a major public health issue with increasing individual and socioeconomic burden. It was recently found that it is the fastest increasing major health condition in terms of the global rank: "Years lived with disability (YLD)" [1]. Although the use of the Kellgren-Lawrence (K/L) score obtained by radiography is a widely accepted method for the diagnosis of OA, several studies have demonstrated the complexity involved in early-stage diagnosis [2–4].

The origin of OA is not exactly clear and the understanding about the etiology of OA is continuously evolving. There are several reports which review the changes observed during progression of OA [5,6]. It is believed that the disease primarily affects the articular cartilage and the associated underlying bone. The prominent histological changes in cartilage are erosion, fibrillation, clefts, chondromalacia, loss of metachromasia, duplication of tidemark and cloning of chondrocytes. The underlying cause of these macroscopic and microscopic structural features is the change in biochemical compositions such as alteration and decrement in content of proteoglycan [7]. Biochemical changes are triggered by expression of enzymes, which are responsible for the breakdown of matrix components. It is believed that expression of such enzymes occurs due to the imbalance in cell homeostasis, *i.e.*, increased catabolic activity by chondrocytes [8]. Several studies also show that progression of OA is associated with increased chondrocyte cell death [6,9,10]. Investigations at the cellular level are, therefore, important in understanding the progression of OA as well as other cartilage disorders.

Raman spectroscopy has recently attracted attention because it shows the potential for minimally invasive, label free, and objective diagnostics at molecular level. The advantage of a label free technique is that it provides the information from the target molecules that are not modified. Therefore, all processes in the target molecules are undisturbed and the cells do not show a stress reaction attributable to an external label [11]. Raman spectroscopy is a laser based vibrational spectroscopic technique and has numerous times successfully demonstrated early detection of neoplasia, differentiation of tumors grades, and been useful in providing sub-cellular information from the living cells [12,13]. However, to the knowledge of author, this is the first report on Raman spectroscopy of osteoarthritic human chondrocytes. The aim of this pilot study is to demonstrate the capabilities of Raman spectroscopy for single cell analysis in order to identify major changes in the spectroscopic signatures of the cells during the progression of OA. In addition, we have demonstrated that based on biochemical compositions at cellular level, Raman spectroscopy is capable to distinguish the chondrocytes isolated from different stages (ICRS Grade-I, II and III) of osteoarthritic cartilage.

2. Results and Discussion

A comparison between the mean (of n = 150) spectra of ICRS Grade-I, II and III is shown in Figure 1. Mean spectra represent the average characteristics of each grade. Typically, Raman spectra obtained from biological samples are complex in nature, consisting of superimposed Raman peaks associated with the various bio-molecular constituents of the cell. Hence, visual interpretation of the mean spectra obtained from a large variety of cells is not straightforward. However a careful qualitative evaluation of the mean spectra can provide an insight into the bio-compositional changes at the cellular level during the progression of OA. In order to confirm that the subtle differences involved in the spectra are from variation of the Raman signals and not from the fluorescence background, the variance between the mean spectra of the three different cell types (Grade I, II and III) at each wavelength positions was plotted (Supplementary Figure S1). The plot of variance *vs.* wavelength expressed major Raman peaks (and not fluorescence background) associated with cellular biochemical components which are described in next section.

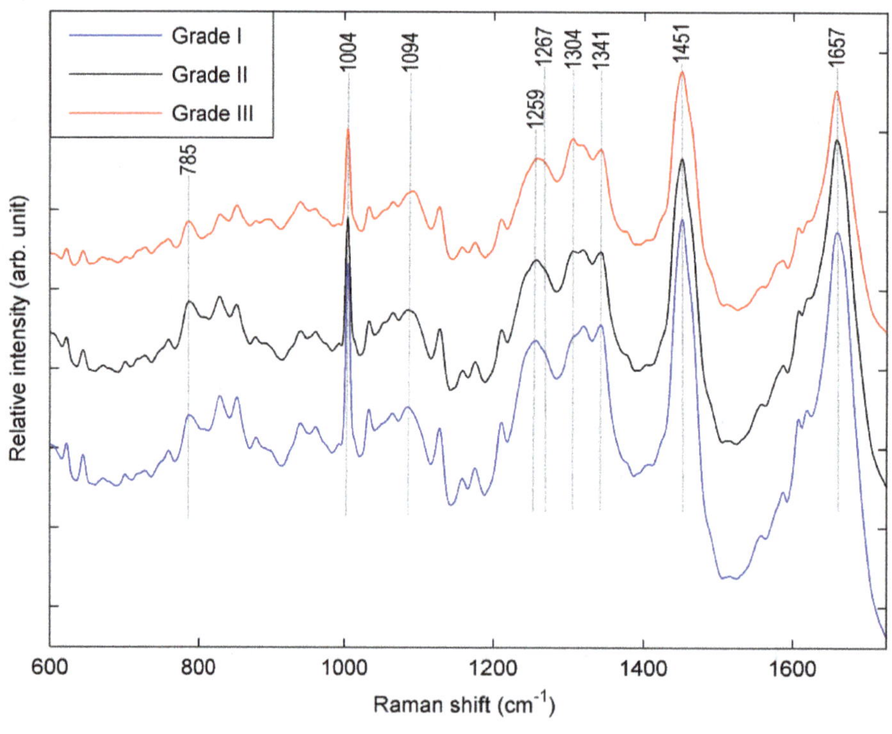

Figure 1. Normalized mean (n = 150) Raman spectra of chondrocytes which were isolated from osteoarthritic cartilage of ICRS Grade-I, II and III. Spectra are vertically offset for clarity.

The main aim of this analysis was to investigate if Raman spectroscopy could be used to identify biomolecular changes at the cellular level and potentially discriminate chondrocytes which were isolated from different stages (ICRS Grade-I, II and III) of osteoarthritic cartilage. The major spectral changes in chondrocytes obtained from different stages of osteoarthritic cartilage can be found by a qualitative analysis of the mean spectra (Figure 1). The Raman spectra clearly indicate changes in protein vibrations. The magnitude of the area (Figure 2a) represented by amide I (1612–1696 cm^{-1}) decreases consistently with an increase in the grade of OA. This observation was supported by similar trend (Figure 2b,c) shown by the content of amide III (1229–1300 cm^{-1}) and phenylalanine (1001–1007 cm^{-1}/peak at 1004 cm^{-1}). These observations indicate a general decrease in protein content inside chondrocytes in higher grades of OA. Moreover, decrease in protein content was observed with decrease in the content of the nucleic acid (780–794 cm^{-1}/peak at 785 cm^{-1}) in cells which may correlate with the enhancement of internucleosomal DNA cleavage [14] during progression of disease (Figure 2e).

As another observation, intensity of the spectral band 1302–1307 cm^{-1}/peak at 1304 cm^{-1} is increasing with progression of OA (Figures 1 and 2d). This peak may be attributed to the lipids [15–19]. In the earlier reports, it was hypothesized that increased lipid deposition may be associated with the advancement of OA [20–23]. Our observation indicates that this peak could be an indicator of lipid deposition and hence, supports the earlier hypothesis that lipid deposition may play a role in progression of OA. However, other alternative interpretation (e.g., amide III and/or nucleobase) besides lipids might also be possible relevant to the peak of 1304 cm^{-1} [15]. Therefore, more experimental studies are necessary to verify the contribution of lipid during progression of the disease.

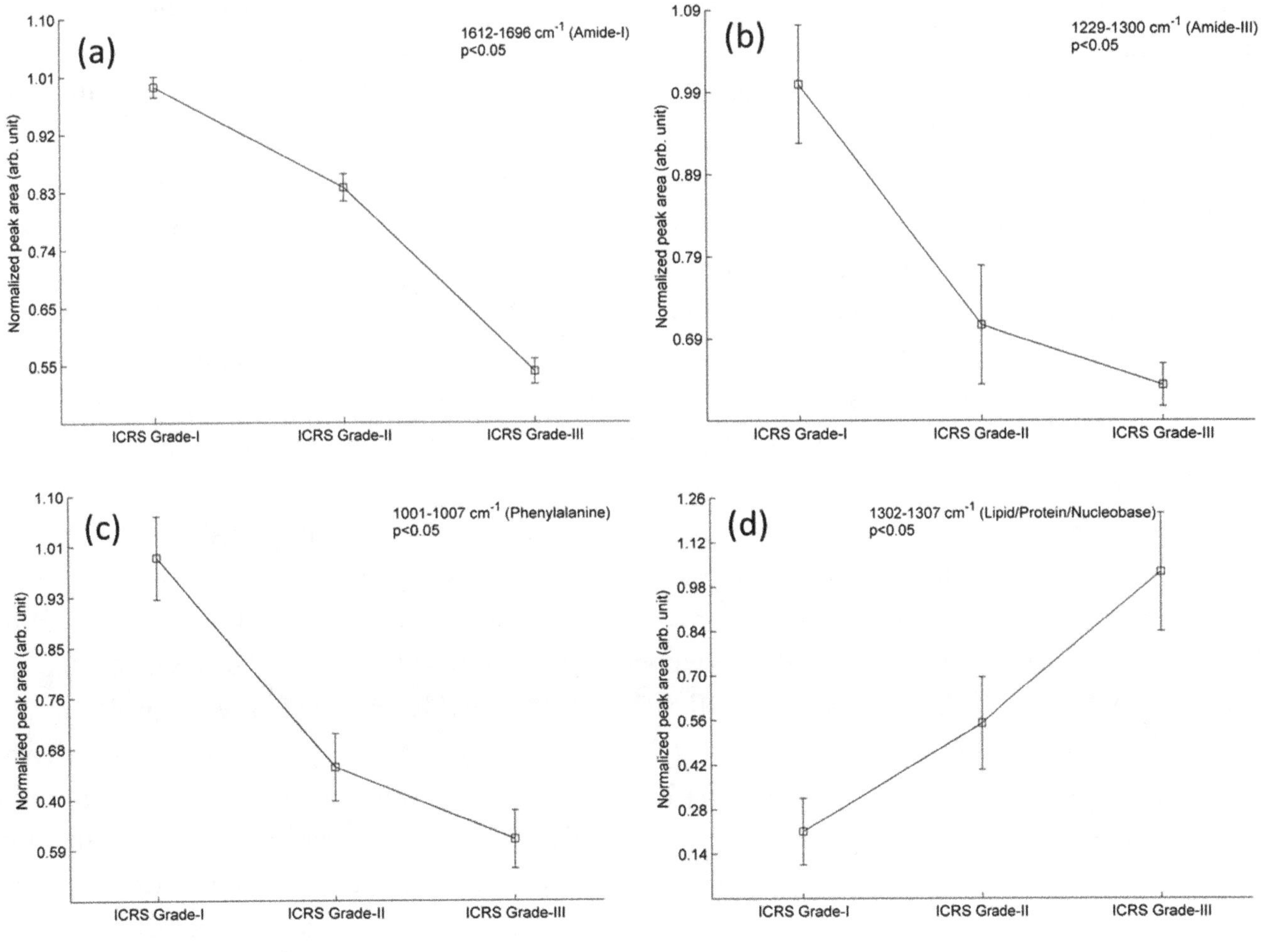

Figure 2. *Cont.*

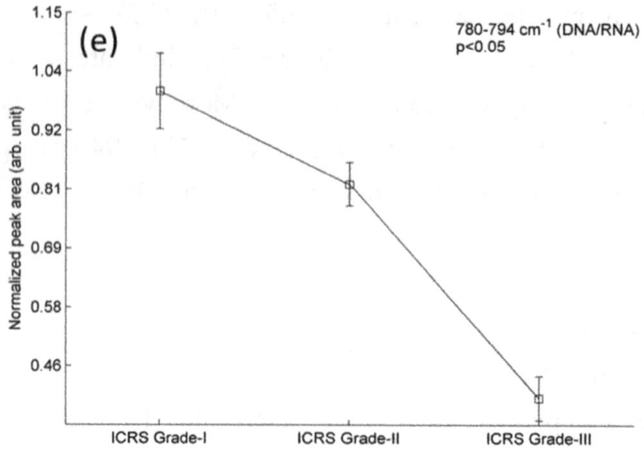

Figure 2. A relative comparison for the content of amide-I (**a**); amide-III (**b**); phenylalanine (**c**); lipid (**d**); and DNA/RNA (**e**) in osteoarthritic chondrocytes (n = 150 cells). These chondrocytes were isolated from osteoarthritic cartilage section of ICRS Grade-I, II and III. Data represents the normalized mean area of Raman band and bar represents the standard error.

In order to test the capability of Raman spectroscopy, spectra were compared objectively by applying multivariate analysis technique, using ICRS as a gold standard. To elucidate and visualize the hidden biochemical information, principal component analysis (PCA) was performed on the total data-set which were acquired from 450 samples (cells). PCA is a common multivariate statistical method that is often used in bioanalytical Raman spectroscopy to reduce dimensionality and to identify combinations of the most important spectral markers which maximizes the data variance and optimizes group separation [24,25]. It is performed over the data sets with the aim of comparing spectral data in terms of similarities and differences in an unsupervised manner [26]. The results of the PCA analysis are shown in Figure 3. The spectra of chondrocytes associated with the three different ICRS grades of OA appeared as distinct clusters when plotted against the three main PCs that explain maximum variance in the data set. The tightness and separation of the clusters illustrates how easily the analytes can be discriminated. In order to quantitatively assess discrimination capability, prediction accuracy was tested subsequently using leave one out full cross-validation technique using Mahalanobis distance as a discriminator. Several reports recommend full cross-validation as a general approach to prediction testing, claiming that this should provide most comprehensive testing of the model [27,28]. The classification matrix summarizes the correct and incorrect classification of the spectra (Table 1). The specificities for ICRS Grade I, II and III were 100.0%, 98.1%, and 90.7% respectively, while, the sensitivities were 98.6%, 82.8%, and 97.5% respectively. The overall predictive efficiency was 92.2%. Additionally, the segmented [27] cross-validation (a segment of 10 spectra at a time) showed similar result in terms of overall predictive efficiency (91.8%). Our analysis shows that the chondrocytes obtained from grade-I osteoarthritic cartilage, were classified with higher accuracy than other grades. Misclassifications mainly occurred between the chondrocytes of grade II and III (Figure 3), which indicates that chondrocytes associated with ICRS grade II and III are more heterogeneous in biochemical compositions than ICRS Grade-I. During progression of OA the quality of articular cartilage degrades and chondrocytes respond in order to cope with physiochemical changes occurred in extracellular matrix [29,30]. In this context, our result indicates that to a greater extent most chondrocytes available in ICRS Grade-I cartilage are in similar stage, but, the partial overlap between the cluster of ICRS Grade-II and Grade-III indicate that all chondrocytes associated with macroscopically higher grade (*i.e.*, Grade-II and III) of osteoarthritic cartilage are not in same stage of degradation.

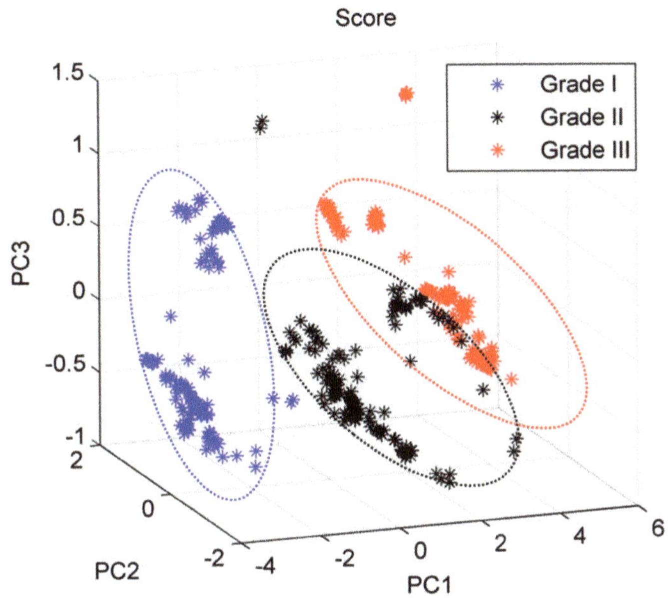

Figure 3. Multivariate-analysis based principal component analysis (PCA) algorithm classifies different osteoarthritic chondrocytes cells associated with ICRS Grade I, II and III into separate clusters.

Table 1. Classification matrix provides the information about classification ability and prediction efficiency of PCA algorithm.

Sample: Chondrocytes	Grade-I	Grade-II	Grade-III
Grade-I (150)	150	0	0
Grade-II (150)	2	145	3
Grade-III (150)	0	30	120

Loading vector is associated to the original Raman spectrum by a variable called PC score, which represents the weight of that particular component against the basis spectrum [31]. Loading vectors (Figure 4) associated with three main principle components (PC-1, PC-2 and PC-3) were plotted. These plots are mainly composed of Raman peaks, which indicate that the major variations in the data are due to Raman signals and not fluorescence background. Loadings associated with PC-1, PC-2 and PC-3 explains 82.30%, 7.06% and 3.75% of the total variance in the data set, respectively. Combined, these three PCs explain 93.11% of the total variation in the data set. The most important spectral features which appeared in the loading plots were tentatively assigned [15,18] to amide-I (1657 cm^{-1}), protein and carbohydrate (1451 cm^{-1}), DNA/RNA (1341 cm^{-1}), lipid and protein (1304 cm^{-1}), amide-III (1267 cm^{-1}), nucleic acids (1094 cm^{-1}), phenylalanine (1004 cm^{-1}) and DNA/RNA (785 cm^{-1}).

However, a few peaks (e.g., 1379 and 1490 cm^{-1}) are remained unassigned for human cells and require further investigation.

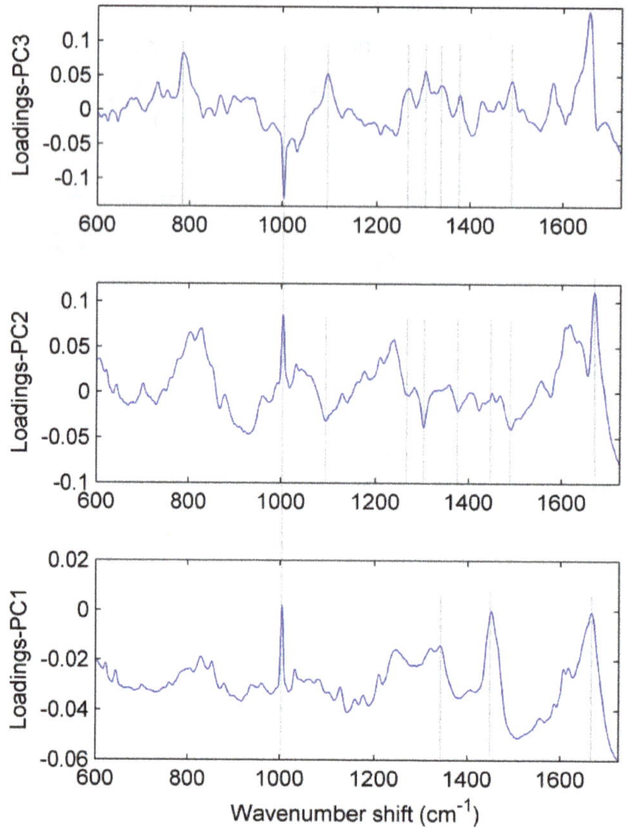

Figure 4. Loading plots relevant to PC1, PC2 and PC3, which are mainly responsible for the discrimination among different clusters of chondrocytes cells associated with different grades of osteoarthritic cartilage.

Fixation of cells may induce spectral changes. Acetone and alcohol (e.g., methanol and ethanol) dissolve lipids. Therefore, other fixatives like formalin and 4% paraformaldehyde was considered as a better option for the fixation. As shown in previous spectroscopic studies using cultured cells that formalin fixation was found to affect the cellular lipid and protein content [32,33]. Ten percent formalin with fixation times of several hours are considered much harsher conditions than 4% paraformaldehyde that could lead to dehydration due to fixation [34]. Although fixation with 4% paraformaldehyde also induces changes in the Raman spectrum of a cell, the induced changes are small compared to the alcohol [34]. Moreover, with longer duration of fixation time, larger spectral changes were observed [34]. Therefore, 4% paraformaldehyde with mild fixation time (~10 min) was chosen to be optimal for the Raman study. However, additional studies are needed to find the fixation induced Raman changes due to the 4% paraformaldehyde especially in the chondrocytes.

Due to erosion and thinning in high grade OA [35–37], it is difficult to isolate the chondrocytes from a given cartilage layer (for example, superficial layer). Therefore, spectra were acquired from a large number of randomly chosen cells that were isolated from the whole volume (bulk) of the osteoarthritic cartilage. Hence, the spectra represent the chondrocyte characteristics of the bulk cartilage rather than a specific layer of cartilage. Thus, the spectral differences are associated with biochemical difference in OA severities (and not only the chondrocyte characteristics associated with different cartilage layer).

Due to ethical considerations, biopsies of normal (healthy) cartilage from human knee were not obtained in this study. Although our study includes the detection of osteoarthritic grades at cellular level, it should be noted that clinical diagnosis of OA is not performed at cellular level. Biopsy of early stage cartilage and then, isolation of chondrocytes for diagnostic purposes is not intended to be performed in clinical practice. Nevertheless, detailed knowledge of OA at cellular level is highly desirable. Enhanced understanding will help in exploring underlying mechanism and thus, in early diagnosis that may lead towards correct treatment and prevention of OA.

3. Materials and Methods

3.1. Human Osteoarthritic Cartilage Sections

The use of human samples was approved by the Regional Committee for Medical Research Ethics (2013/265 REK, Norway). Human articular cartilage samples were obtained from the femoral condyle of osteoarthritic patients undergoing total replacement knee surgery. In order to consider the case of primary OA, only those patients that had not suffered previous knee injury and had not undergone other prior surgery were included in the study. To reduce age-associated variation in the articular cartilage, all patients selected in this study were older than 65 years. It should be pointed out that different grades of osteoarthritic cartilage may be present within one femoral head of the knee joint. The grading of cartilage sections was based on the standard (International Cartilage Repair Society) ICRS classification (Supplementary Table S1) [38,39]. The assignment of ICRS grade to the cartilage sections was performed by two experienced orthopedic surgeons, who were blinded to the classification of each other. Only sections assigned a similar ICRS grade by both orthopedic surgeons were included in this study. In this pilot study, 15 specimens in total, 5 specimens of each ICRS Grade (Grade I, II and III) were included (for details please refer to Supplementary Figure S2). All tissue sections included in this study were obtained from load bearing area (medial) of the femoral condyle of the knee. Only osteoarthritic cartilage specimens of ICRS Grade I, II and III, but not Grade-IV, were included in the study. As described in the ICRS classification scheme, at the location of ICRS Grade IV (*i.e.*, advanced stage of OA), underlying bone is exposed and almost no cartilage present. Therefore, ICRS Grade-IV specimens were not included in this study.

3.2. Isolation of Chondrocytes (Cells)

Isolation of chondrocytes from all osteoarthritic cartilage specimens was performed by enzymatic (collagenase) digestion process. A typical image of osteoarthritic cartilage and isolated single cells is shown in Figure S3. To isolate chondrocytes, firstly, cartilage sections were divided into several pieces (~1–2 mm) using a surgical scalpel. Incubation of trimmed cartilage was then performed for ~18 h in 2–5 mL of Dulbecco's Modified Eagle's Medium (DMEM) containing 0.8 mg/mL collagenase from clostridium histolyticum (C6885; Sigma-Aldrich, Saint Louis, MO, USA) and 50 μg/mL gentamicin (G1397; Sigma-Aldrich). Filtration of the digested cartilage was performed through a 50 μm filter. Centrifugation of the filtered cells was performed at 1500 rmp for 5 min. Pelleted cells were resuspended in phosphate buffer saline (PBS) solution. Centrifugation and resuspension in PBS was repeated second time. Finally, cells were fixed by 4% pararfomaldehyde for the time duration of 10 min. Cells in PBS suspension were stored at 4 °C.

3.3. Raman Spectroscopy

Raman spectra were acquired using a standard confocal Raman microscope (LabRam HR800, HORIBA Jobin Yvon, Villeneuve-d'Ascq, France). The 632.10 nm laser light was focused by an Olympus (Tokyo, Japan) 60×, 1.2 NA water-immersion microscope objective. Cells sedimented in a well on a CaF_2 slide and were covered by a 100 μm quartz cover-slip. 30 Raman spectra were acquired from 30 randomly chosen cells associated with each specimen (Supplementary Figure S2). Therefore, 150 (30 × 5 specimens) spectra were collected from each ICRS grade (Grade I, II, and III). In total, 450 (150 × 3 Grades) spectra obtained from 450 samples (cells) were used in this investigation (Figure S2). Spectra were acquired over the Raman shift region 600–1725 cm^{-1} (fingerprint region). The acquisition time of each spectrum was 90 s. During acquisition of data, the laser power at the surface of the sample was 8 mW. In order to minimize biochemical heterogeneity of different cell compartments (e.g., nucleus, cytoplasm, membrane), during each data acquisition, the laser was focused in the nucleus of the cell. However, due to inherent features such as multiple scattering and the large size of the confocal volume [40,41] the system might also partially examine the perinuclear structures. As described by Pudlas *et al.* [42], the collected Raman signals predominantly represents molecular bonds of the cell nuclei. In order to reduce the fluorescence background and enhance the comparability of the spectra, the raw Raman spectra were pre-processed by removal of cosmic rays (narrow noise spikes), subtraction of background spectra (the corresponding spectra of PBS suspension which were acquired after the change of every focus) and smoothing (Savitzky-Golay filter: 3rd order, 9 point). Spectra were then subjected to extended multiplicative signal correction (EMSC) [43], followed by peak normalization (1004 cm^{-1}). The plots for the normalized mean (n = 150) area of the corresponding Raman band of interest were plotted with standard error. Spectral band assignment of sharp peaks (*i.e.*, 785, 1004 and 1304 cm^{-1}) for the calculation of peak area was performed according to full width at half maxima (FWHM). Statistical comparisons were performed using one-way analysis of variance (ANOVA) and a p-value of less than 0.05 was considered as statistically significant. The data analysis was performed in Matlab (The MathWorks, Natick, MA, USA).

4. Conclusions

In summary, as a proof-of-principle study, we demonstrated Raman spectroscopy as a noninvasive and label free technique that can detect subtle biomolecular changes in osteoarthritic chondrocytes. The relative changes in protein, lipid and RNA content in chondrocytes at different stages of OA may help in exploring the underlying mechanism of the disease at cellular level. Additionally, we have shown that with high specificity and sensitivity, it can discriminate the cells obtained from the different stages of osteoarthritic cartilage. Our analysis might be helpful in the development of Raman spectroscopy technique to monitor OA progression at the cellular level. Further correlation and validation of novel Raman spectral features of chondrocytes using supervised/unsupervised techniques in combination with clinical gold-standard may help Raman spectroscopy to achieve its full potential. Nevertheless, our study encourages further investigations; for example: characterization of proteins [44,45], mapping of single cells [19,46,47], quantitative comparison of biochemical compositions (e.g., DNA/RNA, lipid, collagen, total protein content) [48] in different stage osteoarthritic chondrocytes, which may further reveal hidden features associated with progression of OA at cellular and sub-cellular level.

Acknowledgments

The authors would like to thank Kristin Grendstad Sæterbø for her help in cell laboratory and Astrid Bjørkøy for technical help in the microscope laboratory. Financial support was provided by the Norwegian University of Science and Technology (NTNU).

Author Contributions

Rajesh Kumar and Magnus B. Lilledahl conceived the study and designed the experiment; Rajesh Kumar collected data, performed data analysis, and drafted the manuscript; Kirsten M. Grønhaug and Jon O. Drogset collected osteoarthritic cartilage during knee surgery; Gajendra P. Singh, Kirsten M. Grønhaug, Nils K. Afseth, Catharina de Lange Davies, Jon O. Drogset and Magnus B. Lilledahl contributed in collection of reagents/analysis tools/other materials and contributed in preparation of manuscript with valuable discussions. All authors read and approved the final manuscript.

References

1. Hunter, D.J.; Schofield, D.; Callander, E. The individual and socioeconomic impact of osteoarthritis. *Nat. Rev. Rheumatol.* **2014**, *10*, 437–441.
2. Kornaat, P.R.; Bloem, J.L.; Ceulemans, R.Y.; Riyazi, N.; Rosendaal, F.R.; Nelissen, R.G.; Carter, W.O.; Hellio Le Graverand, M.P.; Kloppenburg, M. Osteoarthritis of the knee: Association between clinical features and MR imaging findings. *Radiology* **2006**, *239*, 811–817.
3. Cicuttini, F.; Hankin, J.; Jones, G.; Wluka, A. Comparison of conventional standing knee radiographs and magnetic resonance imaging in assessing progression of tibiofemoral joint osteoarthritis. *Osteoarthr. Cartil.* **2005**, *13*, 722–727.
4. Buchwald, T.; Niciejewski, K.; Kozielski, M.; Szybowicz, M.; Siatkowski, M.; Krauss, H. Identifying compositional and structural changes in spongy and subchondral bone from the hip joints of patients with osteoarthritis using Raman spectroscopy. *J. Biomed. Opt.* **2012**, *17*, doi:10.1117/1.JBO.17.1.017007.
5. Lee, A.S.; Ellman, M.B.; Yan, D.; Kroin, J.S.; Cole, B.J.; van Wijnen, A.J.; Im, H.J. A current review of molecular mechanisms regarding osteoarthritis and pain. *Gene* **2013**, *527*, 440–447.
6. Horton, W.E., Jr.; Yagi, R.; Laverty, D.; Weiner, S. Overview of studies comparing human normal cartilage with minimal and advanced osteoarthritic cartilage. *Clin. Exp. Rheumatol.* **2005**, *23*, 103–112.
7. Horton, W.E., Jr.; Bennion, P.; Yang, L. Cellular, molecular, and matrix changes in cartilage during aging and osteoarthritis. *J. Musculoskelet. Neuronal Interact.* **2006**, *6*, 379–381.

8. Buck, R.J.; Wirth, W.; Dreher, D.; Nevitt, M.; Eckstein, F. Frequency and spatial distribution of cartilage thickness change in knee osteoarthritis and its relation to clinical and radiographic covariates-data from the osteoarthritis initiative. *Osteoarthr. Cartil.* **2013**, *21*, 102–109.
9. Meachim, G.; Collins, D.H. Cell counts of normal and osteo-arthritic articular cartilage in relation to the uptake of sulphate ($^{35}SO_4$) *in vitro*. *Ann. Rheum. Dis.* **1962**, *21*, 45–50.
10. Zamli, Z.; Adams, M.; Tarlton, J.; Sharif, M. Increased chondrocyte apoptosis is associated with progression of osteoarthritis in spontaneous guinea pig models of the disease. *Int. J. Mol. Sci.* **2013**, *14*, 17729–17743.
11. Dittrich, P.; Jakubowski, N. Current trends in single cell analysis. *Anal. Bioanal. Chem.* **2014**, *406*, 6957–6961.
12. Crow, P.; Stone, N.; Kendall, C.A.; Uff, J.S.; Farmer, J.A.; Barr, H.; Wright, M.P. The use of Raman spectroscopy to identify and grade prostatic adenocarcinoma *in vitro*. *Br. J. Cancer* **2003**, *89*, 106–108.
13. Pascut, F.C.; Goh, H.T.; George, V.; Denning, C.; Notingher, I. Toward label-free Raman-activated cell sorting of cardiomyocytes derived from human embryonic stem cells. *J. Biomed. Opt.* **2011**, *16*, doi:10.1117/1.3570302.
14. Verrier, S.; Notingher, I.; Polak, J.M.; Hench, L.L. *In situ* monitoring of cell death using Raman microspectroscopy. *Biopolymers* **2004**, *74*, 157–162.
15. Aksoy, C.; Severcan, F. Role of vibrational spectroscopy in stem cell research. *Spectrosc. Int. J.* **2012**, *27*, 167–184.
16. Notingher, I.; Bisson, I.; Bishop, A.E.; Randle, W.L.; Polak, J.M.; Hench, L.L. *In situ* spectral monitoring of mRNA translation in embryonic stem cells during differentiation *in vitro*. *Anal. Chem.* **2004**, *76*, 3185–3193.
17. Wu, H.; Volponi, J.V.; Oliver, A.E.; Parikh, A.N.; Simmons, B.A.; Singh, S. *In vivo* lipidomics using single-cell Raman spectroscopy. *Proc. Natl. Acad. Sci. USA* **2011**, *108*, 3809–3814.
18. Notingher, I.; Verrier, S.; Haque, S.; Polak, J.M.; Hench, L.L. Spectroscopic study of human lung epithelial cells (A549) in culture: Living cells *versus* dead cells. *Biopolymers* **2003**, *72*, 230–240.
19. Krafft, C.; Knetschke, T.; Siegner, A.; Funk, R.H.W.; Salzer, R. Mapping of single cells by near infrared Raman microspectroscopy. *Vib. Spectrosc.* **2003**, *32*, 75–83.
20. Gkretsi, V.; Simopoulou, T.; Tsezou, A. Lipid metabolism and osteoarthritis: Lessons from atherosclerosis. *Progress Lipid Res.* **2011**, *50*, 133–140.
21. Tiku, M.L.; Shah, R.; Allison, G.T. Evidence linking chondrocyte lipid peroxidation to cartilage matrix protein degradation. Possible role in cartilage aging and the pathogenesis of osteoarthritis. *J. Biol. Chem.* **2000**, *275*, 20069–20076.
22. Villalvilla, A.; Gómez, R.; Largo, R.; Herrero-Beaumont, G. Lipid transport and metabolism in healthy and osteoarthritic cartilage. *Int. J. Mol. Sci.* **2013**, *14*, 20793–20808.
23. Zhuo, Q.; Yang, W.; Chen, J.; Wang, Y. Metabolic syndrome meets osteoarthritis. *Nat. Rev. Rheumatol.* **2012**, *8*, 729–737.
24. Chan, J.; Fore, S.; Wachsmann-Hogiu, S.; Huser, T. Raman spectroscopy and microscopy of individual cells and cellular components. *Laser Photonics Rev.* **2008**, *2*, 325–349.

25. Barman, I.; Dingari, N.C.; Singh, G.P.; Kumar, R.; Lang, S.; Nabi, G. Selective sampling using confocal Raman spectroscopy provides enhanced specificity for urinary bladder cancer diagnosis. *Anal. Bioanal. Chem.* **2012**, *404*, 3091–3099.
26. Mobili, P.; Londero, A.; Gomez-Zavaglia, A. Multivariate analysis of Raman spectra applied to microbiology: Discrimination of microorganisms at the species level. *Rev. Mex. Fis.* **2010**, *56*, 378–385.
27. Esbensen, K.H. *Multivariate Data Analysis-In Practice*, 5th ed.; CAMO Process AS: Esbjerg, Denmark, 2005.
28. Martens, H.; Naes, T. *Multivariate Calibration*; Wiley: Chichester, UK, 1989.
29. Goldring, M.B. The role of the chondrocyte in osteoarthritis. *Arthritis Rheumatol.* **2000**, *43*, 1916–1926.
30. Van der Kraan, P.M.; van den Berg, W.B. Chondrocyte hypertrophy and osteoarthritis: Role in initiation and progression of cartilage degeneration? *Osteoarthr. Cartil.* **2012**, *20*, 223–232.
31. Ferraro, J.R.; Nakamoto, K.; Brown, C.W. *Introductory Raman Spectroscopy*; Academic Press: Waltham, MA, USA, 2002.
32. Hastings, G.; Wang, R.; Krug, P.; Katz, D.; Hilliard, J. Infrared microscopy for the study of biological cell monolayers. I. Spectral effects of acetone and formalin fixation. *Biopolymers* **2008**, *89*, 921–930.
33. Gazi, E.; Dwyer, J.; Lockyer, N.P.; Miyan, J.; Gardner, P.; Hart, C.; Brown, M.; Clarke, N.W. Fixation protocols for subcellular imaging by synchrotron-based Fourier transform infrared microspectroscopy. *Biopolymers* **2005**, *77*, 18–30.
34. Chan, J.W.; Taylor, D.S.; Thompson, D.L. The effect of cell fixation on the discrimination of normal and leukemia cells with laser tweezers Raman spectroscopy. *Biopolymers* **2009**, *91*, 132–139.
35. Pritzker, K.P.; Gay, S.; Jimenez, S.A.; Ostergaard, K.; Pelletier, J.P.; Revell, P.A.; Salter, D.; van den Berg, W.B. Osteoarthritis cartilage histopathology: Grading and staging. *Osteoarthr. Cartil.* **2006**, *14*, 13–29.
36. Buckland, J. Osteoarthritis: Targeting cartilage erosion in OA. *Nat. Rev. Rheumatol.* **2010**, *6*, 64.
37. Plaas, A.; Li, J.; Riesco, J.; Das, R.; Sandy, J.D.; Harrison, A. Intraarticular injection of hyaluronan prevents cartilage erosion, periarticular fibrosis and mechanical allodynia and normalizes stance time in murine knee osteoarthritis. *Arthritis Res. Ther.* **2011**, *13*, R46.
38. Brittberg, M.; glietti, P.; Gambardella, R.; Hangody, L.; Hauselmann, H.J.; Jakob, R.P.; Levine, D.; Lohmander, S.; Mandelbaum, B.R.; Peterson, L.; *et al.* ICRS Cartilage Injury Evaluation Package. In Proceedings of the 3rd ICRS Meeting, Göteborg, Sweden, 28 April 2000. Available online: http://www.cartilage.org/_files/contentmanagement/ICRS_evaluation.pdf (accessed on 19 March 2015).
39. Kleemann, R.U.; Krocker, D.; Cedraro, A.; Tuischer, J.; Duda, G.N. Altered cartilage mechanics and histology in knee osteoarthritis: Relation to clinical assessment (ICRS Grade). *Osteoarthr. Cartil.* **2005**, *13*, 958–963.

40. Jess, P.R.; Smith, D.D.; Mazilu, M.; Dholakia, K.; Riches, A.C.; Herrington, C.S. Early detection of cervical neoplasia by Raman spectroscopy. *J. Int. Cancer* **2007**, *121*, 2723–2728.
41. Mauch, F.; Lyda, W.; Osten, W. Object depending measurement uncertainty of confocal sensors. In *Fringe 2013*; Osten, W., Ed.; Springer Berlin Heidelberg: Berlin, Germany, 2014; pp. 465–470.
42. Pudlas, M.; Koch, S.; Bolwien, C.; Thude, S.; Jenne, N.; Hirth, T.; Walles, H.; Schenke-Layland, K. Raman spectroscopy: A noninvasive analysis tool for the discrimination of human skin cells. *Tissue Eng. Part C Methods* **2011**, *17*, 1027–1040.
43. Afseth, N.K.; Kohler, A. Extended multiplicative signal correction in vibrational spectroscopy, a tutorial. *Chemom. Intell. Lab. Syst.* **2012**, *117*, 92–99.
44. Maiti, N.C.; Apetri, M.M.; Zagorski, M.G.; Carey, P.R.; Anderson, V.E. Raman spectroscopic characterization of secondary structure in natively unfolded proteins: α-Synuclein. *J. Am. Chem. Soc.* **2004**, *126*, 2399–2408.
45. Thomas, G.J., Jr. Raman spectroscopy of protein and nucleic acid assemblies. *Ann. Rev. Biophys. Biomol. Struct.* **1999**, *28*, 1–27.
46. Klein, K.; Gigler, A.M.; Aschenbrenner, T.; Monetti, R.; Bunk, W.; Jamitzky, F.; Morfill, G.; Stark, Robert W.; Schlegel, J. Label-free live-cell imaging with confocal Raman microscopy. *Biophys. J.* **2012**, *102*, 360–368.
47. Draux, F.; Jeannesson, P.; Beljebbar, A.; Tfayli, A.; Fourre, N.; Manfait, M.; Sule-Suso, J.; Sockalingum, G.D. Raman spectral imaging of single living cancer cells: A preliminary study. *Analyst* **2009**, *134*, 542–548.
48. Swain, R.J.; Stevens, M.M. Raman microspectroscopy for non-invasive biochemical analysis of single cells. *Biochem. Soc. Trans.* **2007**, *35*, 544–549.

3

Simultaneous Characterization of Instantaneous Young's Modulus and Specific Membrane Capacitance of Single Cells using a Microfluidic System

Yang Zhao [1], Deyong Chen [1], Yana Luo [1], Feng Chen [1], Xiaoting Zhao [2], Mei Jiang [2], Wentao Yue [2], Rong Long [3,*], Junbo Wang [1,*] and Jian Chen [1,*]

1 State Key Laboratory of Transducer Technology, Institute of Electronics, Chinese Academy of Sciences, Beijing 100190, China; E-Mails: gaochunzy@gmail.com (Y.Z.); dychen@mail.ie.ac.cn (D.C.); luoyana88@126.com (Y.L.); ccmucf@aliyun.com (F.C.)

2 Department of Cellular and Molecular Biology, Beijing Chest Hospital, Capital Medical University, Beijing 101149, China; E-Mails: zhao_xiaoting@126.com (X.Z.); jmcyf627@sina.com (M.J.); yuewentao@gmail.com (W.Y.)

3 Department of Mechanical Engineering, University of Alberta, Edmonton, AB T6G 2G8, Canada

* Authors to whom correspondence should be addressed; E-Mails: rlong2@ualberta.ca (R.L.); jbwang@mail.ie.ac.cn (J.W.); chenjian@mail.ie.ac.cn (J.C.)

Academic Editor: Kwang W. Oh

Abstract: This paper presents a microfluidics-based approach capable of continuously characterizing instantaneous Young's modulus ($E_{instantaneous}$) and specific membrane capacitance ($C_{specific\ membrane}$) of suspended single cells. In this method, cells were aspirated through a constriction channel while the cellular entry process into the constriction channel was recorded using a high speed camera and the impedance profiles at two frequencies (1 kHz and 100 kHz) were simultaneously measured by a lock-in amplifier. Numerical simulations were conducted to model cellular entry process into the constriction channel, focusing on two key parameters: instantaneous aspiration length ($L_{instantaneous}$) and transitional aspiration length ($L_{transitional}$), which was further translated to $E_{instantaneous}$. An equivalent distribution circuit model for a cell travelling in the constriction channel was used to determine $C_{specific\ membrane}$. A non-small-cell lung cancer cell line 95C (n = 354) was used to evaluate this technique, producing $E_{instantaneous}$ of 2.96 ± 0.40 kPa and $C_{specific\ membrane}$

of 1.59 ± 0.28 μF/cm^2. As a platform for continuous and simultaneous characterization of cellular $E_{instantaneous}$ and $C_{specific\ membrane}$, this approach can facilitate a more comprehensive understanding of cellular biophysical properties.

Keywords: microfluidics; single-cell analysis; cellular biophysics; instantaneous Young's modulus; specific membrane capacitance

1. Introduction

Mechanical properties of the cytoskeleton ($E_{instantaneous}$ and $E_{equilibrium}$) and electrical properties of cell membrane ($C_{specific\ membrane}$) determine the overall cellular biophysical properties [1] and have been correlated with diseases such as malaria and cancer [2,3]. Conventional techniques for cellular mechanical and/or electrical property characterization suffer from limited throughput and cannot quantify mechanical and electrical properties simultaneously [4,5] (e.g., ~20 cells per cell type based on electrorotation [6] and ~10 cells per cell type based on atomic force microscopy [7]).

Advances in microfluidic technologies have enabled mechanical and/or electrical property characterization of single cells in a continuous manner [8–11] (e.g., $E_{instantaneous}$ values from hundreds of A549 cells [12] and $C_{specific\ membrane}$ values from hundreds of H1299 cells [13,14]). However, the majority of reported microfluidic devices were only designed for characterizing either electrical or mechanical properties of a cell, but not both.

Previously, there are three types of devices developed to combine the measurements of cellular mechanical and electrical properties. These approaches were based on principles of microcantilever-based electrodes [15], micropipette aspiration with impedance spectroscopy [16], and constriction channel with impedance spectroscopy [17,18], respectively. Approaches using microcantilever-based electrodes [15] and the combination of micropipette aspiration with impedance spectroscopy [16] have limited throughput and cannot collect data from hundreds of single cells.

In [17,18], single cells were continuously aspirated through a constriction channel (a cross sectional area smaller than cells) while cellular images and single-frequency impedance profiles were obtained simultaneously and translated to cellular mechanical (cellular transition time through the constriction channel) and electrical (impedance amplitude increase during the cellular travelling process within the constriction channel) properties. Although this approach is featured with a higher throughput, which is capable of collecting biophysical data from hundreds of single cells, due to the lack of equivalent mechanical and electrical models, this method cannot interpret raw mechanical and electrical data into intrinsic cellular biophysical parameters ($E_{instantaneous}$ and $C_{specific\ membrane}$), leading to compromised capabilities in cell status evaluation and cell type classification [19].

Motivated by this challenge, in this paper, we present a constriction channel based microfluidic platform for simultaneous characterization of the intrinsic mechanical and electrical properties of single cells in a continuous manner (see Figure 1). In this platform, single cells were aspirated continuously through a constriction channel while the cellular entry process was imaged by a high speed camera and the two-frequency impedance profiles were measured by an impedance analyzer.

Based on equivalent mechanical and electrical models, these raw data were translated to $E_{instantaneous}$ and $C_{specific\ membrane}$, providing a comprehensive understanding of cellular biophysical properties.

Note that this platform is not a simple integration of previous constriction channel based microfluidic devices capable of characterizing either cellular instantaneous Young's modulus ($E_{instantaneous}$) [12] or specific membrane capacitance ($C_{specific\ membrane}$) [13,14]. The simultaneous collection of cellular images and two-frequency impedance data can provide more insights to study cellular behaviors during their entry and traveling processes within the constriction channel and thus provide a more complete characterization of biological cells.

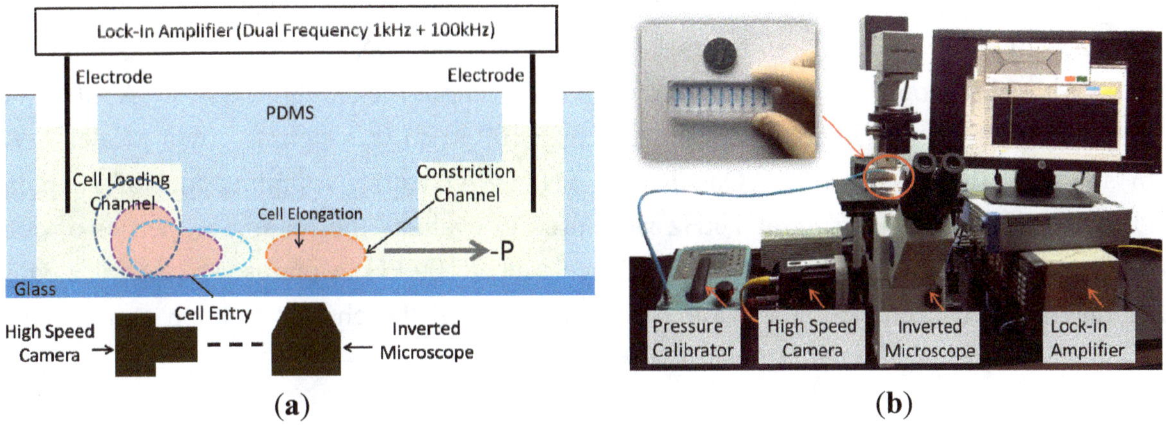

Figure 1. (**a**) Schematic and (**b**) setup of the microfluidic system for continuous characterization of $E_{instantaneous}$ and $C_{specific\ membrane}$ of single cells where cells are aspirated continuously through the constriction channel with cellular entry and traveling images as well as impedance profiles obtained simultaneously and translated to $E_{instantaneous}$ and $C_{specific\ membrane}$ based on proposed equivalent mechanical and electrical models.

2. Materials and Methods

2.1. Materials

All cell-culture reagents were purchased from Life Technologies Corporation (Carlsbad, CA, USA) unless otherwise specified. The materials used during device fabrication were SU-8 photoresist (MicroChem Corp., Newton, MA, USA) and 184 silicone elastomer (Dow Corning Corp., Midland, MI, USA). The non-small-cell lung cancer cell line 95C was cultured at 37 °C in 5% CO_2 in RPMI 1640 medium supplemented with 10% heat-inactivated fetal bovine serum, 100 units/mL penicillin and 100 μg/mL streptomycin.

2.2. Device Fabrication and Operation

The microfluidic device consists of a constriction channel (cross-section area: 10 μm × 10 μm) in polydimethylsiloxane (PDMS) elastomer that was replicated from a double-layer SU-8 mold. The detailed fabrication procedures were described in a previous publication [20]. Briefly, SU-8 5 was spin coated and exposed without development to form the constriction channel with a height of 10 μm. Then SU-8 25 (cell loading channel with a height of 25 μm) was spin coated on top of the first SU-8 layer, exposed with alignment and developed, forming the two-layer mold master. PDMS prepolymers and curing agents (10:1 in weight) were mixed, poured on channel masters and baked in an oven for crosslinking. PDMS channels were then peeled from the SU-8 masters, punched to form reservior holes, and bonded to glass slides after plasma treatment.

During operation the device was first filled with culture medium and a droplet of cell suspension (1×10^6 cells/mL) was pipetted onto the entrance of the cell loading channel. A negative pressure of 500 Pa generated from a pressure calibrator (DPI-610 pressure calibrator, Druck, Billerica, MA, USA) was applied to aspirate cells continuously through the constriction channel while silver electrode wires [16] were inserted into the inlet and the outlet of the PDMS device for impedance profile monitoring. Cell images were taken by an inverted microscope (IX71, Olympus Inc., Tokyo, Japan) connected with a Phantom M320S high speed camera (Phantom Inc., Bublin, OH, USA) at 200 frames per second. Two-frequency impedance data (1 kHz + 100 kHz) was sampled by a Model 7270 DSP lock-in amplifier (Signal Recovery, Oak Ridge, TN, USA) with a sampling rate of 20 points per second. Note that impedance data at 1 kHz were used to evaluate the sealing properties of deformed cells with constriction channel walls and impedance data at 100 kHz were used to quantify equivalent cellular membrane capacitance and cytoplasm resistance. All the characterization experiments were conducted within 30 min after cell trypsinization.

2.3. Equivalent Mechanical and Electrical Models

Numerical simulations were performed using a finite element package ABAQUS (version 6.11, Dassault Systemes Simulia Corp., Providence, RI, USA) to model cellular entry process into the constriction channel with detailed procedures described in a previous publication [12]. Briefly, the channel walls were modeled as rigid surfaces with a geometrical parameter of $D_{channel}$ and the cell was modeled as an incompressible solid with a key mechanical parameter of $E_{instantaneous}$. The friction on cell-wall interface was modeled with a constant friction coefficient f_c.

Figure 2a–d show the simulation results of the cellular entry process including the stage I of instantaneous jump into the channel ($L_{instantaneous}$) (see Figure 2a) and the stage II of cellular creep response (Figure 2b,c), with the ending point at the transitional position ($L_{transitional}$) (Figure 2d). Numerical simulation shows that $L_{instantaneous}/D_{channel}$ and $L_{transitional}/D_{channel}$ are functions of two variables $P_{aspiration}/E_{instantaneous}$ (linear function) and f_c as follows ($P_{aspiration}$ represents aspiration pressure applied to aspirate cells into the constriction channel) [12]:

$$L_{instantaneous}/D_{channel} = \left(44.27f_c^2 - 37.24f_c + 13.70\right) \times P_{aspiration}/E_{instantaneous} - 5.31f_c^2 + 2.84f_c - 0.59$$

$$L_{transitional}/D_{channel} = \left(-60.40f_c^2 + 40.05f_c - 8.68\right) \times P_{aspiration}/E_{instantaneous} + 1.99f_c^2 + 0.03f_c + 1.60$$

An equivalent circuit model (see Figure 2e) was proposed to model the cellular traveling process within the constriction channel with detailed description shown in a previous publication [13]. Briefly, distributed leakage resistors were represented by R_{L1}, R_{L2}, …, R_{Ln}, R_{Ln+1} with n divisions ($n \geq 2$) along the cell elongation direction while cellular electrical components were represented by C_{Ms}, C_{Mp1}, …, C_{Mpn} and R_{C1}, …, R_{Cn+1}, respectively. These distributed equivalent membrane capacitors can be further represented by $C_{specific\ membrane}$ as follows ($L_{cell\ elongation}$ represents the length of an elongated biological cell travelling within the constriction channel) [13]:

$$C_{Ms} = C_{specific membrane} \times D_{channel}^{2}$$

$$C_{Mp1} = C_{Mp2} = \cdots = C_{Mpn} = 4C_{specific membrane} \times L_{cell elongation} \times D_{channel} / n$$

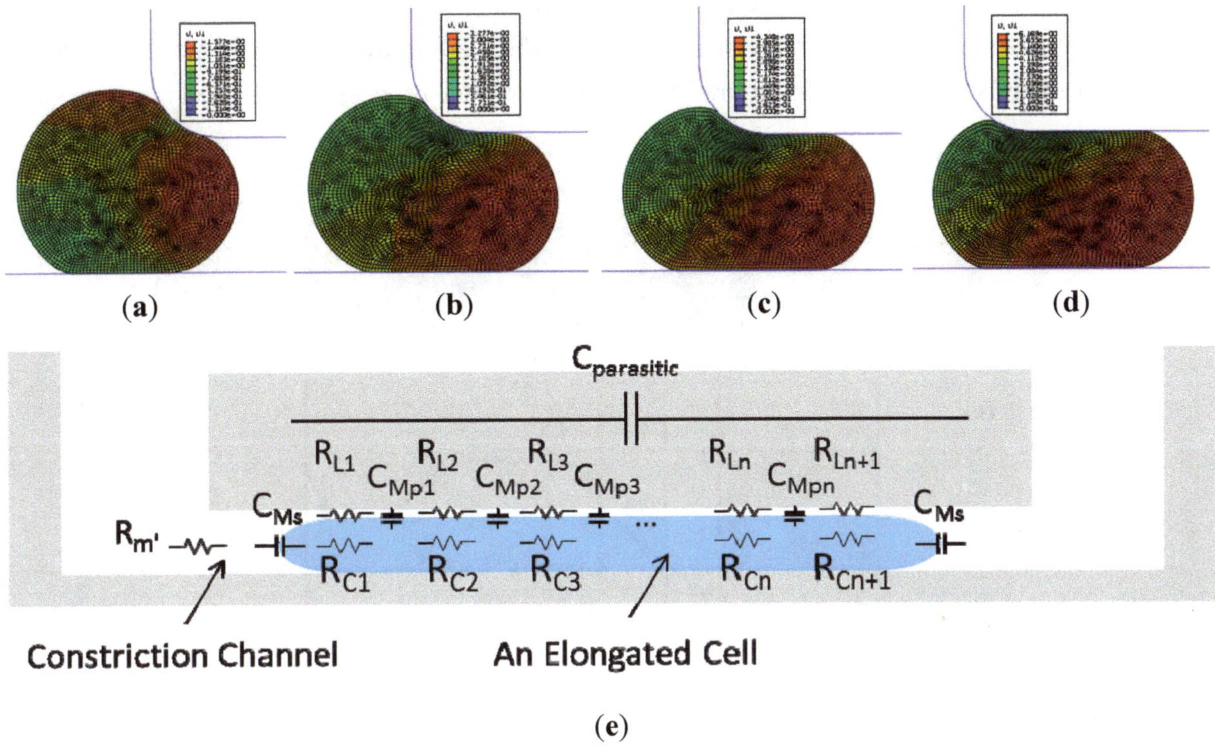

Figure 2. (**a–d**) Numerical simulations to model the cellular entry process including (**a**) initial jump into the channel; (**b**,**c**) cellular creep response and (**d**) creep response termination at the transitional position; (**e**) The proposed equivalent circuit model for a travelling cell within the constriction channel where C_{Ms}, C_{Mp1}, …, C_{Mpn} and R_{C1}, …, R_{Cn+1} represent distributed membrane capacitance and cytoplasm resistance with n divisions along the cell elongation direction. Note that R_{L1}, R_{L2}, …, R_{Ln}, R_{Ln+1} represent distributed leakage resistors, indicating the sealing properties of deformed cells with constriction channel walls.

2.4. Data Analysis

To measure the elongation length of a cell while it is travelling in the constriction channel, we used a previously developed background subtraction technique to process the images captured by the high-speed camera. This technique consists of a sequence of image processing steps adapted to the context of cell elongation (frame differencing, thresholding, particle removal using erosion, and edge detection along the channel) [21].

Based on the equivalent mechanical model in Section 2.3, raw data of cellular aspiration length were translated to intrinsic cellular mechanical parameter of $E_{instantaneous}$ and based on the equivalent electrical model in Section 2.3, raw data of two-frequency impedance values and cell elongation length during its travelling inside the constriction channel were translated to intrinsic cellular electrical parameter of $C_{specific\ membrane}$.

3. Results and Discussion

Figure 3a–d show microscopic pictures of a typical 95C cell's entry process into the constriction channel with raw impedance data of the same cell recorded at 1 kHz and 100 kHz (Figure 3e). Based on the quantified aspiration length (the distance between the leading tip of the cell and the beginning of the constriction channel) *vs.* time (see Figure 3f), the cellular entry process was divided into two stages.

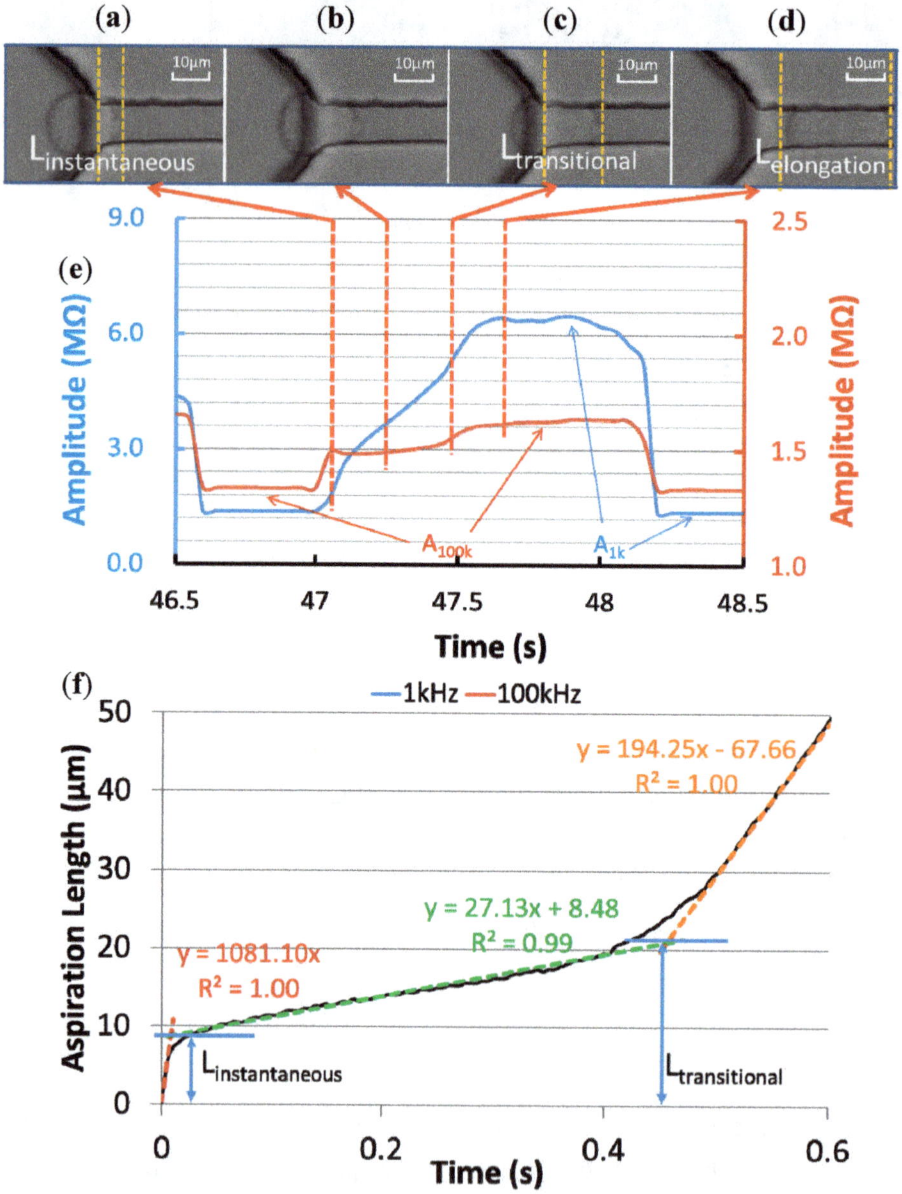

Figure 3. **(a–d)** Microscopic pictures of a 95C cell's entry process into the constriction channel with raw impedance data of the same cell shown in **(e)** and aspiration length *vs.* time shown in **(f)**. Based on raw data processing, four critical parameters $L_{instantaneous}$, $L_{transitional}$, $A_{1\ kHz}$ and $A_{100\ kHz}$ were obtained.

In stage I, an instantaneous jump into the constriction channel was observed (see Figure 3a), characterized by the instantaneous aspiration length ($L_{instantaneous}$) in Figure 3f. An instantaneous amplitude jump at 100 kHz and the initiation of a gradual amplitude increase at 1 kHz were also found in this stage. The 100 kHz electric lines were used to penetrate capacitive cellular membrane portions and in stage I its instantaneous increase in amplitude indicates the existence of an equivalent capacitor representing the leading membrane portion, which was sucked into the constriction channel instantly. The 1 kHz electric lines were used to pass around capacitive cellular membrane portions and in stage I, due to the limited aspiration length of the cell under measurement, no significant increase in amplitude at 1 kHz was observed.

After the instantaneous jump, a gradual increase in aspiration length (see Figure 3b,f) was observed due to the cellular creeping behaviour, and this was defined as stage II. During this stage, there is a gradual increase in amplitude at 1 kHz, confirming a gradual increase of the cellular aspiration length. On the other hand, no significant increase in amplitude at 100 kHz was observed, which is due to the fact that the area of the membrane portion sucked into the constriction channel was much smaller than the membrane portion left outside of the constriction channel and thus the contribution of the equivalent capacitor for the portion outside of the constriction channel was negligible.

With further increase in aspiration length while the cell continues to deform due to viscoelastic relaxation, a transitional position quantified as transitional aspiration length ($L_{transitional}$) was reached (see Figure 3c), which is the end of the creep stage and the cell rapidly enters the channel. From the perspective of impedance measurement, there is a slope (amplitude/time) increase at 1 kHz, confirming the initiation of accelerated cellular entry into the constriction channel.

In addition, a gradual increase in the impedance amplitude at 100 kHz was noticed for this transition status since as the cell started to enter the constriction channel rapidly, both membrane portions for the leading edge and the trailing edge make contribution to the impedance profiles at 100 kHz.

Figure 3d shows the microscopic picture of the cell travelling within the constriction channel and at this stage, stable impedance amplitudes at 1 kHz and 100 kHz were observed. Furthermore, two critical parameters $A_{1\text{ kHz}}$ (the amplitude ratio between the stable impedance amplitude at 1 kHz with travelling cells to the basal impedance amplitude at 1 kHz without cells) and $A_{100\text{ kHz}}$ (the amplitude ratio between the stable impedance amplitude at 100 kHz with travelling cells to the basal impedance amplitude at 100 kHz without cells) were calculated for cellular electrical property quantification.

Figure 4a,b show scatter plots of cellular mechanical data of $L_{instantaneous}/D_{channel}$ *vs.* D_{cell} (a) and $L_{transitional}/D_{channel}$ *vs.* D_{cell} (b), respectively, for 95C cells (n = 354). Quantified $L_{instantaneous}/D_{channel}$ and $L_{transitional}/D_{channel}$ were 0.76 ± 0.15 and 2.00 ± 0.40 where $D_{channel}$ represents the characteristic channel dimension (10 μm). It was noticed that with an increase in cell diameter, there is a corresponding increase for $L_{transitional}/D_{channel}$ where $L_{instantaneous}/D_{channel}$ is not dependent on D_{cell}. This observation was consistent with the previous study where $E_{instantaneous}$ played a major role in determining $L_{instantaneous}/D_{channel}$ while $L_{transitional}/D_{channel}$ is mainly determined by f_c and D_{cell} [12].

Figure 4c,d show scatter plots of cellular electrical data of $A_{1\text{ kHz}}$ *vs.* $L_{cell\ elongation}$ (c) and $A_{100\text{ kHz}}$ *vs.* $L_{cell\ elongation}$ (d), respectively, for 95C cells (n = 354). Quantified $A_{1\text{ kHz}}$ and $A_{100\text{ kHz}}$ were 3.92 ± 1.09 and 1.29 ± 0.04 where $L_{cell\ elongation}$ represents the cellular elongation length during its travelling process within the constriction channel as an indicator of cell sizes. Compared to $A_{1\text{ kHz}}$, $A_{100\text{ kHz}}$ was significantly lower since electric lines at 100 kHz effectively penetrate the cellular membrane portions while electric lines at 1 kHz always travel around capacitive cellular membrane portions. With the increase in $L_{cell\ elongation}$, a corresponding increase in $A_{1\text{ kHz}}$ was observed, further confirming the assumption that electric lines at 1 kHz pass around cells rather than penetrate cellular membrane portions.

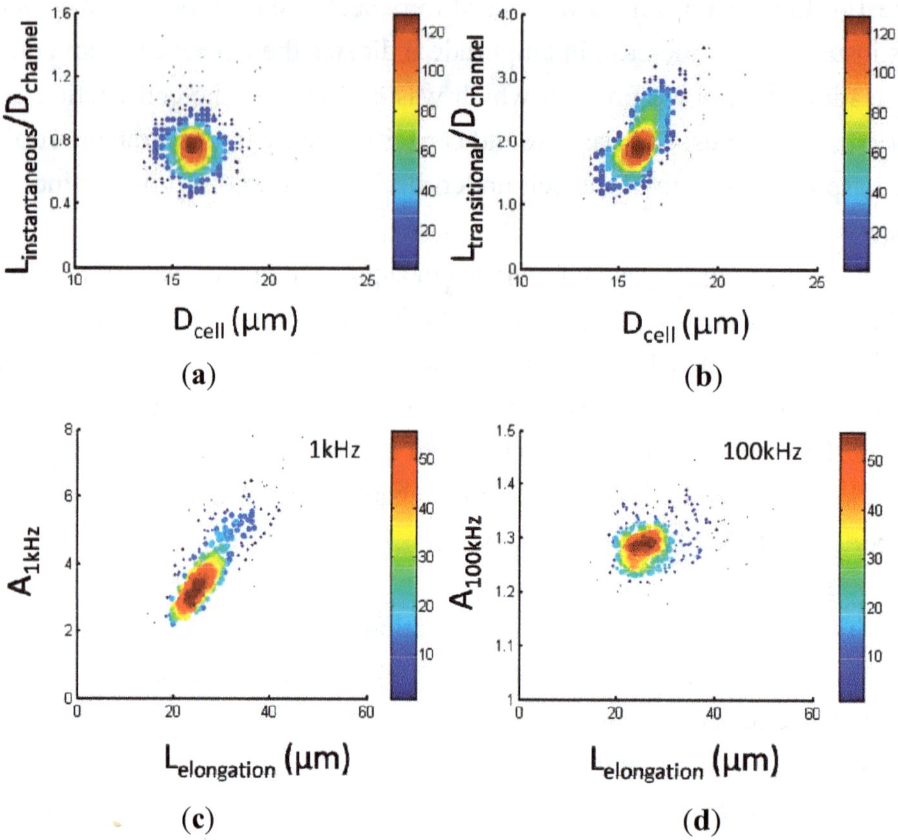

Figure 4. (**a–d**) Scatter plots of raw mechanical data of $L_{instantaneous}/D_{channel}$ *vs.* D_{cell} (**a**) and $L_{transitional}/D_{channel}$ *vs.* D_{cell} (**b**) as well as raw electrical data of A_{1kHz} *vs.* $L_{cell\ elongation}$ (**c**) and A_{100kHz} *vs.* $L_{cell\ elongation}$ (**d**) for 95C cells ($n = 354$).

Figure 5a shows the scatter plot of $C_{specific\ membrane}$ *vs.* $E_{instantaneous}$, which were quantified as 2.96 ± 0.40 kPa ($E_{instantaneous}$) *vs.* 1.59 ± 0.28 μF/cm^2 ($C_{specific\ membrane}$) for 95C cells (n = 354) (see Figure 5b). Figure 5c,d shows distribution percentages of $E_{instantaneous}$ and $C_{specific\ membrane}$ for 95C cells with the peak of $E_{instantaneous}$ within the range of 2.6–2.9 kPa and $C_{specific\ membrane}$ within the range of 1.4–1.6 μF/cm^2. These reported values were consistent with previously reported values [12,14], confirming that the proposed microfluidic platform is capable of characterizing cellular intrinsic mechanical and electrical markers simultaneously.

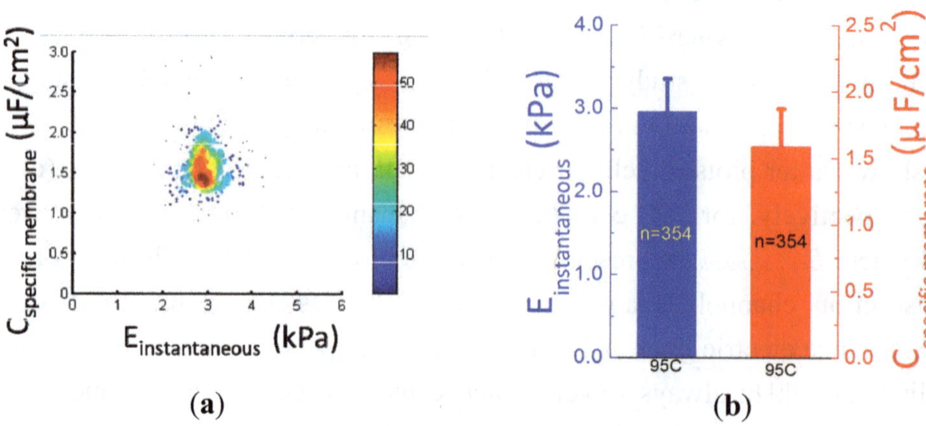

Figure 5. *Cont.*

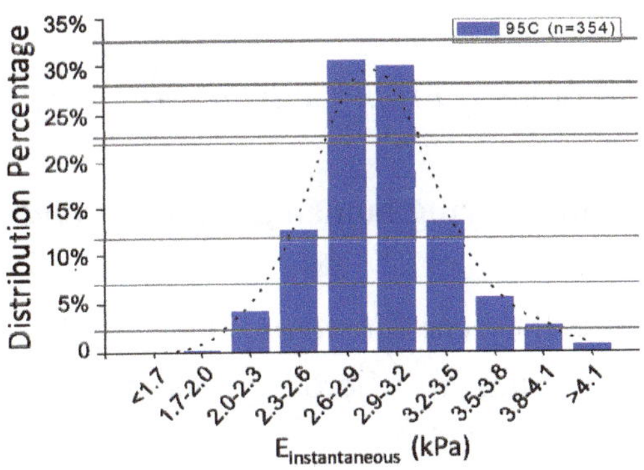

(c)

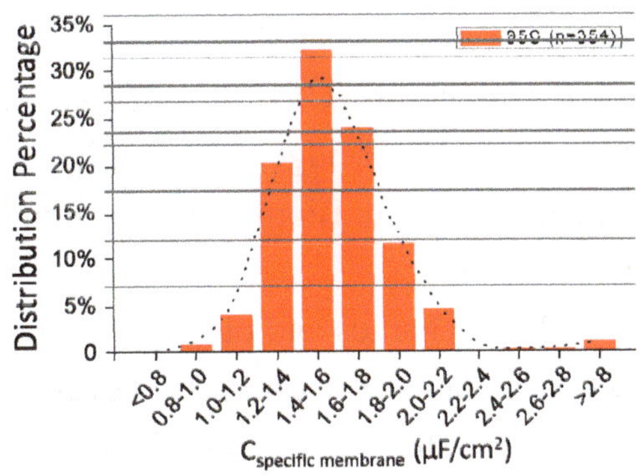

(d)

Figure 5. (**a**) A scatter plot of $C_{specific\ membrane}$ *vs.* $E_{instantaneous}$ for 95C cells (n = 354) with quantified $E_{instantaneous}$ of 2.96 ± 0.40 kPa and $C_{specific\ membrane}$ of 1.59 ± 0.28 μF/cm^2 shown in (**b**). (**c**) and (**d**) Distribution percentages of $E_{instantaneous}$ and $C_{specific\ membrane}$ for 95C cells with the peak of $E_{instantaneous}$ within the range of 2.6–2.9 kPa and $C_{specific\ membrane}$ within the range of 1.4–1.6 μF/cm^2.

4. Conclusions

This paper presented a microfluidic measurement system for characterizing $E_{instantaneous}$ and $C_{specific\ membrane}$ of single cells using constriction channel and impedance spectroscopy. Both microscopic images and impedance profiles confirmed the two-stage cellular entry process into the constriction channel, enabling the quantification of $L_{instantaneous}$ and $L_{transitional}$ as key mechanical parameters, which were then translated to $E_{instantaneous}$. Furthermore, based on the distributed equivalent circuit model for cellular travelling within the constriction channel, impedance profiles at two frequencies were obtained and translated to $C_{specific\ membrane}$. Compared to previous approaches, this technique is capable of characterizing both $E_{instantaneous}$ and $C_{specific\ membrane}$ in a continuous manner, which can potentially lead to a more complete understanding of cellular biophysical properties.

Acknowledgments

The authors would like to acknowledge financial support from National Natural Science Foundation of China (Grant No. 61201077), National Basic Research Program of China (973 Program, Grant No. 2014CB744600), National Natural Science Foundation of China (Grant No. 81261120561 and No. 61431019), National High Technology Research and Development Program of China (863 Program, Grant No. 2014AA093408), Instrument Development Program of Chinese Academy of Sciences (2015-2016) and Beijing NOVA Program.

Author Contributions

Y.Z., Y.L. and X.Z. designed and conducted experiments; F.C. and W.Y. analyzed the data; D.C. and R.L. contributed to equivalent electrical and mechanical models; M.J. conducted cell culture; J.W. and J.C. organized the experiments and drafted the paper.

References

1. Ethier, C.R.; Simmons, C.A. *Introductory Biomechanics: From Cells to Organisms*; Cambridge University Press: Cambridge, UK, 2007.
2. Lee, G.Y.H.; Lim, C.T. Biomechanics approaches to studying human diseases. *Trends Biotechnol.* **2007**, *25*, 111–118.
3. Suresh, S. Biomechanics and biophysics of cancer cells. *Acta Biomater.* **2007**, *3*, 413–438.
4. Kim, D.H.; Wong, P.K.; Park, J.; Levchenko, A.; Sun, Y. Microengineered platforms for cell mechanobiology. *Ann. Rev. Biomed. Eng.* **2009**, *11*, 203–233.
5. Valero, A.; Braschler, T.; Renaud, P. A unified approach to dielectric single cell analysis: Impedance and dielectrophoretic force spectroscopy. *Lab Chip* **2010**, *10*, 2216–2225.
6. Becker, F.F.; Wang, X.B.; Huang, Y.; Pethig, R.; Vykoukal, J.; Gascoyne, P.R.C. Separation of human breast-cancer cells from blood by differential dielectric affinity. *Proc. Natl. Acad. Sci. USA* **1995**, *92*, 860–864.
7. Cross, S.E.; Jin, Y.S.; Rao, J.; Gimzewski, J.K. Nanomechanical analysis of cells from cancer patients. *Nat. Nanotechnol.* **2007**, *2*, 780–783.
8. Mao, X.; Huang, T.J. Exploiting mechanical biomarkers in microfluidics. *Lab Chip* **2012**, *12*, 4006–4009.
9. Zheng, Y.; Sun, Y. Microfluidic devices for mechanical characterisation of single cells in suspension. *Micro Nano Lett.* **2011**, *6*, 327–331.
10. Cheung, K.C.; di Berardino, M.; Schade-Kampmann, G.; Hebeisen, M.; Pierzchalski, A.; Bocsi, J.; Mittag, A.; Tárnok, A. Microfluidic impedance-based flow cytometry. *Cytometry A.* **2010**, *77*, 648–666.
11. Yobas, L. Microsystems for cell-based electrophysiology. *J. Micromech. Microeng.* **2013**, *23*, 083002.
12. Luo, Y.; Chen, D.; Zhao, Y.; Wei, C.; Zhao, X.; Yue, W.; Long, R.; Wang, J.B.; Chen, J. A constriction channel based microfluidic system enabling continuous characterization of cellular instantaneous Young's modulus. *Sens. Actuators B Chem.* **2014**, *202*, 1183–1189.
13. Zhao, Y.; Chen, D.; Luo, Y.; Li, H.; Deng, B.; Huang, S.-B.; Chiu, T.-K.; Wu, M.-H.; Long, R.; Hu, H.; *et al.* A microfluidic system for cell type classification based on cellular size-independent electrical properties. *Lab Chip* **2013**, *13*, 2272–2277.
14. Zhao, Y.; Zhao, X.T.; Chen, D.Y.; Luo, Y.N.; Jiang, M.; Wei, C.; Long, R.; Yue, W.T.; Wang, J.B.; Chen, J. Tumor cell characterization and classification based on cellular specific membrane capacitance and cytoplasm conductivity. *Biosens. Bioelectron.* **2014**, *57*, 245–253.
15. Cho, Y.H.; Yamamoto, T.; Sakai, Y.; Fujii, T.; Kim, B. Development of microfluidic device for electrical/physical characterization of single cell. *J. Microelectromech. Syst.* **2006**, *15*, 287–295.
16. Chen, J.; Zheng, Y.; Tan, Q.; Zhang, Y.L.; Li, J.; Geddie, W.R.; Jewett, M.A.S.; Sun, Y. A microfluidic device for simultaneous electrical and mechanical measurements on single cells. *Biomicrofluidics* **2011**, *5*, 14113.
17. Chen, J.; Zheng, Y.; Tan, Q.; Shojaei-Baghini, E.; Zhang, Y.L.; Li, J.; Prasad, P.; You, L.; Wu, X.Y.; Sun, Y. Classification of cell types using a microfluidic device for mechanical and electrical measurement on single cells. *Lab Chip* **2011**, *11*, 3174–3181.

18. Zheng, Y.; Shojaei-Baghini, E.; Azad, A.; Wang, C.; Sun, Y. High-throughput biophysical measurement of human red blood cells. *Lab Chip* **2012**, *12*, 2560–2567.
19. Zheng, Y.; Nguyen, J.; Wei, Y.; Sun, Y. Recent advances in microfluidic techniques for single-cell biophysical characterization. *Lab Chip* **2013**, *13*, 2464–2483.
20. Zhao, Y.; Chen, D.; Li, H.; Luo, Y.; Deng, B.; Huang, S.B.; Chiu, T.-K.; Wu, M.-H.; Long, R.; Hu, H.; *et al.* A microfluidic system enabling continuous characterization of specific membrane capacitance and cytoplasm conductivity of single cells in suspension. *Biosens. Bioelectron.* **2013**, *43C*, 304–307.
21. Huang, S.B.; Zhao, Z.; Chen, D.Y.; Lee, H.C.; Luo, Y.N.; Chiu, T.K.; Wang, J.; Chen, J.; Wu, M.-H. A clogging-free microfluidic platform with an incorporated pneumatically-driven membrane-based active valve enabling specific membrane capacitance and cytoplasm conductivity characterization of single cells. *Sens. Actuators B Chem.* **2014**, *190*, 928–936.

4

Methods, Challenges and Potentials of Single Cell RNA-seq

Daniel Hebenstreit

The University of Warwick, School of Life Sciences, Coventry CV4 7AL, UK;
E-Mail: d.hebenstreit@warwick.ac.uk

Abstract: RNA-sequencing (RNA-seq) has become the tool of choice for transcriptomics. Several recent studies demonstrate its successful adaption to single cell analysis. This allows new biological insights into cell differentiation, cell-to-cell variation and gene regulation, and how these aspects depend on each other. Here, I review the current single cell RNA-seq (scRNA-seq) efforts and discuss experimental protocols, challenges and potentials.

Keywords: RNA-seq; scRNA-seq; single cell

---

1. Introduction—the Importance of Transcriptional Profiling at Single Cell Resolution

Transcriptome profiling has been a popular tool in molecular biological research for more than a decade. Mostly implemented by microarray technology, it has led to numerous insights and discoveries. These range from cell type specific single factors that were identified in differential expression screens to findings based on large portions of the entire transcriptome, such as disease signatures (e.g., [1]) and the interrelations of epigenetic modifications and gene expression (e.g., [2]).

Following its recent introduction, RNA-sequencing (RNA-seq) [3,4] is rapidly replacing microarrays as the method of choice for the aforementioned endeavors. Besides superior accuracy in the quantification of expression, RNA-seq offers other advantages, such as the possibility to detect novel transcripts, splice variants or allele-specific expression [5]. mRNA-profiling was performed on single cells early [6] and RNA-seq is following suit [7].

The analysis of the transcriptome in individual cells offers a number of advantages compared to cell-averaging experiments. Tissues or other cellular assemblages are heterogeneous even if a single, "traditional" cell type is concerned. This is particularly apparent from the study of immune cell types that are often defined based on the expression of surface markers. Improving experimental technologies reveal continuous expression ranges and fairly unrestricted combinatorial expression of surface markers [8]. This means that boundaries between cell types are blurred and that every individual cell is different. A similar picture emerges with tumors. While tumors have been known to be heterogeneous mixtures of cell types for a long time [9], pioneering studies demonstrate the potential of genome sequencing in individual cells [10]. RNA-seq will thus provide a powerful means to facilitate functional characterization of the differences among the cells in a tumor.

While tumor cell heterogeneity can be attributed to accumulating mutations, even genetically identical cells, under identical conditions, display high variability in their gene and protein expression levels. This is usually referred to as "noise", defined as variance or standard deviation over mean [11]. A number of studies have probed into the origins and mechanisms of noise and found it to be mostly due to the stochastic effects associated with the low numbers of involved molecules [12].

While standard microarray or RNA-seq experiments yield mean expression levels, distributions in single cells demonstrate that only a negligible portion of cells express mRNAs close to the actual mean levels. Depending on the skewness of the distributions, this could mean that cells that express certain mRNAs at "outlier" levels are functionally important, yet remain undetected by traditional experiments. Knowledge about the shape of the distributions can also be used to understand mechanisms that are involved in transcriptional regulation [13].

2. RNA-seq—Basic Experimental Procedures

RNA-seq ideally allows the accurate quantification of mRNA expression levels, covers the entire transcript lengths at equal representation at each position, and retains strand information. Depending on the experimental protocol, particularly for single cell RNA-seq (scRNA-seq), these goals are not simultaneously achievable with current methods.

Sample processing for RNA-seq is largely based on traditional molecular biological protocols that were developed decades ago for common assays such as RT-PCR, northern blots or microarrays. The vast majority of RNA-seq methods thus include the basic steps of poly-(A)+ RNA isolation, fragmentation, reverse transcription, and amplification before the actual sequencing takes place.

The selection of poly-(A)+ RNA is usually performed in order to suppress the "loss" of sequencing reads to structural RNAs such as rRNA and tRNA, which represent the bulk of cellular RNA. One disadvantage of this is that certain protein coding RNAs without the poly-(A) tail, such as replication dependent histones, are lost [14]. Total RNA based protocols do exist [15] and the technological advances in sequencing depth could soon render the removal of RNAs other than mRNA obsolete.

The fragmentation step is carried out in order to produce many short RNA or DNA fragments that represent the original transcript. Various methods, including enzymatical, mechanical, and chemical ones have been used successfully. Depending on the exact protocol, the fragmentation serves to optimize the length for the sequencing machine, generates more reads, thus allowing more accurate quantification, and/or is used to reduce positional biases in the transcript representation.

This bias is largely due to reverse transcription that is used to generate the DNA template for PCR or *in vitro* transcription (IVT), which are both popular means to amplify the material. Reverse transcription is necessary, since the suitability of RNA-dependent RNA polymerases [16] for RNA amplification is unclear. Depending on the type of primer that is used, the reverse transcriptase starts polymerizing internally or at the end of the RNA template. In both cases, a bias is induced that is affected by the processivity of the enzyme. The bias induced by internal random priming worsens with a lower enzyme drop-off (although the total yield gets higher), while the opposite is seen when a fixed primer position (e.g., at the transcript 3' end) is used. In the latter case, a roughly exponentially decreasing coverage is expected [17], which is observed experimentally [18].

These biases can be strongly reduced if the fragmentation step is performed on the original RNA [19]. This may not be desirable, however, as it seems likely to damage the RNA and lead to the loss of some of it. Amplification-free RNA-seq might abrogate the need for reverse transcription and thus remove the aforementioned biases (in addition to PCR biases). One effort to that end has been reported as FRT-seq [20], which has not been tested for scRNA-seq.

3. ScRNA-seq Strategies

Four different scRNA-seq strategies have been explored so far (Figure 1). The first study appeared in 2009 [21]. The authors performed reverse transcription directly in the cell lysate primed from a poly-(dT) oligonucleotide containing an anchor sequence on its 3' end. Avoiding a random primer makes it unnecessary to purify poly-(A)+ RNA. Excess primers are digested with exonuclease after which terminal transferase is used to attach poly-A tails to the 3' ends of the first strands. Another poly-(dT) primer with a different anchor is annealed to the poly-(dA) stretch and is used to initiate second strand synthesis. The library is then PCR-amplified from the two anchor sequences, fragmented, and subjected to the sample preparation specific to the sequencing machine. Importantly, here the second strand synthesis begins from wherever the first strand synthesis stops, introducing a strong bias depending on the drop-off of the reverse transcriptase. Fragmentation after PCR further means that the strand information is lost.

Two other scRNA-seq approaches are similar in their usage of the "template-switching" method. This method, also marketed as "Switching mechanism at the 5' end of the RNA transcript" (SMART) [22] exploits the intrinsic property of the Moloney murine leukemia virus (MMLV) reverse transcriptase to add a few non-templated nucleotides (mainly cytosines) to the 3' end of the cDNA first strand [23]. An oligonucleotide with $(rG)_3$ on its 3' end (and a suitable 5' sequence that will later serve as primer binding site for PCR amplification) anneals to these nucleotides as soon as the first strand is complete, and provides a template-extension. The RT thus switches from an RNA- to a DNA-template (Figure 1). Importantly, addition of the non-templated nucleotides is most efficient at the 5' end of the RNA transcript [22]. Thus, incomplete first-strands are selected against, in contrast to the terminal transferase-based scRNA-seq.

The first SMART-based scRNA-seq protocol was named "Single-cell tagged reverse transcription" (STRT) [24]. The template-switching oligonucleotides feature a barcoding sequence. Samples can thus be pooled after first strand synthesis and can be sequenced in a multiplex reaction. Since the barcode is located at the 5' ends of the transcripts, only this part will be sequenced after fragmentation (and a purification step). This means that there will be a maximum of one sequencing read for each original transcript, which ideally maps to the corresponding transcriptional start sites.

Figure 1. Overview of four scRNA-seq methods. The scheme is a simplification and the focus is the differences between the protocols and the steps that are characteristic to each. The Tang protocol [21,25] includes attachment of adenine nucleotides to the 3' ends of first strands which serve as binding sites for an anchor oligonucleotide, and the RNA and 1st strand primers are removed enzymatically (black crosses). STRT and CEL-seq [23,24,26] include sample pooling based on first strands with barcodes at 5' and 3' regions, respectively, with regards to the original transcripts. CEL-seq uses IVT as first amplification step, which requires direct RNA adapter ligation followed by a second reverse transcription during sequencing sample preparation. Smart-Seq [27] is a simplification of STRT without barcoding. All PCRs are primed from sequences introduced by the flanking oligonucleotides. Arrows indicate primer-initiated DNA polymerization. Nucleotide stretches (such as AAA or TTT) are for illustration only and do not reflect the actual number of nucleotides. The color-coding was chosen to roughly discriminate between the crucial 1st strand-related steps (light blue), protocol-specific steps (dark blue), and the more interchangeable sequencing sample preparation (burgundy).

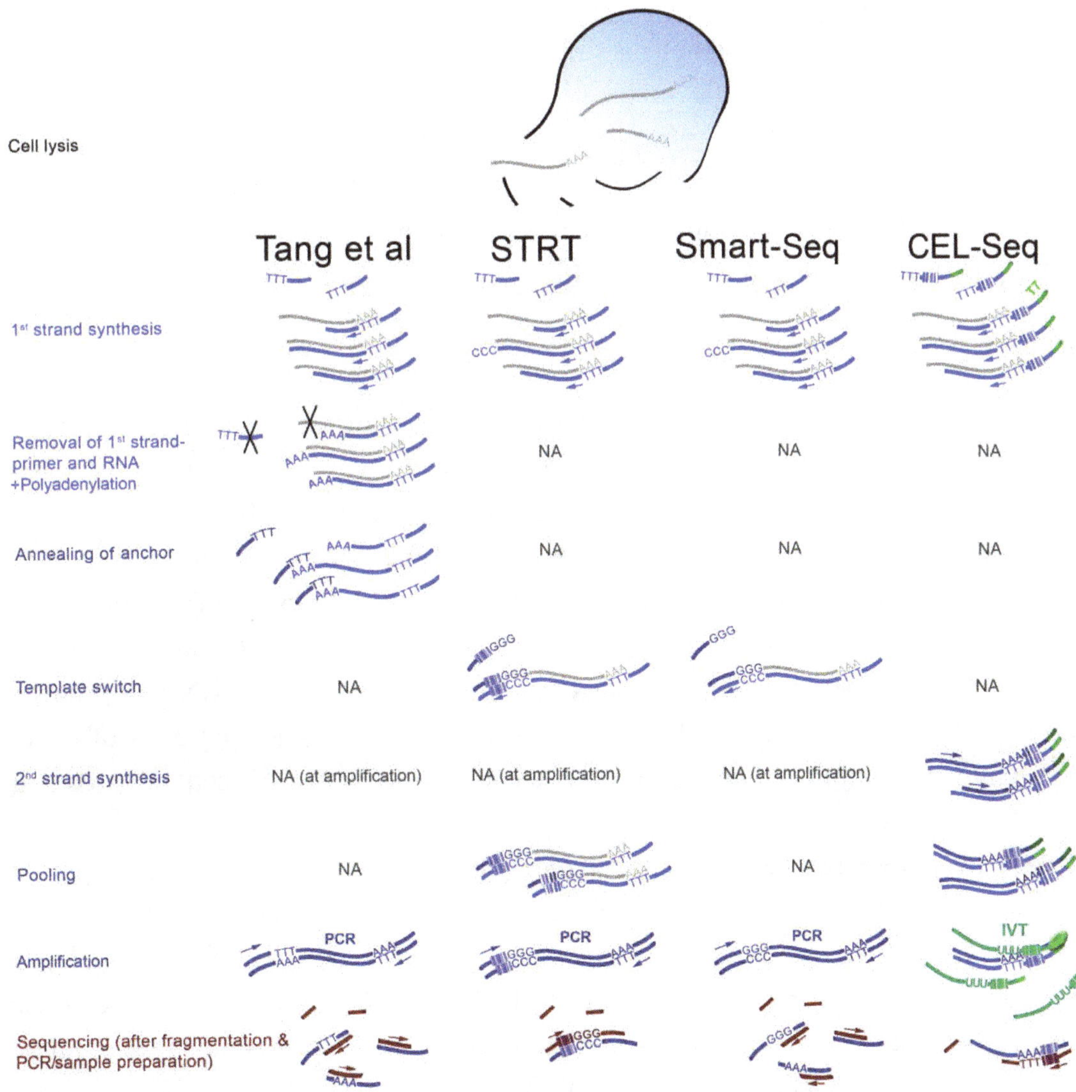

Although it was published later, the second SMART-based scRNA-seq, "Smart-Seq", is essentially a simplified version of the first one [27]. No barcoding was used, and fragments were not purified according to their position or orientation. Similar to the Tang protocol [21], oligo-(dT) priming was used directly in the cell lysate to circumvent purification of poly-(A)+ RNA. Although strand-specificity was thus lost, the authors report a markedly improved coverage across the transcript lengths. This protocol, in slightly altered form, was successfully implemented to profile the transcriptome of single neurons [28].

CEL-seq provides the latest addition among the tested scRNA-seq protocols [26]. This strategy is unique in relying on IVT instead of PCR for sample enrichment. The IVT is initiated from a T7 promoter that is introduced with a barcoded oligo-(dT) 1st strand primer. Its advantage lies in the linear mode of amplification, which, unlike PCR, does not exponentially deplete of sequences that are difficult to process [29]. The ability of T7 polymerase to repeatedly bind to the promoter and polymerize means that substantial amplification is achieved in the same reaction, without the need for denaturing and primer re-annealing. CEL-seq is demonstrated to outperform STRT in sensitivity and accuracy [26]. Curiously, the protocol converts the mRNA to DNA, to RNA, and again to DNA (Figure 1).

Common to all four scRNA-seq methods is the usage of two rounds of amplification (including the sequencing technology-specific sample preparation). It should be further noted that the SMART protocol does not lead to better transcript coverage in the sense that the full-length cDNA is synthesized more often. Instead, incomplete cDNAs are less likely to be carried on in the protocol. The downside of this is that transcripts, which could at least have partially contributed to the information gained by the experiment, are discarded.

4. Findings from the First scRNA-seq Studies

It is difficult to directly compare the technical performance of the existing scRNA-seq schemes based on their original descriptions, aside from the obvious differences such as strand specificity or transcript coverage. The means to assess performance were largely different in the three works (comparison with microarrays, spike-in controls, titration of the RNA starting mass, correlation with standard RNA-seq, *etc.*) and differences in parameters such as PCR cycle numbers might have strong effects on the final results. A specific effort to evaluate all four methods against each other under identical conditions might be desirable, as it has been done for conventional RNA-seq protocols [30]. To a certain degree, the different methods are modular and allow recombining parts of the protocols. For instance, template switching can be used for 1st strand synthesis in CEL-seq, or barcoding may be used with the Tang protocol.

The potency of scRNA-seq to reveal novel biological insights is explored in the original and follow-up studies of these. One widely used advantage of RNA-seq over microarrays is the possibility to detect splice variants, which has been demonstrated on single cell basis for all three PCR-based scRNA-seq protocols. The Smart-Seq and STRT works further illustrate how global gene expression profiles can be used to categorize cells into traditional cell types, while individual cells might exhibit closer relatedness to cell types other than the one they are classified as [24,27]. The usefulness of CEL-seq is demonstrated for cell lineage analysis in *C. Elegans* [26]. Since the exact number of cells is known with certainty in scRNA-seq experiments, spike-in controls can be used to estimate the absolute levels of mRNA expression. Islam *et al.* use this to reveal a 20-fold higher RNA-content in mouse embryonic fibroblasts (MEFs) compared to embryonic stem cells (ESCs) [24]. In a first application of their scRNA-seq protocol, Tang et al. investigated the transition of inner cell mass (ICM) cells to ESCs in mouse embryos [31]. This revealed a high cell-to-cell variation for genes expressed at intermediate levels, suggesting that they have a higher propensity for dynamic transcriptional regulation among cells of the same traditional cell type. In addition, many probable target genes of pluripotency factors such as Nanog, Sox2, Oct4, or Esrrb appeared to be regulated in a cell autonomous manner [31]. This demonstrates the usefulness of scRNA-seq and the novel biological insights it generates.

5. Outlook—Challenges, Future Issues and Emerging Technologies

The first few studies on scRNA-seq have proven the feasibility of this type of experimental assay. The true strength of the method will probably become fully apparent when it is implemented to study the characteristics of distributions of mRNA expression levels among individual cells. This is likely to yield decisive new insights into transcriptional regulation [13].

To fully exploit this potential, several requirements must be met. Precise parameter estimation for mathematical models of transcriptional regulation is usually based on single-molecule fluorescent *in situ* hybridization (smFISH) of at least a few hundred cells each under different conditions (e.g., [32]). This necessitates an upscaling of scRNA-seq, which will be probably most easily implemented by multiplexing as in STRT and CEL-seq. A further issue that is currently unsolved is the quantitative accuracy of scRNA-seq. Spike-in controls will help to understand how sequencing read numbers translate into the original numbers of transcripts. Yet, this does not appear to be a trivial issue, as synthetic RNA seems to behave differently from endogenous RNA in the STRT protocol, for instance [24]. Approaches such as smFISH or NanoString technology provide means to calibrate scRNA-seq based on absolute numbers of endogenous RNA [33,34]. To achieve accurate calibration, this will have to be done for a large number of different types/lengths of mRNA, though.

One means to monitor amplification bias is to introduce a different type of barcode that identifies individual molecules [35]. Multiple occurrences of the same barcode indicate that the corresponding sequenced fragments were derived from only one original molecule. This has highlighted the biases introduced by PCR and has been shown to increase RNA-seq accuracy [36,37]. It has been used only downstream of cDNA production though, and does not protect against complete loss of molecules. It will thus be further beneficial to have mathematical models that describe the sample preparation process, as it is clear that the SMART protocols deplete of incomplete first strands. In addition, CEL-seq probably induces non-trivial biases during the multiple strand conversions. These effects have to be taken account of and thus precisely estimated, if the number of sequencing reads is to be used to determine the exact original transcript numbers.

Another promising strategy is clearly to minimize the number of processing steps. An effort in this vein is direct sequencing technology, as implemented by the Helicos system. This allows sequencing of cDNAs without sample preparation [38] and makes it possible to abolish reverse transcription altogether by direct sequencing of RNA [39,40]. One limitation is the lower number of reads that are currently obtained with this technology [5]. Other upcoming technologies such as the method developed by Pacific Biosciences enable single molecule sequencing and are potentially adaptable for single cell experiments [41,42].

Particularly promising are nanopore sequencing technologies, which represent an entirely new approach. Here, the DNA or RNA strand is threaded through (biological or solid-state) pores of nanometer dimension that are inserted into membranes. Bases that pass through the pores are detected electronically or optically [43].

Further help for sample preparation will be microfluidics systems such as provided by RainDance Technologies or Fluidigm. These have been mainly designed for targeted amplification and highly parallelized PCRs (e.g., [44]), but have considerable potential for adaption for scRNA-seq. Fluidigm notably offers a machine ("C1") that captures single cells and performs reverse transcription and amplification.

Acknowledgments

I would like to thank Arthur Wuster for reading the manuscript and making helpful suggestions.

References and Notes

1. McKinney, E.F.; Lyons, P.A.; Carr, E.J.; Hollis, J.L.; Jayne, D.R.; Willcocks, L.C.; Koukoulaki, M.; Brazma, A.; Jovanovic, V.; Kemeny, D.M.; *et al.* A cd8+ t cell transcription signature predicts prognosis in autoimmune disease. *Nat. Med.* **2010**, *16*, 586–591, 581 following 591.
2. Wei, G.; Wei, L.; Zhu, J.; Zang, C.; Hu-Li, J.; Yao, Z.; Cui, K.; Kanno, Y.; Roh, T.Y.; Watford, W.T.; *et al.* Global mapping of h3k4me3 and h3k27me3 reveals specificity and plasticity in lineage fate determination of differentiating cd4+ t cells. *Immunity* **2009**, *30*, 155–167.
3. Morin, R.; Bainbridge, M.; Fejes, A.; Hirst, M.; Krzywinski, M.; Pugh, T.; McDonald, H.; Varhol, R.; Jones, S.; Marra, M. Profiling the hela s3 transcriptome using randomly primed cdna and massively parallel short-read sequencing. *Biotechniques* **2008**, *45*, 81–94.
4. Sultan, M.; Schulz, M.H.; Richard, H.; Magen, A.; Klingenhoff, A.; Scherf, M.; Seifert, M.; Borodina, T.; Soldatov, A.; Parkhomchuk, D.; *et al.* A global view of gene activity and alternative splicing by deep sequencing of the human transcriptome. *Science* **2008**, *321*, 956–960.
5. Ozsolak, F.; Milos, P.M. Rna sequencing: Advances, challenges and opportunities. *Nat. Rev. Genet.* **2011**, *12*, 87–98.
6. Eberwine, J.; Yeh, H.; Miyashiro, K.; Cao, Y.; Nair, S.; Finnell, R.; Zettel, M.; Coleman, P. Analysis of gene expression in single live neurons. *Proc. Natl. Acad. Sci. USA* **1992**, *89*, 3010–3014.
7. Tang, F.; Lao, K.; Surani, M.A. Development and applications of single-cell transcriptome analysis. *Nat. Methods* **2011**, *8*, S6–S11.
8. Bendall, S.C.; Simonds, E.F.; Qiu, P.; Amir el, A.D.; Krutzik, P.O.; Finck, R.; Bruggner, R.V.; Melamed, R.; Trejo, A.; Ornatsky, O.I.; *et al.* Single-cell mass cytometry of differential immune and drug responses across a human hematopoietic continuum. *Science* **2011**, *332*, 687–696.
9. Shackleton, M.; Quintana, E.; Fearon, E.R.; Morrison, S.J. Heterogeneity in cancer: Cancer stem cells versus clonal evolution. *Cell* **2009**, *138*, 822–829.

10. Navin, N.; Kendall, J.; Troge, J.; Andrews, P.; Rodgers, L.; McIndoo, J.; Cook, K.; Stepansky, A.; Levy, D.; Esposito, D.; *et al.* Tumour evolution inferred by single-cell sequencing. *Nature* **2011**, *472*, 90–94.
11. Paulsson, J. Models of stochastic gene expression. *Phys. Life Rev.* **2005**, *2*, 157–175.
12. Maheshri, N.; O'Shea, E.K. Living with noisy genes: How cells function reliably with inherent variability in gene expression. *Annu. Rev. Biophys. Biomol. Struct.* **2007**, *36*, 413–434.
13. Munsky, B.; Neuert, G.; van Oudenaarden, A. Using gene expression noise to understand gene regulation. *Science* **2012**, *336*, 183–187.
14. Marzluff, W.F.; Wagner, E.J.; Duronio, R.J. Metabolism and regulation of canonical histone mrnas: Life without a poly(a) tail. *Nat. Rev. Genet.* **2008**, *9*, 843–854.
15. Tariq, M.A.; Kim, H.J.; Jejelowo, O.; Pourmand, N. Whole-transcriptome rnaseq analysis from minute amount of total rna. *Nucleic Acids Res.* **2011**, *39*, e120.
16. Ortin, J.; Parra, F. Structure and function of rna replication. *Annu. Rev. Microbiol.* **2006**, *60*, 305–326.
17. Von Hippel, P.H.; Fairfield, F.R.; Dolejsi, M.K. On the processivity of polymerases. *Ann. NY Acad. Sci.* **1994**, *726*, 118–131.
18. Fox-Walsh, K.; Davis-Turak, J.; Zhou, Y.; Li, H.; Fu, X.D. A multiplex rna-seq strategy to profile poly(a+) rna: Application to analysis of transcription response and 3' end formation. *Genomics* **2011**, *98*, 266–271.
19. Mortazavi, A.; Williams, B.A.; McCue, K.; Schaeffer, L.; Wold, B. Mapping and quantifying mammalian transcriptomes by rna-seq. *Nat. Methods* **2008**, *5*, 621–628.
20. Mamanova, L.; Andrews, R.M.; James, K.D.; Sheridan, E.M.; Ellis, P.D.; Langford, C.F.; Ost, T.W.; Collins, J.E.; Turner, D.J. Frt-seq: Amplification-free, strand-specific transcriptome sequencing. *Nat. Methods* **2010**, *7*, 130–132.
21. Tang, F.; Barbacioru, C.; Wang, Y.; Nordman, E.; Lee, C.; Xu, N.; Wang, X.; Bodeau, J.; Tuch, B.B.; Siddiqui, A.; *et al.* Mrna-seq whole-transcriptome analysis of a single cell. *Nat. Methods* **2009**, *6*, 377–382.
22. Zhu, Y.Y.; Machleder, E.M.; Chenchik, A.; Li, R.; Siebert, P.D. Reverse transcriptase template switching: A smart approach for full-length cdna library construction. *Biotechniques* **2001**, *30*, 892–897.
23. Islam, S.; Kjallquist, U.; Moliner, A.; Zajac, P.; Fan, J.B.; Lonnerberg, P.; Linnarsson, S. Highly multiplexed and strand-specific single-cell rna 5' end sequencing. *Nat. Protoc.* **2012**, *7*, 813–828.
24. Islam, S.; Kjallquist, U.; Moliner, A.; Zajac, P.; Fan, J.B.; Lonnerberg, P.; Linnarsson, S. Characterization of the single-cell transcriptional landscape by highly multiplex rna-seq. *Genome Res.* **2011**, *21*, 1160–1167.
25. Tang, F.; Barbacioru, C.; Nordman, E.; Li, B.; Xu, N.; Bashkirov, V.I.; Lao, K.; Surani, M.A. Rna-seq analysis to capture the transcriptome landscape of a single cell. *Nat. Protoc.* **2010**, *5*, 516–535.
26. Hashimshony, T.; Wagner, F.; Sher, N.; Yanai, I. Cel-seq: Single-cell rna-seq by multiplexed linear amplification. *Cell Rep.* **2012**, *2*, 666–673.

27. Ramskold, D.; Luo, S.; Wang, Y.C.; Li, R.; Deng, Q.; Faridani, O.R.; Daniels, G.A.; Khrebtukova, I.; Loring, J.F.; Laurent, L.C.; *et al.* Full-length mrna-seq from single-cell levels of rna and individual circulating tumor cells. *Nat. Biotechnol.* **2012**, *30*, 777–782.
28. Qiu, S.; Luo, S.; Evgrafov, O.; Li, R.; Schroth, G.P.; Levitt, P.; Knowles, J.A.; Wang, K. Single-neuron rna-seq: Technical feasibility and reproducibility. *Frontiers Genet.* **2012**, *3*, 124.
29. RNA Sample Amplification. Available online: http://www.ipc.nxgenomics.org/newsletter/no7.htm (accessed on 11 October 2012).
30. Levin, J.Z.; Yassour, M.; Adiconis, X.; Nusbaum, C.; Thompson, D.A.; Friedman, N.; Gnirke, A.; Regev, A. Comprehensive comparative analysis of strand-specific rna sequencing methods. *Nat. Methods* **2010**, *7*, 709–715.
31. Tang, F.; Barbacioru, C.; Bao, S.; Lee, C.; Nordman, E.; Wang, X.; Lao, K.; Surani, M.A. Tracing the derivation of embryonic stem cells from the inner cell mass by single-cell rna-seq analysis. *Cell Stem Cell* **2010**, *6*, 468–478.
32. So, L.H.; Ghosh, A.; Zong, C.; Sepulveda, L.A.; Segev, R.; Golding, I. General properties of transcriptional time series in escherichia coli. *Nat. Genet.* **2011**, *43*, 554–560.
33. Hebenstreit, D.; Fang, M.; Gu, M.; Charoensawan, V.; van Oudenaarden, A.; Teichmann, S.A. RNA sequencing reveals two major classes of gene expression levels in metazoan cells. *Mol. Syst. Biol.* **2011**, *7*, 497.
34. Geiss, G.K.; Bumgarner, R.E.; Birditt, B.; Dahl, T.; Dowidar, N.; Dunaway, D.L.; Fell, H.P.; Ferree, S.; George, R.D.; Grogan, T.; *et al.* Direct multiplexed measurement of gene expression with color-coded probe pairs. *Nat. Biotechnol.* **2008**, *26*, 317–325.
35. Fu, G.K.; Hu, J.; Wang, P.H.; Fodor, S.P. Counting individual DNA molecules by the stochastic attachment of diverse labels. *Proc. Natl. Acad. Sci. USA* **2011**, *108*, 9026–9031.
36. Kivioja, T.; Vaharautio, A.; Karlsson, K.; Bonke, M.; Enge, M.; Linnarsson, S.; Taipale, J. Counting absolute numbers of molecules using unique molecular identifiers. *Nat. Methods* **2012**, *9*, 72–74.
37. Shiroguchi, K.; Jia, T.Z.; Sims, P.A.; Xie, X.S. Digital rna sequencing minimizes sequence-dependent bias and amplification noise with optimized single-molecule barcodes. *Proc. Natl. Acad. Sci. USA* **2012**, *109*, 1347–1352.
38. Lipson, D.; Raz, T.; Kieu, A.; Jones, D.R.; Giladi, E.; Thayer, E.; Thompson, J.F.; Letovsky, S.; Milos, P.; Causey, M. Quantification of the yeast transcriptome by single-molecule sequencing. *Nat. Biotechnol.* **2009**, *27*, 652–658.
39. Ozsolak, F.; Platt, A.R.; Jones, D.R.; Reifenberger, J.G.; Sass, L.E.; McInerney, P.; Thompson, J.F.; Bowers, J.; Jarosz, M.; Milos, P.M. Direct rna sequencing. *Nature* **2009**, *461*, 814–818.
40. Ozsolak, F.; Milos, P.M. Single-molecule direct rna sequencing without cdna synthesis. *Wiley Interdiscip. Rev. RNA* **2011**, *2*, 565–570.
41. Eid, J.; Fehr, A.; Gray, J.; Luong, K.; Lyle, J.; Otto, G.; Peluso, P.; Rank, D.; Baybayan, P.; Bettman, B.; *et al.* Real-time DNA sequencing from single polymerase molecules. *Science* **2009**, *323*, 133–138.
42. Uemura, S.; Aitken, C.E.; Korlach, J.; Flusberg, B.A.; Turner, S.W.; Puglisi, J.D. Real-time trna transit on single translating ribosomes at codon resolution. *Nature* **2010**, *464*, 1012–1017.

43. Niedringhaus, T.P.; Milanova, D.; Kerby, M.B.; Snyder, M.P.; Barron, A.E. Landscape of next-generation sequencing technologies. *Anal. Chem.* **2011**, *83*, 4327–4341.
44. Buganim, Y.; Faddah, D.A.; Cheng, A.W.; Itskovich, E.; Markoulaki, S.; Ganz, K.; Klemm, S.L.; van Oudenaarden, A.; Jaenisch, R. Single-cell expression analyses during cellular reprogramming reveal an early stochastic and a late hierarchic phase. *Cell* **2012**, *150*, 1209–1222.

5

Protein Expression Analyses at the Single Cell Level

Masae Ohno, Peter Karagiannis and Yuichi Taniguchi *

Laboratory for Single Cell Gene Dynamics, Quantitative Biology Center, RIKEN Address, 6-2-3 Furuedai, Suita, Osaka 565-0874, Japan

* Author to whom correspondence should be addressed; E-Mail: taniguchi@riken.jp

---

Abstract: The central dogma of molecular biology explains how genetic information is converted into its end product, proteins, which are responsible for the phenotypic state of the cell. Along with the protein type, the phenotypic state depends on the protein copy number. Therefore, quantification of the protein expression in a single cell is critical for quantitative characterization of the phenotypic states. Protein expression is typically a dynamic and stochastic phenomenon that cannot be well described by standard experimental methods. As an alternative, fluorescence imaging is being explored for the study of protein expression, because of its high sensitivity and high throughput. Here we review key recent progresses in fluorescence imaging-based methods and discuss their application to proteome analysis at the single cell level.

Keywords: single cell analysis; protein expression; stochastic gene expression; burst; single molecule fluorescence microscopy; proteome analysis

---

1. Introduction

Gene expression describes how the genome sequence is decoded into proteins via an intermediate product, mRNA. Despite the relationship between genes, RNA, and proteins, their copy numbers show significant differences. The copy number of the genome in a single cell is constant: at the G1 phase, one haploid has one copy and one diploid has two. In contrast, the copy numbers of mRNAs and proteins fluctuate with time and vary from cell to cell, including even isogenic cell populations, owing to stochastic molecular reactions during gene expressions, such as the stochastic binding of RNA polymerases to the genome and ribosomes to the mRNA [1]. Thus, a single genome sequence can cause a variety of cellular phenotypes by generating different mRNA and protein copy numbers. Although a stochastic mechanism for gene expression may intuitively seem detrimental to cell stability, this property is thought crucial for many biological processes, such as differentiation and evolution [2,3]. One approach to studying this stochasticity and its impact on cell function is single cell gene expression analysis methods.

Because of their numbers, quantifying how many molecules are in a single cell is a tremendous challenge. Single cell transcriptome analysis techniques based on mRNA sequencing and cDNA microarray have made great inroads on this goal [4,5]. Single cell proteome analysis, however, has proven more difficult. Nevertheless, increasing improvements in the sensitivity of techniques such as mass spectroscopy have made this task more achievable [6,7].

Another approach gaining popularity for the characterization of protein expressions in a single cell is fluorescence imaging [1,8–12]. Here, the protein amount can be quantified by imaging proteins that are efficiently labeled with a fluorescent protein. Fluorescence-based methods enable time-lapse measurements [8], because it is not necessary to decompose the observed cell for analysis, which is unlike the case in mass spectroscopy. In addition, fluorescence imaging can measure the localization of the proteins with high resolution at the micrometer level and realize single molecule sensitivity [9,10]. Thus, it offers molecular counting in a single cell with precise quantification of the stochastic gene expression. Currently, the number of protein types that can be imaged at the same time is limited, but several methods for system-wide characterization have been proposed [11,12]. In this review, we highlight recent methods for single cell protein expression analysis based on fluorescence imaging.

2. Significance of Single Cell Protein Expression Analyses

For genome-wide analyses in a single-cell, gene expression is routinely determined by measuring mRNA. Recent advances in amplification techniques have greatly aided the detection and quantification of mRNA compared with proteins. Because protein levels are predominantly determined by corresponding mRNA levels, mRNA measuring techniques have been used to estimate protein levels. In fact, this correlation is common across taxa, from bacteria to mammals [13].

However, at the same time, several reports have suggested that the protein copy number is determined more by translational control than by transcriptional control, arguing these indirect observations of protein abundance can be misleading. Schwanhäusser *et al.*, calculated the copy numbers and half-lives of mRNA and proteins for over 5000 genes simultaneously in mouse fibroblast cells by pulse labelling followed by mass spectrometry and RNA-sequencing [14]. By combining the measured concentrations and degradation rates of the mRNA and proteins into a mathematical model, they predicted the rates of transcription and translation. The authors concluded that the protein level was controlled more by the translation rate than the transcription rate, demonstrating the importance of measuring not just the transcriptome, but also the proteome when studying cell function.

Accumulating data also suggest that protein concentrations generally cover a higher dynamic range than mRNA concentrations across organisms (Figure 1a). For example, in *E. coli*, protein copy numbers range from 10^{-1} to 10^{4} per cell, while mRNA copy numbers range far less, from 10^{-3} to 10^{1} per cell [12]. In baker's yeast, cells were seen to contain 10^{1} to 10^{5} copies of protein, but approximately only 10^{-1} to

10^2 copies of mRNA [15]. Similarly, in mouse fibroblasts, the copy numbers of protein and mRNA were measured as 10^2 to 10^8 and 1 to 10^4, respectively [14]. The difference in the dynamic range of protein and mRNA copy numbers indicates there exists an amplification step post-transcription.

Figure 1. (**a**) Relationship between proteins and mRNA levels across organisms. Data for *E. coli*, *S. cerevisiae*, and mouse fibroblast were adapted from references 12, 15, and 14, respectively. (**b**) Correlation plot of mRNA and protein levels of a TufA gene tagged with YFP in single *E. coli* cells. (**c**) Among the 129 genes that were high copy and analyzed, no correlation was observed between mRNA and protein levels in a single *E. coli* cell. Data taken from [12].

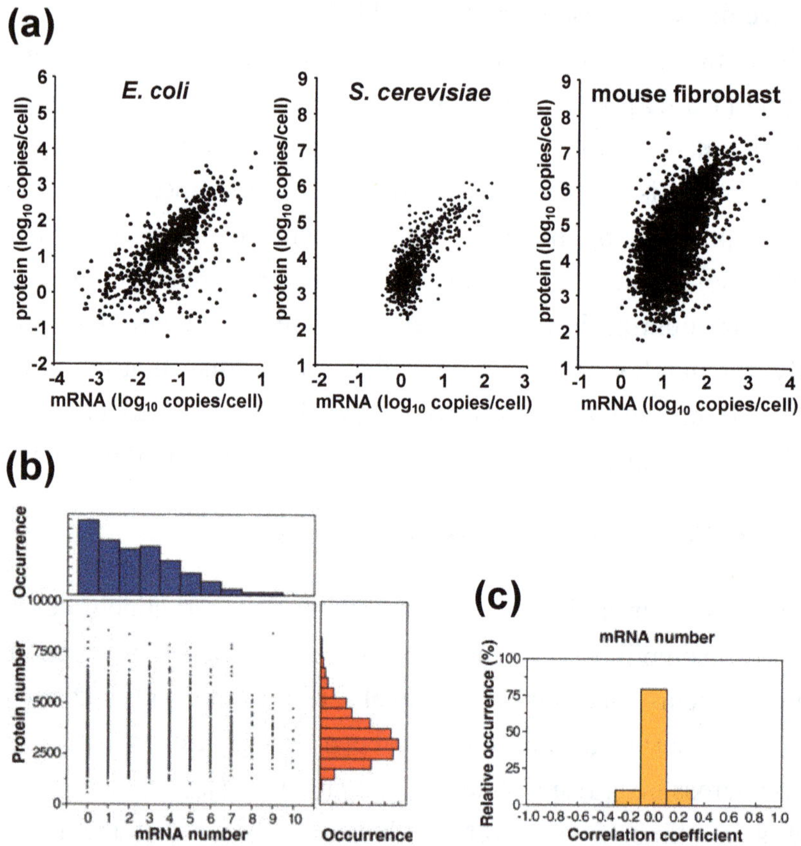

Those studies compared protein and mRNA abundances from the same gene in the same species. Others have examined orthologs across diverse species. Comparisons of nematode and fly showed that the protein abundances of orthologous genes correlate better than do the protein abundance and mRNA abundance within the same organism or the mRNA abundances across organisms [16]. Similar results have been obtained by comparing the protein and mRNA abundances of orthologs across seven organisms that range from bacteria to human [17]. In addition, a recent report indicated that the protein levels in the lymphoblastoid cells of three different primates (human, chimpanzee, and rhesus macaque) are more conserved than are the mRNA levels [18]. That report further suggested that protein expression levels evolve under greater constraints than respective mRNA levels.

The above studies measured mRNA and protein abundances using bulk measurements. On the other hand, Taniguchi *et al.* quantified the abundances simultaneously in a single-cell. They conducted fluorescence *in situ* hybridization (to detect mRNA) in combination with single molecule fluorescence microscopy to measure an *E. coli* fluorescent-protein tagged protein library. They measured 129 highly expressed genes (an example of a correlation plot is shown in Figure 1b) and showed that protein and mRNA copy numbers are uncorrelated at the single-cell level (Figure 1c) [12]. The absence of correlation between protein and mRNA copy numbers in a single-cell can be explained by the different lifetimes of the proteins and mRNAs, as bacterial mRNAs are short-lived (0.6–36 min), whereas their proteins have a long lifetime exceeding cell division (~150 min). Therefore, the abundances of mRNA can only reflect gene expression states over a short period of no more than half an hour, while that of proteins can reflect the accumulated results of gene expression for a period that goes beyond the duration of the cell cycle. This result offers a cautionary note for single-cell transcriptome analysis and suggests the importance of single cell proteome analysis.

3. Stochastic Nature in Gene Expression

Stochasticity in gene expression is often dominated by "bursting", a phenomenon that describes how multiple gene expression products are generated during intermittent active periods that are separated by inactive periods. According to the central dogma, two types of bursts exist, transcriptional bursts and translational bursts. Regarding transcriptional bursts, many mRNA molecules are produced in a short period rather than with a constant probability per unit time. On the other hand, in translational bursts, many protein molecules are produced from a single mRNA when the rate of protein translation is greater than that of mRNA degradation.

Experimentally, transcriptional bursts have been observed in single *E. coli* cells (Figure 2a) [19,20], eukaryotes [21], and other higher eukaryotes [22]. Real-time observation of mRNA expression in living cells has demonstrated that genes are transcribed in a discontinuous fashion. To visualize mRNA molecules, the MS2 system has been used [19–22]. This approach exploits the specific interactions between the MS2 hairpin RNA and a complex that includes a fluorescent protein and the MS2 capsid protein. Additionally, luminescence imaging has revealed the transcription kinetics of the short-lived luciferase reporter system [23], and RNA FISH experiments allow us to analyze variations in the number of mRNA molecules that result from transcriptional bursts without the use of a genetic tag [24–27].

There are several possible mechanisms that explain the transcription burst phenomena [28]. One is promoter on/off switching. The binding and unbinding of regulatory molecules, such as histones and transcription factors, may also be a cause. A recent *in vitro* assay has indicated that DNA supercoiling in the promoter region may also be responsible [29]. In addition, global factors that affect gene activities at the whole gene level, such as cell-cycle-dependent regulation of the promoter activity and the cooperative recruitment of RNA polymerases, are also considered possible causes [28].

Similarly, the observation of translational bursts has been reported. One study monitored the production of single proteins in single *E. coli* cells by imaging newly synthesized fluorescent-fused proteins under the control of the lac promoter (Figure 2b) [10]. The same study demonstrated that proteins were generated in bursts from a single mRNA, with each burst generating an average of 4.2 proteins. The same lab acquired similar results when applying the enzymatic amplification method on the same lac promoter and another reporter gene, β-galactosidase, in *S. cerevisiae* [9].

Figure 2. Stochastic gene expression. (**a**) Experimental observations of transcriptional bursts. Expressions of mRNA molecules in single *E. coli* cells are visualized in real-time by labeling the MS2-GFP system [19]. Each green spot consists of one or more mRNA molecules (**left**). Time trace of mRNA expression in single cells (**right**). (**b**) Experimental observations of translational bursts. The expression of proteins labeled with Venus in single *E. coli* cells are visualized in real-time. Venus molecules are photobleached immediately after the image acquisition so that only newly synthesized proteins are imaged each time [10]. Each yellow spot represents single protein molecules (**top**). Corresponding time courses (**bottom**). (**c**) Model for stochastic gene expression. The promoter transitions between an inactive 'off' state and an active "on" state with rates k_{on} and k_{off}. During the on state, mRNAs are transcribed at the rate of k_1, and proteins are translated at the rate k_2. mRNAs and proteins are degraded at the rates γ_1 and γ_2, respectively. The states of degraded mRNAs and proteins are indicated as ϕ. Images are taken from [10,19].

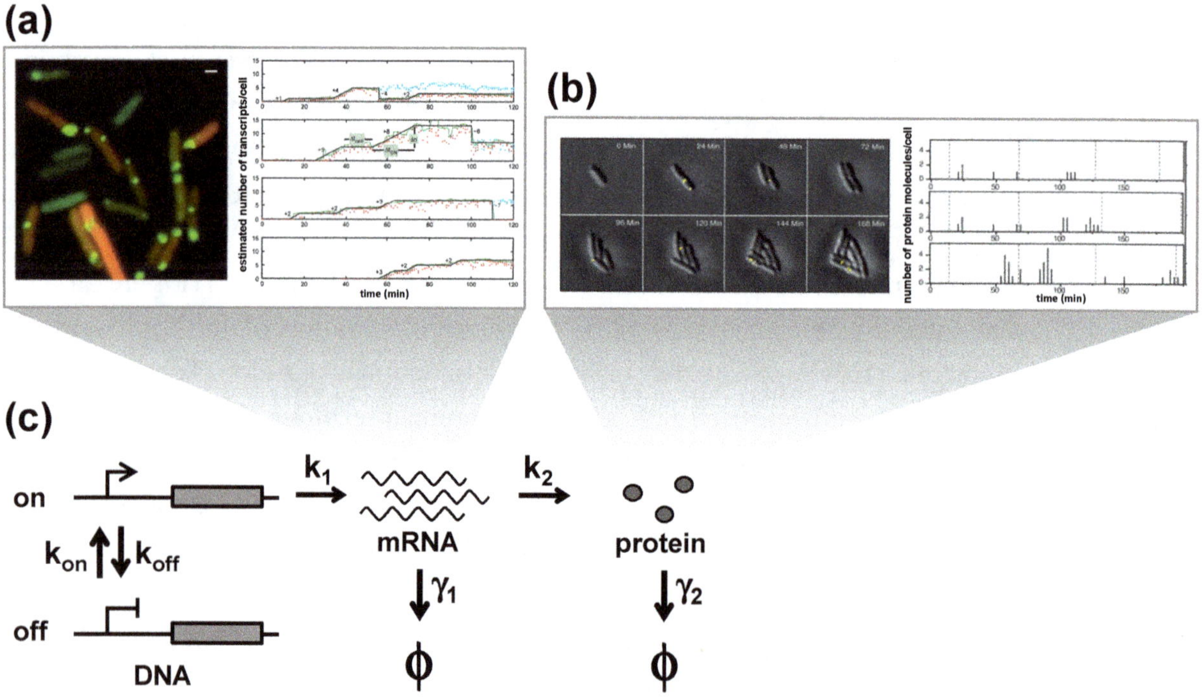

Bursting behaviors can be described with a simple kinetic model, where transcription, translation, promoter switching, and the degradation of mRNA and proteins, are assumed as first-order reactions (Figure 2c). This model was first introduced by Peccoud and Ycart [30] and is now broadly used to understand quantitative gene expression at the single molecule level. The rates of transition between the on and off states are indicated in the figure as k_{on} and k_{off}, respectively. The rate of transcription in the on state is described as k_1, and the rate of translation as k_2. The rates of degradation for mRNA and protein are defined as γ_1 and γ_2, respectively. Transcriptional bursting occurs during the on state. The number of mRNAs produced in a single transcriptional burst, or transcriptional burst size, is given by k_1/k_{off}. The transcriptional burst frequency changes with the duration of the inactive period (k_{on}). In contrast, the translational burst size, which denotes the number of proteins per single mRNA, is given by k_2/γ_1. This means that bursts of protein expression occur when k_2 is greater than γ_1, which is the case for most genes in budding yeast [31].

Because γ_1 is much greater than γ_2, the steady-state distribution of mRNA and protein copy numbers among a cell population can be given by solving the master equations for this model [31]. The probability density function of having m mRNAs at steady-state is:

$$P_m = \frac{m_s^m e^{-m_s}}{m!} \cdot \frac{\Gamma(\zeta_{on} + m)\Gamma(\zeta_{on} + \zeta_{off})}{\Gamma(\zeta_{on} + \zeta_{off} + m)\Gamma(\zeta_{on})} \cdot {}_1F_1\left(\zeta_{off}, \zeta_{on} + \zeta_{off} + m; m_s\right)$$

where $m_s = k_1/\gamma_1$, $\xi_{on} = k_{on}/\gamma_1$, and $\xi_{off} = k_{off}/\gamma_1$. ${}_1F_1(a,b;z)$ is the confluent hypergeometric function of the first kind. In contrast, the probability of having n proteins is given by:

$$P_n = \frac{\Gamma(\alpha + n)\Gamma(\beta + n)\Gamma(k_{on} + k_{off})}{\Gamma(n+1)\Gamma(\alpha)\Gamma(\beta)\Gamma(k_{on} + k_{off} + n)} \cdot \left(\frac{b}{1+b}\right)^n \cdot \left(1 - \frac{b}{1+b}\right)^\alpha \cdot {}_2F_1\left(\alpha + n, k_{on} + k_{off} - \beta, k_{on} + k_{off} + n; \frac{b}{1+b}\right)$$

where $a = k_1/\gamma_2$, $b = k_2/\gamma_1$, $\alpha = (1/2)(a + k_{on} + k_{off} + \phi)$, $\beta = (1/2)(a + k_{on} + k_{off} - \phi)$ and $\phi = (a + k_{on} + k_{off})^2 - 4ak_{on}$. ${}_2F_1(a,b,c;z)$ is the Gaussian hypergeometric function. Assuming that DNA is always active at steady state ($k_{off} \rightarrow 0$), this equation becomes a negative binomial distribution [32]. Moreover, assuming a continuous variable for n leads to a gamma distribution [33], which has been used to analyze the distributions of protein concentrations (but not protein copy numbers) in single *E. coli* cells, and minimizes any effects on the analysis caused by abrupt changes in protein numbers before and after cell division [12]. In fact, it has been shown that in *E. coli* the distribution of almost all protein concentrations can be described by a gamma distribution.

In the above equation, the protein noise, defined by $\eta^2 = \langle n^2\rangle/\langle n\rangle^2$, satisfies:

$$\eta^2 = \frac{1}{\langle n\rangle} + \frac{\gamma_2}{\gamma_1}\frac{1}{\langle m\rangle} + \frac{\gamma_2}{\gamma_2 + k_{on} + k_{off}}\eta_D^2$$

where $\langle m\rangle$ and $\langle n\rangle$ are the mean numbers of mRNA and protein, respectively, and η_D^2 is the noise when DNA is in its active state and equals k_{off}/k_{on}. This equation demonstrates that the noise in protein expression levels can be influenced by any rates of translation, transcription or promoter switching, which is consistent with an expanded model [34].

The above model attributes intrinsic noise in protein copy numbers to the stochasticity inherent in gene expression. Additionally, a previous work has shown the existence of another noise source, extrinsic noise, due to fluctuations in other cellular components [1]. Subsequent work has indicated that extrinsic noise is inherited over a timescale of about one cell cycle, while intrinsic noise decays rapidly [8]. Extrinsic noise can be modeled in the above simple kinetic model by adding static or slowly varying heterogeneities in individual rate constants among cells [12]. In the *E. coli* case where the intrinsic noise distribution is described as a gamma distribution with two adjustable parameters, $P_n(a_E, b_E)$, the total noise distribution can be expressed as:

$$P_{n_{total}} = \int_0^\infty \int_0^\infty P_n(a_E, b_E) f(a_E) g(b_E) da_E db_E$$

where it is assumed that a_E and b_E exist with distributions of $f(a_E)$ and $g(b_E)$, respectively. Thus, the noise, η_{total}^2, is given by:

$$\eta_{total}^2 = \frac{\langle b \rangle + \langle b \rangle \eta_b^2}{\langle n \rangle} + \eta_a^2 + \eta_a^2 \eta_b^2 + \eta_b^2$$

where $\eta_a^2 = \langle a_E^2 \rangle / \langle a_E \rangle^2$ and $\eta_b^2 = \langle b_E^2 \rangle / \langle b_E \rangle^2$. The extrinsic noise in the last three terms predicts the existence of an unavoidable, global noise floor independent of the mean protein abundance, $\langle n \rangle$, which is consistent with previously reported experimental observations [12].

Taken together, adding to the simple kinetic model described in Figure 2 the assumption of static heterogeneities in kinetic rate constants among cells can describe both intrinsic and extrinsic noise at the single cell level. These mathematical descriptions quantitatively connect the steady-state distributions of mRNA and protein copy numbers among cells with fluctuations of mRNA and protein copy numbers over time within a cell lineage. More importantly, the models are conducive for understanding the complexity of gene expression behaviors at the single cell level.

4. Single Molecule Detection of Protein Expression

As noted above, fluorescence imaging can provide quantitative measurements of protein expression levels in single cells, which are essential to understanding cell dynamics and cell heterogeneity. One common approach to quantifying protein copy numbers in a single living cell is the use of a genetic tag. Typically, the coding sequence of a fluorescent protein is genetically inserted downstream of a promoter or is fused to or replaced with an open reading frame of the protein of interest so that the target promoter or protein expression coincides with the expression of the fluorescent protein. Protein expression levels can thus be monitored in real-time with fluorescence microscopy by evaluating integrated intensities within the cells. An alternative way to study protein expression levels is to label proteins with fluorescence antibodies in fixed cells. This method, unfortunately, does not perform real-time monitoring of protein levels and is highly affected by the recognition ability of the antibody. However, it also has no need for *a priori* genetic manipulation and causes no artifactual perturbation in the genetic tagging, which makes it suitable for many purposes in both scientific research and clinical diagnosis.

Concentrated efforts on new technologies have led to the detection of single molecules in single cells by single-molecule fluorescence microscopy. Such sensitivity promises new insights on gene expression, including the ability to detect the copy numbers of proteins, which can then be used to characterize protein localization and diffusion with high spatiotemporal resolution. For example, Choi *et al.*, monitored stochastic phenotype switching in the carbohydrate metabolism of single *E. coli* cells [35]. Below we introduce several methods for single molecule detection.

Yu *et al.*, visualized gene expressions of single living *E. coli* cells at the single-molecule level (Figure 3a) [10], finding they occur in bursts. To investigate the expression of the lac promoter, a complex that includes the membrane protein Tsr and YFP-variant fluorescence protein Venus [36] was inserted downstream. Tsr was used to localize the fluorescence probe to the membrane, as otherwise the complex could diffuse into the cytoplasm leaving its signal to be overwhelmed by the autofluorescence of the cell. Venus was included for several reasons. First, the maturation time of Venus to emit fluorescence (~7 min) is much faster than the cell division time (about 150 min in typical *E. coli* cells in minimal media). Second, the brightness of Venus is sufficient to detect single molecules. Third, the relatively higher wavelength of excitation for Venus (514 nm, *cf.* 488 nm for GFP) can minimize photo-excitation caused by cellular autofluorescence.

Figure 3. Selected methods for single cell protein expression analysis. (**a**) Single molecule detection of protein expression at the membrane. The inclusion of a membrane protein (Tsr) into the gene assures the protein of interest will be localized at the membrane, while the inclusion of a fluorescent protein (Venus) into the gene assures detection of the membrane localization [10]. (**b**) Stroboscopic detection of single fluorescent proteins in single *E. coli* cells [37]. (**c**) Counting the number of FliM-Dendra2 in a *E. coli* cell with PALM/STORM. Bright-field image, super-resolution Flim-Dendra2 overlay image and diffraction-limited Flim-Dendra2 image are shown [41]. (**d**) Microfluidic system for monitoring single reporter molecules with the enzymatic assay. Schematic diagram of fluorogenic enzymatic reactions is shown on the bottom [9]. (**e**) Single-cell analysis chip. Schematic diagram of the molecule detection section illuminated by a focused excitation laser is shown on the top right. Examples of recorded frames are shown on the bottom right [43]. Images are taken from [9,10,37,41] and [43].

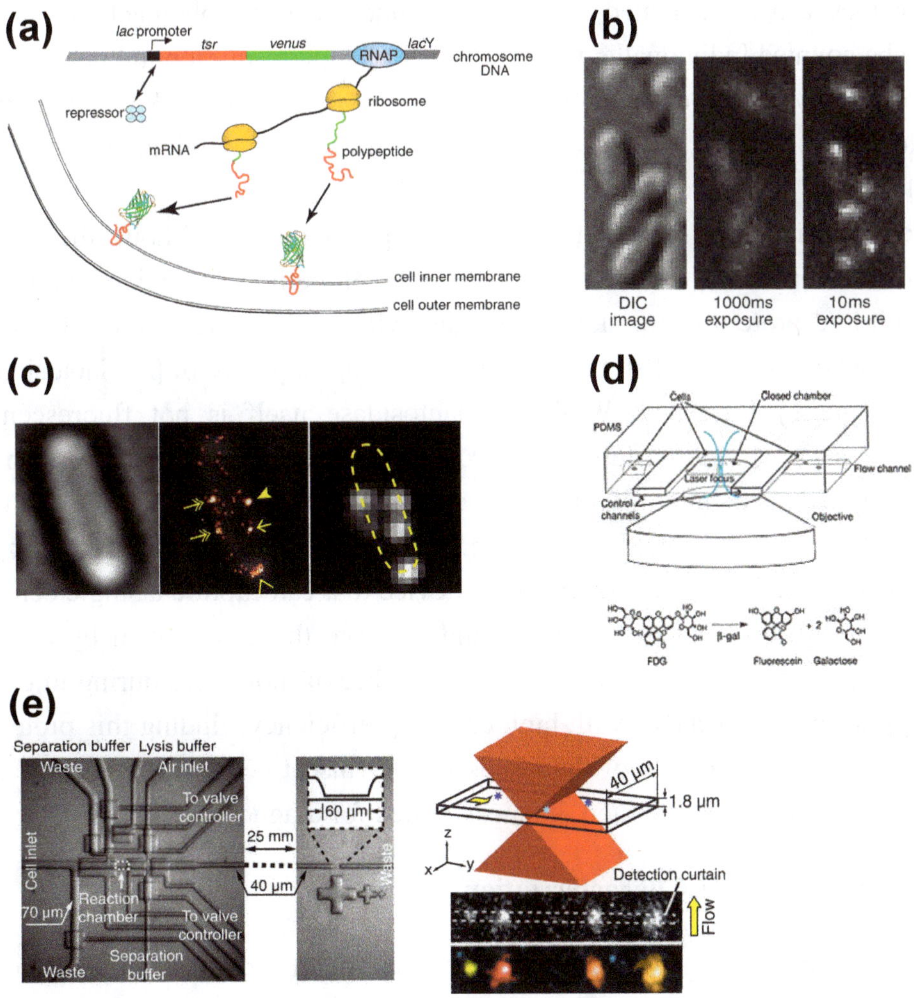

However, the inclusion of Tsr or any other membrane protein is a problem if one is interested in protein localization. As an alternative, Elf *et al.*, have proposed a stroboscopic method (Figure 3b) [37].

This method uses short laser pulses (~1 ms) to excite the Venus probe. Because of the short lifetime of the pulse, even diffusing fluorescent protein molecules can be visualized as spots because the diffusion is nearly within the diffraction limit of light. Using this method, they characterized the diffusion states of the lac repressor interacting with the lac operator in living *E. coli* cells.

When quantifying gene expression noise, *i.e.*, measuring the distributions of protein copy numbers in a cell population, the above techniques are suitable only when investigating low copy numbers. For high copy numbers, a different strategy should be considered, because the many resulting fluorescent spots are likely to overlap with one another, which compromises the measurement. Taniguchi *et al.*, designed a technique that quantifies gene expression noise for any protein abundance [12]. In this method, distributions of the number of fluorescent probes can be analyzed by deconvolving the measured fluorescence from the cell autofluorescence.

However, even that approach does not allow for the investigation of high protein abundance at the single cell level. Recently, methods based on superresolution microscopy, such as photoactivated localization microscopy (PALM) [38] or stochastic optical reconstruction microscopy (STORM) [39], have been proposed for this purpose (Figure 3c) [40–42]. Here, the protein of interest is genetically tagged with a photo-activatable protein such as Dendra2 or mMaple3. The advantage of photo-activatable proteins is that they can be activated incrementally under a weak photoactivation condition, which allows them to be counted individually. One drawback of superresolution microscopy, however, is that it is not suitable for living cell analysis because of the relatively high risk of photodamage due to the long acquisition time. Still, it is arguably the most effective technique for counting proteins in individual cells.

Fluorescence imaging can be used to study copy numbers without the use of exogenous fluorescence proteins as well, which reduces the risk of both phototoxicity and autofluorescence. Cai *et al.*, monitored protein expressions in single living cells using a microfluidic device and the β-galactosidase assay (Figure 3d) [9]. In this assay, a coding sequence of β-galactosidase is genetically inserted into the gene of interest. While β-galactosidase itself is not fluorescent, it hydrolyzes fluorogenic substrates to produce a fluorescent product. To monitor protein expressions in a single cell, single cells are trapped inside microfluidic chambers and their fluorescence levels are measured.

Huang *et al.* developed a method for protein counting that needs no genetic tagging nor editing (Figure 3e) [43]. This method uses a microfluidic device that can capture a single cell, lyse it, label the proteins with fluorescently-labeled antibodies, and separate the cell content by electrophoresis. The protein number can be determined by counting the number of molecules during migration. The group measured the β_2 adrenergic receptor with high counting efficiency, finding this protein can vary from just a few thousand to several tens of thousands in SF9 insect cells. Although this approach cannot analyze the dynamics of gene expression, it is well suited for diagnostic purposes.

5. Platforms for System-Wide Characterization

Because any given cell function can be controlled by the expression of thousands of genes, systems-wide characterization is needed. Such a task is not trivial for fluorescence-based methods, because the broad excitation/emission spectrum seen for many fluorescent probes limits the number of probe species that can be analyzed independently and consequently the number of molecule types that can be analyzed. For mRNA analysis, Levsky *et al.* offered a solution by proposing combinatorial labeling, which labels each mRNA species with different combinations of several different dyes [44]. This barcoding strategy increases the number of unique fluorescence spectra, and is especially advantageous when using superresolution microscopy to discriminate individual mRNA molecules [45,46]. However, applying this strategy to protein analysis is much more challenging, because the copy number of most proteins is much higher and the proteins themselves are harder to discriminate due to them localizing at higher densities.

Figure 4. High-throughput single molecule fluorescence microscopy. (**a**) Chromosomal YFP fusion *E. coli* library. In each library strain, a YFP is translationally fused to the C-terminus of a protein in its native chromosomal position. (**b**) Microfluidic chip for imaging 96 library strains. Each strain is immobilized on a polylysine-coated coverslip and is scanned with a single molecule fluorescence microscope. (**c–e**) Three examples from a library of *E. coli* strains are shown. Images to the left are representative fluorescent images overlaid on phase-contrast images; histograms to the right are the distribution of copy numbers. Lines are curve fits to gamma distributions with parameters *a* and *b*. Taken from [12].

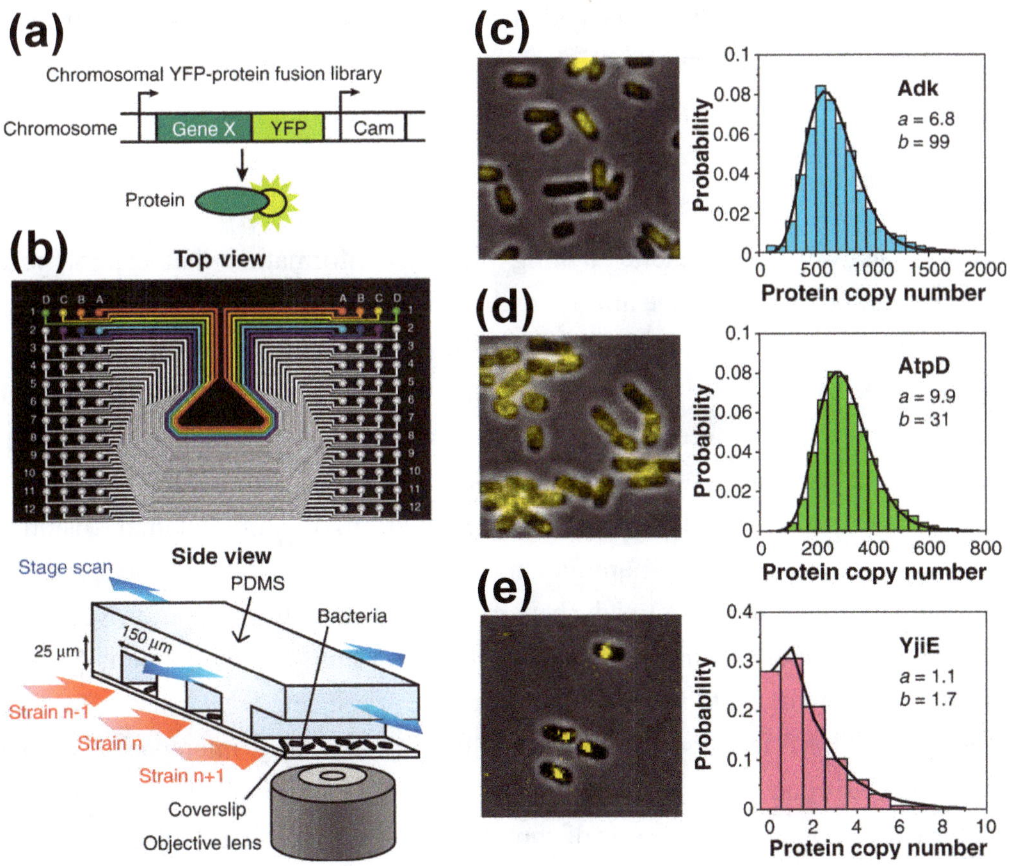

One standard strategy for system-wide quantification of protein expression levels based on fluorescence imaging is to use a cell strain library. A library is typically a collection of many recombinant cell strains, in which a fluorescent protein is chromosomally fused to different genes [47–49]. The cell strains are measured using a high-throughput system from which system-wide information can be acquired. This method cannot measure the proteome in the same single cell, but can take advantages of fluorescence microscopy, which means it can be used to analyze dynamics with single molecule sensitivity. Such cell strain libraries have been constructed for many organisms, including *E. coli* [12], *S. cerevisiae* [47,50], *C. elegans* [48], and human H1299 lung carcinoma cells [51]. Furthermore, correlation analysis of all gene pairs in the entire gene set can describe the entire gene network in a single cell.

To perform high-throughput analysis of many cell strains, high-throughput flow cytometry is preferred, as this technology automates the delivery of cell strains to a flow cytometer via an autosampling device [11,52]. With this system, approximately seven strains at >50,000 cells per strain can be measured per minute. Newman *et al.* have successfully measured >2500 GFP-tagged yeast strains at single-cell resolution [11]. Unfortunately, the flow cytometry method is inferior in sensitivity and resolution compared to standard cellular imaging. An alternative approach is to use an automated microscope system. The system described above by Taniguchi *et al.* can measure about two strains per minute and ~4000 *E. coli* cells per strain [12] (Figure 4). Accordingly, it was used to measure, 1018 Venus-tagged *E. coli* strains at single-cell resolution and single molecule sensitivity.

A protein chip approach that does not need any gene recombination has also been proposed [53]. Shi *et al.* developed the single-cell barcode chip, which is a microfluidics system in which single cells are captured and lysed in 2 nL chambers and the released proteins are assayed using an array of antibodies patterned inside the chambers. With this chip, 11 different protein species associated with the phosphoinositide 3-kinase signaling pathway were quantified.

6. Future Perspectives

Single live cell analysis with genetic tagging provides information that cannot be acquired by studying ensemble cell populations. The ability to characterize the dynamics and heterogeneity of gene expression will be instrumental in correlating stochastic gene expressions with biological phenomena, as has already been seen in bacterial persisters [54] and the translocation of transcription factors [55]. Similarly, correlation analysis between a large number of genes in single cells is expected to illuminate the network pathways and modules of gene regulations [56].

One critical obstacle in live cell analysis is the creation of chromosomal fusion cell strains, especially in higher eukaryote and mammalian cells that have low efficiency of homologous recombination. However, progress is already being made, as random integration via a retrovirus was used to create a chromosomal fusion strain library of human lung carcinoma cells [57]. For efficient genetic engineering, technologies that use site-specific engineered nucleases such as truncated transcription activator-like effector nucleases and clustered regularly interspaced short palindromic repeat RNA-guided Cas9 nuclease have been investigated [58]. It is expected that these technologies will accelerate the generation of complete sets of chromosomal fusion strains in mouse and human cells.

Another obstacle is imaging thicker cells or tissues (*E. coli*, *cf.* 0.5 μm; yeast, 5 μm; mammals >10 μm). In these cases, 3D stacking of images is done using different focal positions. To reduce the strong out-of-focus background that prevents quantitative analyses, 3D sectioning of stacked images is recommended. Confocal microscopy may be one option for this purpose, but it has relatively low sensitivity and the out-of-focus exposure of the excitation laser causes higher photodamage to the sample. In this regard, sheet illumination microscopy may be preferred [59]. This method combines light-sheet illumination with orthogonal camera-based detection to achieve selective plane illumination that minimizes the photodamage. Using similar systems, the imaging of single molecule fluorophores has been achieved in living mammalian cells and tissues [60,61].

Meanwhile, developments in single-cell analytic methods not requiring genetic tags will greatly advance diagnostic and clinical work. To analyze protein levels and localizations with fluorescence imaging, fluorescent antibodies are commonly used to label the proteins because of their high specificity. Protein interactions in single cells can also be probed using fluorescent antibodies. For example, multiplex proximity ligation assay was used to analyze the co-existence of two proteins bound to different antibodies that were conjugated with unique tag sequences and rolling circle amplification [62]. These single-cell technologies can visualize protein molecules and complexes at a single-cell level in tissue sections and are sufficiently sensitive for clinical samples.

Alternatively, mass spectroscopy may not be as sensitive as fluorescent-dependent systems, but it offers to simultaneously measure vast kinds of molecules with high spectral resolution. Recently, single-cell mass cytometry has been reported [63]. Here, cells labeled with antibodies coupled to transition element isotopes are nebulized into single cell droplets, and the metal tags are quantified by time-of-flight mass spectrometry. This method has been used to characterize human born marrow cells with as many as 34 parameters, which is far more parameters than other systems have reported [64]. The high sensitivity is especially advantageous when only having access to a minimal amount of samples from patients, such as biopsies and blood tests. These technologies are expected to reveal differences in phenotypic states of single cells.

Considering the exponential progress in single cell proteomics over the last decade, it is easy to anticipate new innovations that will make currently impractical experiments feasible in the near future. This progress will result in large datasets that are built from single-cell resolution and single-molecule sensitivity imaging on a systems-wide scale. These datasets will present new challenges, as investigators will need appropriate tools to analyze their content. The accumulating information from these techniques is allowing scientists to reveal more and more about the mechanisms and general rules of single-cell gene expressions, which are the basis for heterogeneous single cell activities and complex biological behaviors.

Acknowledgments

We thank Y. Ogawa and K. Nishimura for discussions. We acknowledge support from the Grant-in-Aid for Young Scientists (No. 24687022), the Grant-in-Aid for Scientific Research on Innovative Areas (No. 23115005), the Grant-in-Aid for Exploratory Research (No. 26650055), the Mochida Memorial Foundation for Medical and Pharmaceutical Research, and the Takeda Science Foundation.

Author Contributions

All authors wrote portions of this review.

References

1. Elowitz, M.B.; Levine, A.J.; Siggia, E.D.; Swain, P.S. Stochastic gene expression in a single cell. *Science* **2002**, *297*, 1183–1186.
2. Eldar, A.; Elowitz, M.B. Functional roles for noise in genetic circuits. *Nature* **2010**, *467*, 167–173.
3. Raj, A.; van Oudenaaden, A. Nature, nurture, or chance: Stochastic gene expression and its consequences. *Cell* **2008**, *135*, 216–226.
4. Tang, F.; Barbacioru, C.; Wang, Y.; Nordman, E.; Lee, C.; Xu, N.; Wang, X.; Bodeau, J.; Tuch, B.B.; Siddiqui, A.; *et al.* mRNA-Seq whole-transcriptome analysis of a single cell. *Nat. Methods* **2009**, *6*, 377–382.
5. Tietjen, I.; Rihel, J.M.; Cao, Y.; Koentges, G.; Zakhary, L.; Dulac, C. Single-Cell transcriptional analysis of neuronal progenitors. *Neuron* **2003**, *38*, 161–175.
6. Wu, M.; Singh, A.K. Single-Cell protein analysis. *Curr. Opin. Biotechnol.* **2012**, *23*, 83–88.
7. Mellors, J.S.; Jorabchi, K.; Smith, L.M.; Ramsey, J.M. Integrated microfluidic device for automated single cell analysis using electrophoretic separation and electrospray ionization mass spectrometry. *Anal. Chem.* **2010**, *82*, 967–973.
8. Rosenfeld, N.; Young, J.W.; Alon, U.; Swain, P.S.; Elowitz, M.B. Gene regulation at the single-cell level. *Science* **2005**, *307*, 1962–1965.
9. Cai, L.; Friedman, N.; Xie, X.S. Stochastic protein expression in individual cells at the single molecule level. *Nature* **2006**, *440*, 358–362.
10. Yu, J.; Xiao, J.; Ren, X.; Lao, K.; Xie, X.S. Probing gene expression in live cells, one protein molecule at a time. *Science* **2006**, *311*, 1600–1603.
11. Newman, J.R.; Ghaemmaghami, S.; Ihmels, J.; Breslow, D.K.; Noble, M.; DeRisi, J.L.; Weissman, J.S. Single-Cell proteomic analysis of S. cerevisiae reveals the architecture of biological noise. *Nature* **2006**, *441*, 840–846.
12. Taniguchi, Y.; Choi, P.J.; Li, G.W.; Chen, H.; Babu, M.; Hearn, J.; Emili, A.; Xie, X.S. Quantifying *E. coli* proteome and transcriptome with single-molecule sensitivity in single cells. *Science* **2010**, *329*, 533–538.
13. De Sousa Abreu, R.; Penalva, L.O.; Marcotte, E.M.; Vogel, C. Global signatures of protein and mRNA expression levels. *Mol. Biosyst.* **2009**, *5*, 1512–1526.
14. Schwanhäusser, B.; Busse, D.; Li, N.; Dittmar, G.; Schuchhardt, J.; Wolf, J.; Chen, W.; Selbach, M. Global quantification of mammalian gene expression control. *Nature* **2011**, *473*, 337–342.
15. Lu, P.; Vogel, C.; Wang, R.; Yao, X.; Marcotte, E.M. Absolute protein expression profiling estimates the relative contributions of transcriptional and translational regulation. *Nat. Biotechnol.* **2007**, *25*, 117–124.
16. Schrimpf, S.P.; Weiss, M.; Reiter, L.; Ahrens, C.H.; Jovanovic, M.; Malmström, J.; Brunner, E.; Mohanty, S.; Lercher, M.J.; Hunziker, P.E.; *et al.* Comparative functional analysis of the Caenorhabditis elegans and Drosophila melanogaster proteomes. *PLoS Biol.* **2009**, 7, e48.
17. Laurent, J.M.; Vogel, C.; Kwon, T.; Craig, S.A.; Boutz, D.R.; Huse, H.K.; Nozue, K.; Walia, H.; Whiteley, M.; Ronald, P.C.; *et al.* Protein abundances are more conserved than mRNA abundances across diverse taxa. *Proteomics* **2010**, *10*, 4209–4212.

18. Khan, Z.; Ford, M.J.; Cusanovich, D.A.; Mitrano, A.; Pritchard, J.K.; Gilad, Y. Primate transcript and protein expression levels evolve under compensatory selection pressures. *Science* **2013**, *342*, 1100–1104.
19. Golding, I.; Paulsson, J.; Zawilski, S.M.; Cox, E.C. Real-Time kinetics of gene activity in individual bacteria. *Cell* **2005**, *123*, 1025–1036.
20. Le, T.T.; Harlepp, S.; Guet, C.C.; Dittmar, K.; Emonet, T.; Pan, T.; Cluzel, P. Real-Time RNA profiling within a single bacterium. *Proc. Natl. Acad. Sci. USA* **2005**, *102*, 9160–9164.
21. Chubb, J.R.; Trcek, T.; Shenoy, S.M.; Singer, R.H. Transcriptional pulsing of a developmental gene. *Curr. Biol.* **2006**, *16*, 1018–1025.
22. Yunger, S.; Rosenfeld, L.; Garini, Y.; Shav-Tal, Y. Single-Allele analysis of transcription kinetics in living mammalian cells. *Nat. Methods* **2010**, *7*, 631–633.
23. Suter, D.M.; Molina, N.; Gatfield, D.; Schneider, K.; Schibler, U.; Naef, F. Mammalian genes are transcribed with widely different bursting kinetics. *Science* **2011**, *332*, 472–474.
24. So, L.H.; Ghosh, A.; Zong, C.; Sepúlveda, L.A.; Segev, R.; Golding, I. General properties of transcriptional time series in Escherichia coli. *Nat. Genet.* **2011**, *43*, 554–560.
25. Zenklusen, D.; Larson, D.R.; Singer, R.H. Single-RNA counting reveals alternative modes of gene expression in yeast. *Nat. Struct. Mol. Biol.* **2008**, *15*, 1263–1271.
26. Raj, A.; Peskin, C.S.; Tranchina, D.; Vargas, D.Y.; Tyagi, S. Stochastic mRNA synthesis in mammalian cells. *PLoS Biol.* **2006**, *4*, e309.
27. Paré, A.; Lemons, D.; Kosman, D.; Beaver, W.; Freund, Y.; McGinnis, W. Visualization of individual Scr mRNAs during Drosophila embryogenesis yields evidence for transcriptional bursting. *Curr. Biol.* **2009**, *19*, 2037–2042.
28. Sanchez, A.; Golding, I. Genetic determinants and cellular constraints in noisy gene expression. *Science* **2013**, *342*, 1188–1193.
29. Chong, S.; Chen, C.; Ge, H.; Xie, X.S. Mechanism of transcriptional bursting in bacteria. *Cell* **2014**, *158*, 314–326.
30. Peccoud, J.; Ycart, B. Markovian modeling of gene-product synthesis. *Theor. Popul. Biol.* **1995**, *48*, 222–234.
31. Shahrezaei, V.; Swain, P.S. Analytical distributions for stochastic gene expression. *Proc. Natl. Acad. Sci. USA* **2008**, *105*, 17256–17261.
32. Paulsson, J.; Ehrenberg, M. Random signal fluctuations can reduce random fluctuations in regulated components of chemical regulatory networks. *Phys. Rev. Lett.* **2000**, *84*, 5447–5450.
33. Friedman, N.; Cai, L.; Xie, X.S. Linking stochastic dynamics to population distribution: an analytical framework of gene expression. *Phys. Rev. Lett.* **2006**, *97*, 168302.
34. Paulsson, J. Summing up the noise in gene networks. *Nature* **2004**, *427*, 415–418.
35. Choi, P.J.; Cai, L.; Frieda, K.; Xie, X.S. A stochastic single-molecule event triggers phenotype switching of a bacterial cell. *Science* **2008**, *322*, 442–446.
36. Nagai, T.; Ibata, K.; Park, E.S.; Kubota, M.; Mikoshiba, K.; Miyawaki, A. A variant of yellow fluorescent protein with fast and efficient maturation for cell-biological applications. *Nat. Biotechnol.* **2002**, *20*, 87–90.

37. Elf, J.; Li, G.W.; Xie, X.S. Probing transcription factor dynamics at the single-molecule level in a living cell. *Science* **2007**, *316*, 1191–1194.
38. Betzig, E.; Patterson, G.H.; Sougrat, R.; Lindwasser, O.W.; Olenych, S.; Bonifacino, J.S.; Davidson, M.W.; Lippincott-Schwartz, J.; Hess, H.F. Imaging intracellular fluorescent proteins at nanometer resolution. *Science* **2006**, *313*, 1642–1645.
39. Rust, M.J.; Bates, M.; Zhuang, X. Sub-Diffraction-Limit imaging by stochastic optical reconstruction microscopy (STORM). *Nat. Methods* **2006**, *3*, 793–795.
40. Durisic, N.; Laparra-Cuervo, L.; Sandoval-Alvarez, A.; Borbely, J.S.; Lakadamyali, M. Single-Molecule evaluation of fluorescent protein photoactivation efficiency using an *in vivo* nanotemplate. *Nat. Methods* **2014** , *11*, 156–162.
41. Lee, S.H.; Shin, J.Y.; Lee, A.; Bustamante, C. Counting single photoactivatable fluorescent molecules by photoactivated localization microscopy (PALM). *Proc. Natl. Acad. Sci. USA* **2012**, *109*, 17436–17441.
42. Wang, S.; Moffitt, J.R.; Dempsey, G.T.; Xie, X.S.; Zhuang, X. Characterization and development of photoactivatable fluorescent proteins for single-molecule-based superresolution imaging. *Proc. Natl. Acad. Sci. USA* **2014**, *111*, 8452–8457.
43. Huang, B.; Wu, H.; Bhaya, D.; Grossman, A.; Granier, S.; Kobilka, B.K.; Zare, R.N. Counting low-copy number proteins in a single cell. *Science* **2007**, *315*, 81–84.
44. Levsky, J.M.; Shenoy, S.M.; Pezo, R.C.; Singer, R.H. Single-cell gene expression profiling. *Science* **2002**, *297*, 836–840.
45. Lubeck, E.; Cai, L. Single-cell systems biology by super-resolution imaging and combinatorial labeling. *Nat. Methods* **2012**, *9*, 743–748.
46. Lubeck, E.; Coskun, A.F.; Zhiyentayev, T.; Ahmad, M.; Cai, L. Single-cell *in situ* RNA profiling by sequential hybridization. *Nat. Methods* **2014**, *11*, 360–361.
47. Huh, W.K.; Falvo, J.V.; Gerke, L.C.; Carroll, A.S.; Howson, R.W.; Weissman, J.S.; O'Shea, E.K. Global analysis of protein localization in budding yeast. *Nature* **2003**, *425*, 686–691.
48. Sarov, M.; Murray, J.I.; Schanze, K.; Pozniakovski, A.; Niu, W.; Angermann, K.; Hasse, S.; Rupprecht, M.; Vinis, E.; Tinney, M.; *et al*. A genome-scale resource for *in vivo* tag-based protein function exploration in C. elegans. *Cell* **2012**, *150*, 855–866.
49. Sigal, A.; Danon, T.; Cohen, A.; Milo, R.; Geva-Zatorsky, N.; Lustig, G.; Liron, Y.; Alon, U.; Perzov, N. Generation of a fluorescently labeled endogenous protein library in living human cells. *Nat. Protoc.* **2007**, *2*, 1515–1527.
50. Ghaemmaghami, S.; Huh, W.K.; Bower, K.; Howson, R.W.; Belle, A.; Dephoure, N.; O'Shea, E.K.; Weissman, J.S. Global analysis of protein expression in yeast. *Nature* **2003**, *425*, 737–741.
51. Cohen, A.A.; Geva-Zatorsky, N.; Eden, E.; Frenkel-Morgenstern, M.; Issaeva, I.; Sigal, A.; Milo, R.; Cohen-Saidon, C.; Liron, Y.; Kam, Z.; *et al*. Dynamic proteomics of individual cancer cells in response to a drug. *Science* **2008**, *322*, 1511–1516.

52. Bar-Even, A.; Paulsson, J.; Maheshri, N.; Carmi, M.; O'Shea, E.; Pilpel, Y.; Barkai, N. Noise in protein expression scales with natural protein abundance. *Nat. Genet.* **2006**, *38*, 636–643.
53. Shi, Q.; Qin, L.; Wei, W.; Geng, F.; Fan, R.; Shin, Y.S.; Guo, D.; Hood, L.; Mischel, P.S.; Heath, J.R. Single-cell proteomic chip for profiling intracellular signaling pathways in single tumor cells. *Proc. Natl. Acad. Sci. USA* **2012**, *109*, 419–424.
54. Wakamoto, Y.; Dhar, N.; Chait, R.; Schneider, K.; Signorino-Gelo, F.; Leibler, S.; McKinney, J.D. Dynamic persistence of antibiotic-stressed mycobacteria. *Science* **2013**, *339*, 91–95.
55. Cai, L.; Dalal, C.K.; Elowitz, M.B. Frequency-Modulated nuclear localization bursts coordinate gene regulation. *Nature* **2008**, *455*, 485–490.
56. Stewart-Ornstein, J.; Weissman, J.S.; El-Samad, H. Cellular noise regulons underlie fluctuations in Saccharomyces cerevisiae. *Mol. Cell* **2012**, *45*, 483–493.
57. Sigal, A.; Milo, R.; Cohen, A.; Geva-Zatorsky, N.; Klein, Y.; Alaluf, I.; Swerdlin, N.; Perzov, N.; Danon, T.; Liron, Y.; *et al*. Dynamic proteomics in individual human cells uncovers widespread cell-cycle dependence of nuclear proteins. *Nat. Methods* **2006**, *3*, 525–531.
58. Kim, H.; Kim, J.S. A guide to genome engineering with programmable nucleases. *Nat. Rev. Genet.* **2014**, *15*, 321–334.
59. Huisken, J.; Swoger, J.; Bene, F.D.; Wittbrodt, J.; Stelzer, E.H. Optical sectioning deep inside live embryos by selective plane illumination microscopy. *Science* **2004**, *305*, 1007–1009.
60. Ritter, J.G.; Veith, R.; Veenendaal, A.; Siebrasse, J.P.; Kubitscheck, U. Light sheet microscopy for single molecule tracking in living tissue. *PLoS One* **2010**, *5*, e11639.
61. Gebhardt, J.C.; Suter, D.M.; Roy, R.; Zhao, Z.W.; Chapman, A.R.; Basu, S.; Maniatis, T.; Xie, X.S. Single-molecule imaging of transcription factor binding to DNA in live mammalian cells. *Nat. Methods* **2013**, *10*, 421–426.
62. Leuchowius, K.J.; Clausson, C.M.; Grannas, K.; Erbilgin, Y.; Botling, J.; Zieba, A.; Landegren, U.; Söderberg, O. Parallel visualization of multiple protein complexes in individual cells in tumor tissue. Parallel visualization of multiple protein complexes in individual cells in tumor tissue. *Mol. Cell. Proteomics* **2013**, *12*, 1563–1571.
63. Bandura, D.R.; Baranov, V.I.; Ornatsky, O.I.; Antonov, A.; Kinach, R.; Lou, X.; Pavlov, S.; Vorobiev, S.; Dick, J.E.; Tanner, S.D. Mass cytometry: Technique for real time single cell multitarget immunoassay based on inductively coupled plasma time-of-flight mass spectrometry. *Anal. Chem.* **2009**, *81*, 6813–6822.
64. Bendall, S.C.; Simonds, E.F.; Qiu, P.; Amir, E.A.D.; Krutzik, P.O.; Finck, R.; Bruggner, R.V.; Melamed, R.; Trejo, A.; Ornatsky, O.I.; *et al*. Single-cell mass cytometry of differential immune and drug responses across a human hematopoietic continuum. *Science* **2011**, *332*, 687–696.

6

Single Cell Electrical Characterization Techniques

Muhammad Asraf Mansor [1] **and Mohd Ridzuan Ahmad** [2,*]

[1] Faculty of Biosciences and Medical Engineering, Universiti Teknologi Malaysia, 81310-UTM Skudai, Johor, Malaysia; E-Mail: asraf@biomedical.utm.my

[2] Faculty of Electrical Engineering, Universiti Teknologi Malaysia, 81310-UTM Skudai, Johor, Malaysia

* Author to whom correspondence should be addressed; E-Mail: ridzuan@fke.utm.my

Academic Editor: Fan-Gang Tseng

Abstract: Electrical properties of living cells have been proven to play significant roles in understanding of various biological activities including disease progression both at the cellular and molecular levels. Since two decades ago, many researchers have developed tools to analyze the cell's electrical states especially in single cell analysis (SCA). In depth analysis and more fully described activities of cell differentiation and cancer can only be accomplished with single cell analysis. This growing interest was supported by the emergence of various microfluidic techniques to fulfill high precisions screening, reduced equipment cost and low analysis time for characterization of the single cell's electrical properties, as compared to classical bulky technique. This paper presents a historical review of single cell electrical properties analysis development from classical techniques to recent advances in microfluidic techniques. Technical details of the different microfluidic techniques are highlighted, and the advantages and limitations of various microfluidic devices are discussed.

Keywords: conversional patch clamp; electrical properties; electrorotation; impedance flow cytometry; microelectrical impedance spectroscopy (μEIS); single cell analysis (SCA)

---

1. Introduction

Study of the cell has emerged as a distinct new field, and acknowledged to be one of the fundamental building blocks of life. Moreover, the cells have unique biophysical and biochemical properties to maintain and sense the physiological surrounding environment to fulfill its specific functions [1,2]. Cellular biophysical properties analysis, such as the electrical, mechanical, optical and thermal characterization of cells, provides critical knowledge to diagnostics, clinical science and pharmaceutical industry [1]. Biophysical properties of cells provide early signals of disease or abnormal condition to the human body, which make it them valuable as potential markers for identifying cancers [3–7], bacteria [8–10], toxin detection [11] and the status of tissues [12,13]. Furthermore rapid growing technologies (e.g., conventional patch-clamp, dual nanoprobe-ESEM (environmental scanning electron microscope) and microfluidics) to investigate the biophysical properties of cells have been invented and developed by the researchers in the last decades. The technologies are continually improved make substantial contributions to biology and the clinical research community [14,15].

Single cell analysis (SCA) has become a trend and major topic to engineers and scientists in the last 20 years to develop the experimental tools and technologies able to carry out single cell measurement. In addition, in depth analysis and more fully described activities of cell differentiation and cancer can only be accomplished with single cell analysis [16]. In conventional methods of cellular analysis, population based studies have been utilized for cellular processes such as metabolism, motility, cell growth and proliferation. Population methods use averages of cell properties to measure and predict the biophysical and biochemical parameters of cell. However, this method suffers from inaccurate measurements and often overlooks the essential information available in the cell due to the heterogeneity of cells (e.g., specific gene expression levels) [17]. For this reason, single cell studies have been emphasized to provide biologists and scientists to peer into the molecular machinery of individual cells. Single cell analysis has also been essential to our understanding of some fundamental questions, such as what makes single cells different biophysically, biochemically and functionally. Single cell analysis has been a key in probing of cancer [4,18], and thus helps doctors to develop a prognosis and design a treatment plan for particular patients.

Electrical properties of cells provide some insight and vital information to aid the understanding of complex physiological states of the cell. Cells that experience abnormalities or are infected by bacteria may have altered ion channel activity [19], cytoplasm conductivity and resistance [20,21] and deformability [22]. For instance, red blood cells (RBCs) infected by Plasmodium falciparum, which cause malaria in humans, reduce deformability of the RBC by producing cytoadherence-related neoantigens that increase the rigidity and internal viscosity of the cytomembrane [23,24]. Each RBC which experiences the deformation process, has difference resistance, where the average resistance value of normal RBCs and rigidified RBCs are 14.2 and 19.6 Ω respectively [23]. Since electrical properties of cells have several advantages in cells analysis such as counting, separating, trapping and characterizing of single cell, development of suitable devices for single cell electrical analysis in term of accuracy prediction, portable, and user friendly are very important. In this review, we present an overview of classical technique and microfluidics technique in single cell electrical properties analysis.

2. Classical Platforms

The classical technique for a cell's electrical properties analysis was originated in 1791, when Luigi Galvani conducted the first experiment for measuring electrical activity in animals, which is evoking muscular contractions in frog nerve muscle preparations by electrical stimulation with metal wires [25]. From that study, tools for analyzing a cell's electrical properties have development over the years. Conventional patch clamp and probing were have been the classical platform tools for characterization of single cell electrical properties.

2.1. Conventional Patch Clamp

The patch clamp technique is unique in enabling high-resolution recording of the ionic currents flowing through a cell's plasma membrane. Since the introduction of the patch-clamp technique by Neher and Sakmann in 1976, patch-clamp was adopted by researchers in cellular and molecular biology research areas for studying and providing valuable information of biological cell electrical properties [26,27]. The patch-clamp technique is also capable to analyze ionic currents in the cell membrane under conditions of complete control over transmembrane voltage and ionic gradients. Figure 1a illustrates the basic principle of patch clamp technique. A glass micropipette is used as a probe to suck a cell membrane into a micropipette to form a high electrical resistance or also called as giga-seals (e.g., normally between 10 and 100 GΩ [28]. Thus, the ion current that flows through the pipette (containing an electrode) is measured through an amplifier. Patch clamp can be operated in two modes, which are voltage and current mode. Voltage mode is used to measure voltage specific activity of an ionic channel, while current mode is used to measure the potential change in membrane when a current pulse has been injected into the cell [29]. Furthermore, the patch clamp technique has five basic measurement configurations such as cell-attached patch (CAP), whole-cell (WC), inside-out patch (IOP), outside-out patch (OOP) and permeabilized-patch WC-configuration (ppWC) [30,31]. More detail on the working principle of the patch clamp technique has been described [32,33].

The work of Hamill *et al.* sparked an approach for obtaining information about the characteristics and distribution of ion channels in living cells [28]. They used frog muscle fibres and rat myoballs as cell samples to detail several variants of this technique to create complete electrical isolation of the patched membrane for a variety of cells. This whole cell configuration is the most often utilized mode of the patch clamp technique. Zhang *et al.* combined the whole-cell patch clamp with fluorescence ratio imaging for measuring the electric properties of a cell membrane [34]. Fluorescence dye was used to monitor the transmembrane potential change of the cell in the long term without seriously perturbing the intracellular milieu. Both techniques combined have been successfully used to distinguish between differentiated and undifferentiated N1E-115 neuroblastoma cells according to the values of the resting potentials.

The conventional patch clamp technique has several disadvantages. First, the patch clamp technique is time consuming process [29,35]. The entire dish of cells needs to be replaced after the extracellular fluid has been manipulated, before continue the recording. Second, the quality of the cell and suspension must remain in good condition for channel expression to be homogenous [29]. Third, an experienced operator is required to move the glass pipette over the single cell for measuring current and voltage changes across the membrane through ion channels without damaging the whole cell. Other issues arise such as recoding quality and temperature control. Nevertheless, the patch clamp technique offers high sensitivity (pA resolution) and allows low noise measurement of the currents passing through the low conductance (pS) ion channels [25]. The evolution of upgraded modifications of the patch clamp technique can be found elsewhere [36].

2.2. Nanoprobe

Nanoprobes could potentially be used to perform single cell's electrical characterization. The nanoprobe capable to measure direct electrical properties of single cell and quantitatively determine the viability of single cells. M. R. Ahmad *et al.* developed a dual nanoprobe integrated with nanomanipulator units inside environmental scanning electron microscope (ESEM) to perform electrical probing on single cells for novel single cell viability detection [37]. Figure 1b illustrated the working principle of dual nanoprobe for single cell electrical measurement. Based on Ohm's law, current flow passing through the intracellular area of the cell was measured when a dual nanoprobe penetrated the intracellular area. ESEM was used for high resolution observation while preserving the cell's native state even when the cell is moving out of its buffer [38]. This technique successfully differentiated the live and dead cells of W303 wild yeast cells based on the electrical properties of the cell [37]. Recently, electrostatic force microscopy (EFM) was utilized to quantify the electric polarization response of single bacterial cells with high accuracy and reproducibility [39]. They demonstrated effective dielectric constants obtained from the different bacterial types (*Salmonella typhimurium*, *Escherichia coli*, *Lactobacillus sakei* and *Listeria innocua*), which were well correlated with the hydration state of bacteria. Figure 1c illustrated the working procedure of effective dielectric constant measurement using EFM. The electric polarization force between a bacterium and a nanometric-conducting tip mounted on a force-sensing cantilever was measured at different positions. Topographic images were used to obtain the geometry of the bacterium and finite element numerical simulations of a homogeneous bacterium were utilized to measure the effective dielectric constants of the cell [39,40]. A simple single cell electrical model was used in order to measure electrical properties of yeast cells. However, this technique requires a skilled operator to perform the measurement and is time consuming. The device is bulky system, which can only be performed in a restricted area, e.g., clean room [40].

3. Microfluidics Platforms

An advance in microfabrication technique, such as soft lithography, creates new opportunities for producing structures at micrometer scale inexpensively and rapidly [41]. For this reason, we have witnessed rapid development of microfluidics system for more than a decade ago for biology and medical research [1,14,42]. Microfluidics systems are a science and technology of manipulating fluids at the submillimetre length scale in the microscale fluidic channel. Microfluidics recognized as micro total analysis systems (μTASs) [43] or lab-on-a-chip (LoC) technologies have attracted attention because of the potential to improve diagnostics and biology research. Microfluidic systems have shown a potential to become widely adopted in modern clinical diagnosis and biology research (e.g., DNA analysis [44] and cell analysis [45]) because they are reproducible, have low power consumption, less sample and reagent consumption, are economical, amenable to modifications and can be integrated with other technologies [46,47]. The ability of microfluidics system to perform early cancer detection and address some problems in cellular analysis, make them suitable to replace the classical technique in single cell electrical analysis. Several microfluidic systems have been developed for single cell electrical properties analysis, such as electrorotation, impedance flow cytometry and microelectrical impedance spectroscopy (μEIS).

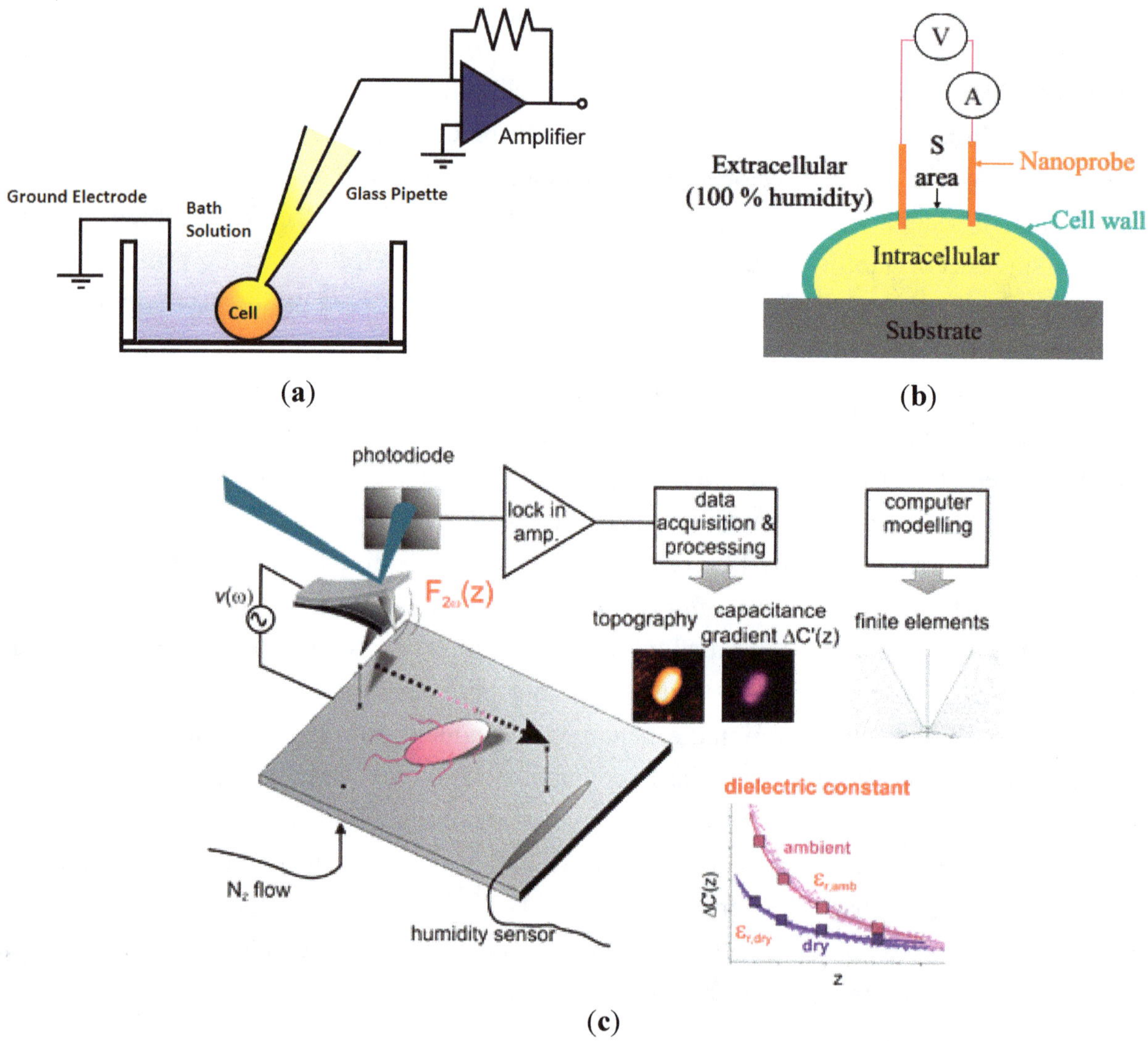

Figure 1. (**a**) Schematic diagram of conventional patch clamp technique; (**b**) Single cell electrical measurement using dual nanoprobes incorporated with ESEM. Reprinted with permission from [37]; (**c**) Schematic of measurement of the effective dielectric constant of a single bacterium using electrostatic force microscopy. Reprinted with permission from [39].

3.1. Electrorotation

A cell shows a rotated ability when it is placed into a rotating electric field within a medium with a non-uniform electric field. Analysis of these phenomena called an electrorotation (ROT), is commonly used for measuring the dielectric properties of cells without invasion. ROT measurement theory is based on rotational speed of cells/particles when the cell and the suspending medium have different electric polarizability, by referring to the frequency of a rotational electric field. This electric field is generated by quadrupole (arranged in a crisscross pattern) electrodes and each electrode is connected to an AC signal with a 90° phase difference from each other.

The quadrupole electrodes connected to sine wave was a famous design in ROT technique [48–50]. Figure 2a shows a working principle of ROT, four electrodes were energized by sinusoidal signal generator created rotating electrical field, E. Laser tweezers were used to drag a single cell to the center of a four-electrode chamber, then a single cell, P will rotate in either the same direction (co-field) or in the opposite direction (anti-field) to the rotating field [51]. The direction was taken by the cell depending to the dielectric properties of the cell and suspending medium along with the frequency of the electric field. The dielectric properties of a single cell can then be extracted by utilizing Maxwell's mixture theory, to associate the complex permittivity of the suspension to the complex permittivity of the cell [49]. More detail on theory and working principle of electrorotation can be found in other articles [52–54] and a book [55].

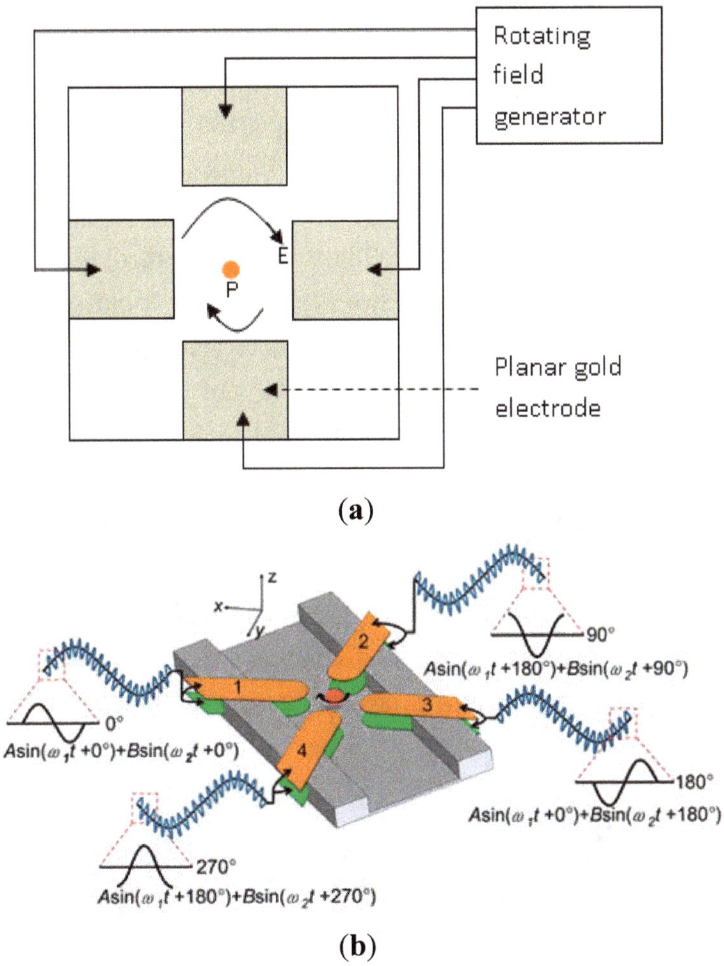

Figure 2. (**a**) An illustration of the working principle of electrorotation to analyse single cells; (**b**) The electrorotation (ROT)-microchip incorporated with the 3D octode. nQDEP (negative quadrupole dielectrophoresis) signal, Asin ($\omega_1 t$ + 0°) and Asin ($\omega_1 t$ + 180°) are used for a single cell trapping, while the ROT signals, Bsin($\omega_2 t$ + 0°), Bsin ($\omega_2 t$ + 90°), Bsin ($\omega_2 t$ + 180°) and Bsin ($\omega_2 t$ + 270°) are used to simultaneously generate torque. Reprinted with permission from [56].

In the ROT technique, the amplitude of the electric field remains unchanged because the cells are only rotated at a certain position in an electric field [57]. Therefore, it is suitable for fitting the rotation spectra at frequency range from 1 kHz to around 200 MHz to determine the intrinsic electrical properties of single cells such as cytoplasm conductivity, cytoplasm permittivity and specific membrane capacitance [48,58,59]. Electrorotation spectra are referred to cellular rotation rate *versus* frequency of the applied field. Jun Yang *et al.* [48] used frequency range between 1 kHz and 120 MHz, to fitting the rotation spectra in order to extract dielectric properties (membrane capacitance) of four main leukocyte subpopulations, *i.e.*, T- and B-lymphocytes, monocytes, and granulocytes. From this experiment, ROT was capable to characterize the dielectric properties of cell subpopulations within a cell mixture. In addition, ROT was utilized to determine the cell viability at real time assessment [51,60,61]. C. Dalton demonstrated that electrorotation technique can be used to determine the viability of two intestinal parasites, *i.e.*, *Giardia Intestinalis* and *Cyclospora Cayetanensis* [51]. An ellipsoidal two-shell model [58] was utilized to analyse the data and estimate the electrical parameter value.

Recently, the concept of negative quadrupole dielectrophoresis (nQDEP) and ROT signals superposed on each other in electrorotation technique was reported (Figure 2b). An accurate ROT spectrum was measured without any other disturbances because repulsive force by the nQDEP signal is stronger [56]. Specific membrane capacitance and cytoplasm conductivity of human leukocyte subpopulations (T-lymphocytes, B-lymphocytes, granulocytes, and monocytes) and metastatic cancer cell lines (SkBr3 and A549) were well achieved. Although the electrorotation technique is powerful tool, capable of extracting the electrical properties of the cell, such as cytoplasm conductivity and membrane permittivity, ROT technique has several drawbacks. Time consumption is a major factor for why the ROT technique has been unable to enter the modern clinical disease diagnosis as an analysis tool. G. De Gasperis *et al.* and M. Cristofanilli *et al.*, utilized ROT technique to analyze single cells and it took approximately 30 min to test a single cell [62,63]. These reports indicate that electrorotaion is a slow technique. Electrorotation also requires a skilled operator to position a single cell in the middle of a rotating electric field and also count the number of revolutions made by particles [64]. Nevertheless, electrorotation is a noninvasive technique which allows it to be used in sequential investigations. ROT also operates at a single-organism level and does not require extensive cell preparations [59]. Table 1 shows a summary of a microfluidic device using electrorotation technique for single cell electrical characterization.

Table 1. Microfluidic electrorotation device for single cell electrical analysis.

Authors	Techniques	Experimental Samples	Frequency	Dielectric Parameter	Summary	Reference
X.B. Wang *et al.* (1994)	Four electrode in phase quadrature	DS19	10 kHz–100 Mhz	Specific membrane capacitance DS19 (1.82 ± 0.24 μF/cm^2) DS19-HMBA (1.6 ± 0.25 μF/cm^2)	Specific membrane capacitance was determined by the complexity of surface features.	[65]
F.F. Becker *et al.* (1995)	Four electrode in phase quadrature	MDA231, T lymphocytes and Erythrocytes	1 kHz–1 GHz	Specific membrane capacitance MDA231 (26 ± 4.2 mF/m^2) T lymphocytes (11 ± 1.1 mF/m^2) Erythrocytes (9 ± 0.80 mF/m^2)	Specific membrane capacitance, cytoplasm conductivity, and cytoplasm permittivity values were reported.	[66]
R. Hoizel (1997)	Four electrode in phase quadrature	Yeast cells	100 Hz–1.6 GHz	Membrane capacitance yeast (0.76 μF/cm^2)	Specific capacitance of plasma membrane, periplasmic space and outer wall region values were reported.	[67]
J. Yang *et al.* (1999)	Four electrode in phase quadrature	Leukocyte (WBCs)	10 kHz–120 Mhz	Specific membrane capacitance T-lymphocytes (10.5 ± 3.1 mF/m^2) B-lymphocytes (12.6 ± 3.5 mF/m^2) Monocytes (15.3 ± 4.3 mF/m^2) Granulocytes (11.0 ± 3.2 mF/m^2)	Four main leukocyte subpopulations were discriminate based on their electrical properties.	[48]
C. Dalton (2001)	Four electrode in phase quadrature	Giardia intestinalis and Cyclospora cayetanensis	20–400 kHz	Membrane conductivity Giardia intestinalis 2 ± 0.81 μS·m^{-1} (viable) & 10 ± 0.2 μS·m^{-1} (nonviable)	Viable and nonviable Giardia intestinalis was differentiated based on dielectric parameter value.	[51]
M. Cristofanilli *et al.* (2002)	Four electrode in phase quadrature	MCF/neo,MCF/HER2-11 and MCF/HER2-18	10 kHz–100 MHz	Specific membrane capacitance MCF/neo (2.09027 μF/cm^2) MCF/HER2-11 (1.70481 μF/cm^2) MCF/HER2-18 (2.5684 μF/cm^2)	Specific membrane capacitance of breast cancer cell lines was reported.	[63]
S. Han (2013)	Four electrode in phase quadrature	Leukocyte (WBCs), SkBr3 and A549	10 kHz–10 MHz	Specific membrane capacitance T lymphocytes (7.01 ± 0.91 mF/m^2) B lymphocytes (10.33 ± 1.6 mF/m^2) Granulocytes (9.14 ± 1.06 mF/m^2) Monocytes (11.77 ± 2.12 mF/m^2) SkBr3 (14.83 ± 1.74 mF/m^2) A549 (16.95 ± 2.93 mF/m^2)	Specific membrane capacitance and cytoplasm conductivity of WBCs and cancer cells was determined using a single-shell dielectric model.	[56]

3.2. Impedance Flow Cytometry

Flow cytometry is a fundamental and powerful analytical tool in cell biology and cellular disease diagnosis for many years. Flow cytometry has an ability to address some problems in single cell analysis such as identifiying, counting and sorting cells [68,69]. Based on laser-induced fluorescence detection in flow cytometry for single-cell studies within cell populations of relatively large sizes [70], flow cytometry creates an ideal scenario to analyze single cell electrical properties from a cell population. Coulter [71] developed the first flow cytometry tool having capability to measure the electrical properties of single particles, which is known as the microfluidic Coulter counter. A Coulter counter measures the changing of DC resistance between two electrically isolated fluid-filled chambers when microparticles act as an insulating layer at DC pass through a small connecting orifice. Figure 3 illustrates the working principle of the Coulter counter, where two large electrodes are placed on connecting chamber. When a particles or biological cells flow through a sensing aperture which has current flow, it will displace the conductive fluid and alters the resistance. The current flow was decreased as a particle passes through and for this reason, individual cells can be counted and sized [72]. The Coulter counter is limited to counting cells and classifying cell types based on size due to challenging of selecting electrode design and channel geometry [73]. The Microfluidic Coulter counter is incapable of characterizing electrical properties of cell. In order to determine the electrical properties of cells, Sohn *et al.* [74] developed flow cytometry based on capacitance principle to measure the DNA content of fixed eukaryotic cells. Electrical properties of individual cells were referred to distinct peaks measured by a capacitance bridge at 1 kHz frequency.

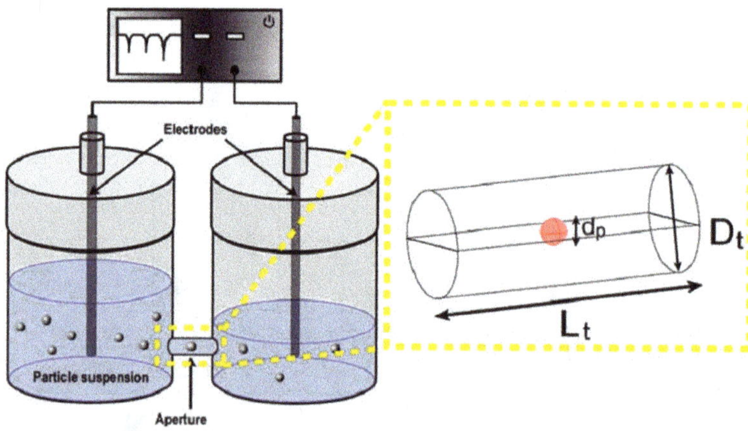

Figure 3. Schematic diagram of the Coulter counter working principle. Reprinted with permission from [72].

Gawad *et al.* [75] developed a significant device in single cell impedance technology, which is known as the impedance flow cytometry (IPC). This device used coplanar electrodes to measure clear differentiation of beads and also erythrocytes and ghost cells (ghosts are RBCs that have been lysed in hypotonic buffer, leaving behind a membrane sack filled with ionic solution). As shown in Figure 4a, three microelectrodes were fabricated on the bottom of a microfluidic channel. An AC voltage was supplied to energise the electrodes for generating a non-uniform electric field within the channel. The impedance value within channel was changed, when a single cell was flowing through the detection area. This impedance value was used to characterize the electrical properties of single cell. However, this electrode configuration may affect impedance measurement when single cell was at variation position. To address this issue, K. Cheung *et al.* [76] designed parallel facing electrodes in a microfluidic channel (Figure 4b). One pair of parallel electrodes was used to detect cells and measure electric current fluctuation, whereas the other one was acted as a reference. Then, the difference between the two signals was measured. The device has the ability to measure electrical properties of normal RBCs and glutaraldehyde-fixed RBCs. More details for the derivation of the electric field distribution for two different electrode configurations, based on Schwarz–Christoffel Mapping (SCM) have been described [77]. In addition, a similar system (parallel facing electrodes) was used by Kampmann *et al.* [78] to monitor frequency effect during conducted measurement processes. The result showed that the cell can be accurately sized at around 500 kHz, where low frequency behaviour is dominated by the electrical double layer (EDL). Meanwhile, at intermediate frequencies behaviour is dominated by the membrane capacitance and at high frequencies, the cell cytoplasm becomes important. High frequency (8.7 Mhz) measurements were used to detect infection of RBCs with the parasite Babesia bovis based on the changes in the electrical properties of the cell cytoplasm [79]. Recently, an impedance flow cytometry that covers frequency range from DC up to 500 Mhz was developed by Niels Haandbæk *et al.* [80]. The device has a capability of dielectric characterization of subcellular components of yeast cells, such as vacuoles and cell nuclei, and can be used for discriminating wild-type yeast from a mutant.

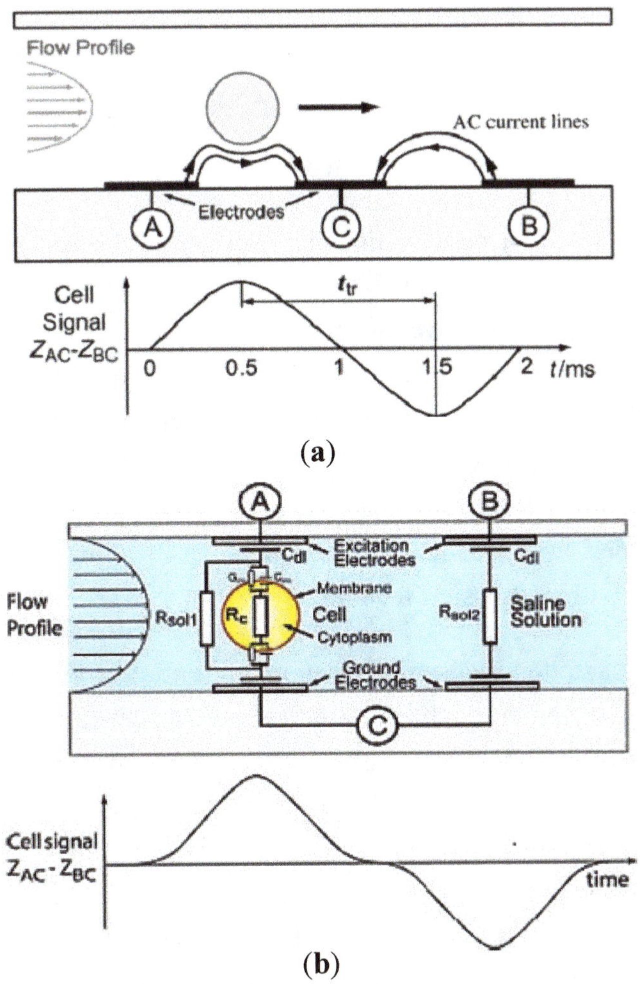

Figure 4. *Cont.*

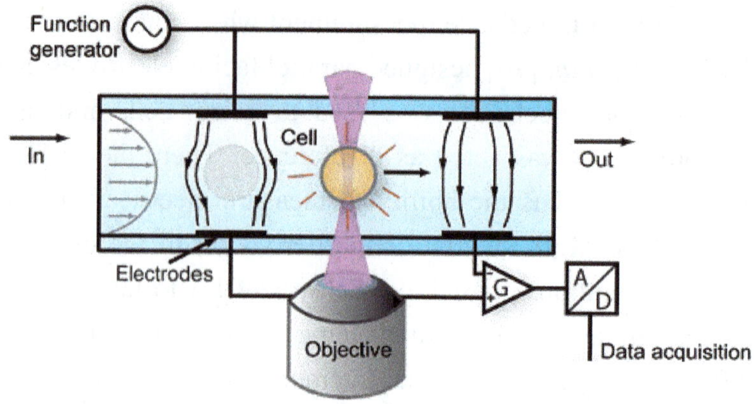

(c)

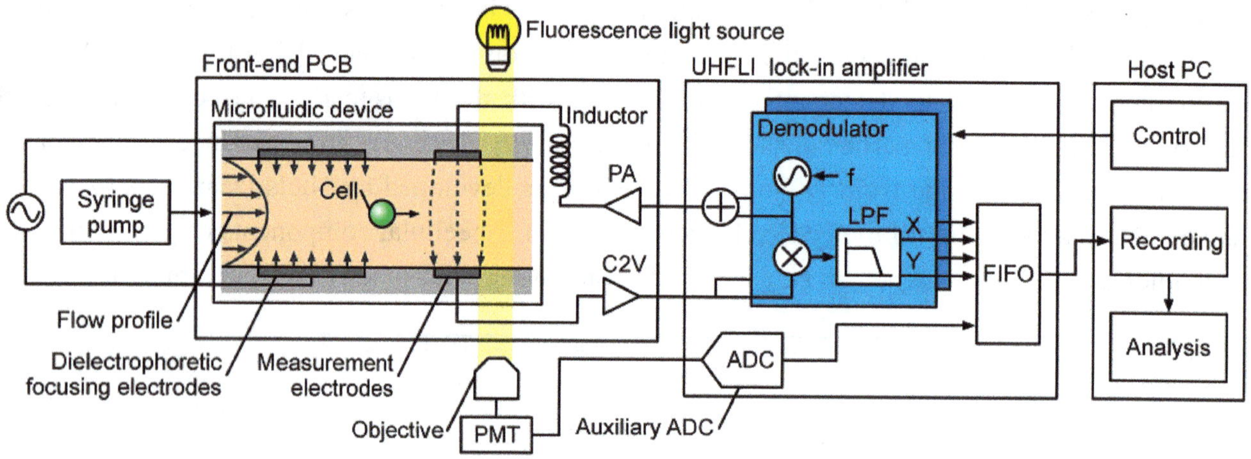

(d)

Figure 4. **(a)** Illustration of a particle flowing over three electrodes inside a microfluidic channel, and a typical impedance signal for a single particle. Reprinted with permission from [75]; **(b)** A single cell flowing over one pair of electrode and second pair used as reference is shown. Reprinted with permission from [76]; **(c)** Schematic diagram of the micro impedance cytometer system, including the confocal-optical detection. Reprinted with permission from [81]; **(d)** Schematic of the complete microfluidic cytometer. The lock-in amplifier drives the series resonance circuit, formed by the discrete inductor and the impedance between the measurement electrodes, with an alternating current (AC) signal at a frequency close to resonance. Reprinted with permission from [82].

Holmes *et al.* [81] demonstrated measurement and differential of single cells at a high speed level by using microfluidic flow cytometry with an attached fluorescence measurement unit (Figure 4c). The device accurately identified T-lymphocytes, monocytes and neutrophils of WBC and a full three-part differential count of whole human blood was achieved. Despite single cells being measured, the data represented the average of the population and was not accurately measured on an individual cell. In addition, similar research groups developed an integrated microfluidic impedance flow cytometry system with haemoglobin concentration measurement unit [83] and RBC lysis [84]. In order to increase the signal-to-noise ratio relative to a single-phase, unfocused stream, while to avoid large shear forces on cells, Mikael Evander [85] developed a microfluidic impedance cytometer that utilizes dielectrophoretic focusing technique. This technique was used to center cells in a fluid stream, thus forms the core of a two-phase flow. Then, this flow will pass between electrodes for analysis of cells at various frequencies from range 280 kHz to 4 MHz. As a result, this technique is able to distinguish between red blood cells and platelets and between resting and activated platelets.

A label-free cell cytometry based on electrophysiological response to stimulus was reported [86]. This method recorded a cell's functionality rather than its expression profile or physical characteristics. In order to distinguish different cells types, they used nature electrically excitable cells that are activated by sufficient transmembrane electric fields. During this activation, the extracellular field potential (FP) signal from cells was produced and detected by electrode inside microchannel. Human induced pluripotent stem cell-derived cardiomyocyte (iPSC-CM) clusters from undifferentiated iPSC clusters were differentiated by using these signals. A contactless measurement method to perform single cell impedance cytometry using a disposable biochip integrated with a printed circuit board that has reusable electrodes was reported [87]. The device can detect and measure impedance of biological cells in a real biological sample (e.g., whole blood (sheep)) and also significantly reduces the manufacturing costs. Recently, a microfluidic impedance cytometer, incorporated with an electrical resonator was reported. This device is high sensitive and capable to measure at high frequencies. Figure 4d showed microfluidic system integrated with a resonant circuit which consists of a discrete inductor in series with the impedance between the measurement electrodes [82]. The cells detection principle is based on the resonance- enhanced phase shift of the measurement current induced by cells or particles passing through the microfluidic channel. Discrimination based on the differences in dielectric properties of *E. coli* and B. subtilis was well achieved. T. Sun *et al.* extends impedance measurements from one dimension to two or three dimensions by utilized electrical impedance tomography (EIT) [88]. A circular 16-electrode array with equal spacing was fabricated and images of Physarum polycephalum were reconstructed by measuring the voltages across sequential electrode-pair combinations. Human fibroblast cells were used to differentiate between an environment of growth medium with and without cells using EIT [89]. Table 2 shows a summary of microfluidic impedance flow cyometry techniques for single cell electrical characterization.

Table 2. Microfluidic impedance flow cytometry device for single cell electrical analysis.

Authors	Techniques	Experimental Samples	Frequency	Summary	Reference
K. Cheung *et al.* (2005)	Parallel facing electrodes	RBCs, ghost RBCs and fixed RBCs	602 kHz and 10 MHz	Controlled RBCs, ghost RBCs and fixed RBCs were distinguished using impedance opacity.	[76]
G. Benazzi *et al.* (2007)	Coplanar electrodes	Algae	327 kHz and 6.03 MHz	Three populations of algae were distinguished on the basis of impedance measurement.	[90]
C. Kuttel *et al.* (2007)	Coplanar electrodes	Babesia bovis infected RBCs	8.7 MHz	The real part and imaginary part of the impedance signal were used for cell type classification.	[79]
G. Schade-Kampmann *et al.* (2008)	Parallel facing electrodes	Jurkat cell, yeast cell and 3T3-L1	624 kHz and 1–15 MHz	Various cell lines, human monocytes and *in vitro*-differentiated dendritic cells and macrophages, viable and apoptotic Jurkat cells were discriminated. Yeast cell growth was also monitored using impedance measurement.	[78]
Y. Katsumoto *et al.* (2008)	Parallel facing electrodes	rabbit erythrocytes and human erythrocytes	10 kHz–100 MHz	Specific membrane capacitance and cytoplasm conductivity values were determined from their dielectric dispersion using new numerical method based on rigorous electric-field simulation combined with three-dimensional modeling of an erythrocyte.	[91]
D. Holmes *et al.* (2009)	Parallel facing electrodes	WBCs	573 kHz and 1.7 MHz	Microfluidic impedance flow cytometry was incorporated with fluorescence detection.	[81]
K.C. Cheung *et al.* (2010)	Parallel facing electrodes	Macrophage, MCF-7, RN22, blood cells and yeast	0.5–15 MHz	Macrophage differentiation, cell viability, blood cells, and RN22 with altered membrane potential and intercellular calcium concentration were distinguished.	[92]
C. Bernabini (2011)	Parallel facing electrodes + hydrodynamic focus	*E. coli* and 1 & 2 μm beads	503 kHz	A focusing technique mitigated the clogging issue and increased sensitivity.	[93]
J. Chen *et al.* (2011)	Constriction channel	MC-3T3	100 Hz–1 MHz	Specific membrane capacitance and cytoplasm conductivity values were determined using a simple equivalent circuit models.	[94]

Table 2. *Cont.*

Authors	Techniques	Experimental Samples	Frequency	Summary	Reference
X.J. Han (2012)	Parallel facing electrodes	RBCs and WBCs	573 kHz–1.7 MHz	The functions of blood dilution, RBCs lysis, and hemoglobin detection were integrated.	[84]
G. Mernier (2012)	Liquid electrodes + DEP focusing	Yeast Cells	500 kHz–15 Mhz	DEP was applied to reduce measurement variations by focusing particles in the middle of the channel.	[95]
Y. Zheng (2013)	Constriction channel + 7 fequencies measurement	AML-2 and HL-60	1–400 kHz	Specific membrane capacitance and cytoplasm conductivity values were determined at speed of 5–10 cell·s^{-1}.	[96]
F.B. Myers (2013)	Electrophysiological cytometry	Pluripontent stem cells	N/A	Clusters of undifferentiated human-induced pluripotent stem cells (iPSC) were identified from iPSC-derived cardiomyocyte (iPSC-CM) clusters.	[86]
Haandbæk *et al.* (2014)	Parallel facing electrodes + resonant circuit	*E. coli* and B. subtilis	89.2 and 87.2 MHz	Discrimination based on the differences in dielectric properties of *E. coli* and B. subtilis.	[82]

RBC: red blood cell; WBC: white blood cell.

3.3. Micro Electrical Impedance Spectroscopy (μ-EIS)

Micro electrical impedance spectroscopy (μ-EIS) is a technique where dielectric properties in a frequency domain of a cell is measured to characterize and differentiate the various types of cell. Mainly this technique analyzed the current response when a single cell was trapped in a trapping system where an alternating current (AC) was applied across the trapping zone. A trapping system is a major contribution and significant part in μ-EIS device. For this reason, development of a trapping system is very crucial and varieties of the trapping system have been developed, such as hydrodynamic traps, negative pressure traps and DEP traps.

First development of micro electrical impedance spectroscopy (μ-EIS) was reported in 2006 [97]. They developed microfluidic device which utilized the negative pressure to capture the single cell into the analysis cavity (Figure 5a). This device was used to measure the electrical impedance of human breast cancer cell lines of different pathological stages (MCF-7, MDA-MB-231, and MDA-MB-435) [18]. However this device has a disadvantage to monitor the cell capturing process using a microscope because the contrast difference in the silicon nitride membrane composing the cell traps area and the surroundings. The same group, Cho *et al.* [4] developed an array of horizontal cell traps of an μ-EIS device to overcome the limitation of the previous device. Negative pressure was used to capture single cells and impedance measurement was performed to obtain the electrical impedance spectra of metastatic head and neck cancer (HNC) cell lines. This device also can minimize the leakage current due to the position of cells formed in direct contact between cells and electrodes.

Furthermore, the concept of vertical trapping system in μ-EIS has been used to monitor the dynamic change of single cell electrical properties over a period of time [98,99]. Hydrodynamic trapping system (e.g., micropillars) within a microfluidic channel was developed by Jang *et al.* [21]. Figure 5b showed the micropillars structure inside microfluidic channel and capable to capture physically single cells. A single human cervical epithelioid carcinoma (HeLa) was successfully captured by the micro pillars and its impedance was measured. Mondal *et al.* performed impedance measurement of HeLa cell based on two geometry structures of micropillars trapping system, namely, parallel and elliptical geometry [100]. Malleo *et al.* demonstrated a hydrodynamic trapping device which has a differential electrode arrangement that measures multiple signals from multiple trapping sites. Measurements was performed by recording the current from two electrode pairs, one empty (reference) and one containing HeLa cells [101]. The device continuously monitored the toxin activity at the single cell level.

Recently, the concept of dielectrophoresis (DEP) for trapping system was reported [102]. The non-uniform electric field distribution between the top and bottom electrodes caused the red blood cells (RBCs) to experience positive dielectrophoresis at 80 kHz frequency [103]. As a result, the red blood cells have been trapped inside microwells, thus the impedance of RBCs was measured [102]. Another DEP trapping technique was developed by Tsai *et al.* to capture a single HeLa cell, then impedance measurement was performed [104]. Figure 5c illustrated trapping system using DEP [104]. Despite that microelectrical impedance spectroscopy (μ-EIS) has several advantages such as label free, real time measurement and non invasive, μ-EIS also has some drawbacks. For example, μ-EIS requires theoretical model for data analysis [105] and time consuming (trapping and releasing process take time to be completed) [106]. Table 3 shows a summary of microelectrical impedance spectroscopy technologies for analyzing the single cell's electrical properties.

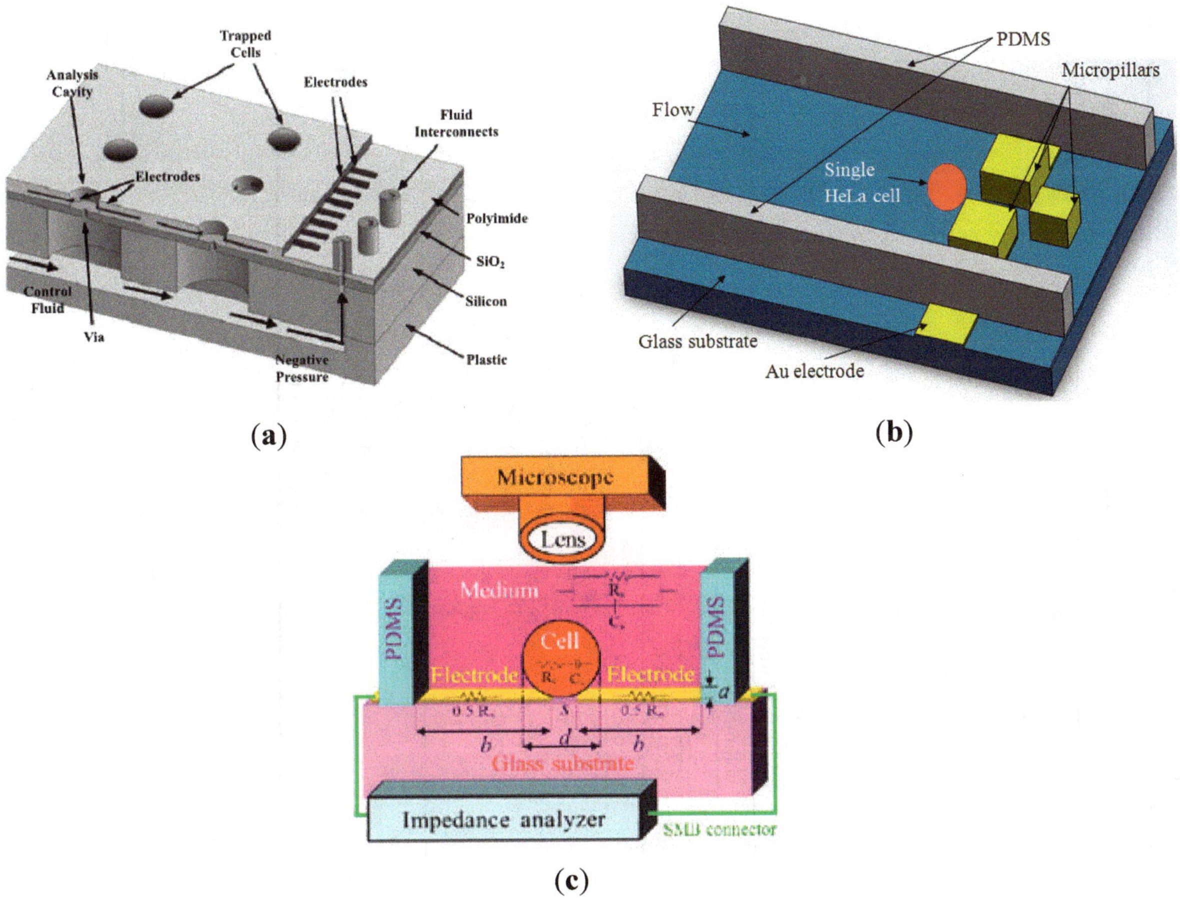

Figure 5. (**a**) Illustrated a micro electrical impedance spectroscopy system using multielectrode configurations within an analysis cavity. Reprinted with permission from [97]; (**b**) Shown 3D schematic of the μ-EIS device incorporated with micropillars structure for capture the single cells; (**c**) Schematic diagram of cell measurement using DEP cell trapping technique. Reprinted with permission from [104].

Table 3. Microelectrical impedance spectoscopy device for single cell electrical analysis.

Authors	Techniques	Experimental Samples	Frequency	Dielectric Parameter	Summary	Reference
A. Han *et al.* (2003)	Vertical hole	MCF-7,MCF-10A, MCF-MB-231 and MDA-MB-435	100 Hz–3 MHz	specific membrane capacitance MCF-10A (1.94 ± 0.14 μF/cm^2) MCF-7 (1.86 ± 0.11 μF/cm^2) MDA-MB-231 (1.63 ± 0.17 μF/cm^2) MDA-MB-435 (1.57 ± 0.12 μF/cm^2)	Impedance spectra were shown to be significantly different between the normal cell lines and each of the cancer cell lines.	[97]
L.S. Jang *et al.* (2007)	Micropillars	Hela	1 Hz–100 kHz	cell membrane C_c 2.5×10^{-12} F cytoplasm R_c 6×10^7 Ω	A circuit model was developed to obtained and calculate electrical parameters of HeLa cells.	[21]
S.B. Cho *et al.* (2007)	Vertical hole	L929	1 Hz–100 MHz	N/A	A culture of L929 cells and the toxicity effect on impedance measurement were monitored on the micro hole. Cell growth and the membrane integrity can monitored without any labelling.	[98]
Y. Cho *et al.* (2009)	Parallel lateral trapping holes	686LN and 686LN-M4e	40 Hz–10 MHz	N/A	The phase part of impedances could be used to differentiate the poorly metastatic cell line from the highly metastatic cell line.	[4]

Table 3. *Cont.*

Authors	Techniques	Experimental Samples	Frequency	Dielectric Parameter	Summary	Reference
D. Malleo *et al.* (2010)	Hydraulic trapping	Hela	300 kHz	N/A	Effect of a surfactant and a pore-forming toxin on captured cells was monitored by referring the impedance value of captured cells.	[101]
C.L. Kung *et al.* (2011)	DEP trapping	Hela	1 Hz–100 kHz	N/A	An alternating current electrothermal effect (ACET) and a negative dielectrophoresis (nDEP) force was utilized to trap cells.	[107]
C.M. Kurz *et al.* (2011)	Vertical hole	Arpe-19	1 kHz	N/A	The subtoxic effect of cells was measured by monitoring impedance signals over time.	[99]
Y. Zhao *et al.* (2014)	Constriction channel + impedance measurement	95D and 95D CCNY-KD	1 and 100 kHz	specific membrane capacitance 95D (1.8–2.0 μF/cm^2) 95D CCNY-KD (1.4–1.6 μF/cm^2)	Specific membrane capacitance and cytoplasm conductivity were determined.	[7]
P. Shah *et al.* (2014)	pDEP trapping	CCL-149 (Rat lung epithelial cells)	1 Hz–10 MHz	impedance in absence 1.51 MΩ impedance in the presence of cell 17 MΩ	Impedance spectrum used to monitoring in absence and in the presence of a single cell in microwell.	[108]
S.-B. Huang, *et al.* (2014)	Constriction channel with an incorporated pneumatically driven + impedance measurement	CCL-185	1 and 100 kHz	specific membrane capacitance 2.17 ± 0.58 μF/cm^2 cytoplasm conductivity 0.74 ± 0.20 S/m	A pneumatically driven membrane-based active valve was utilized for unblocking cell aggregates at the entrance constriction channel.	[20]

4. Discussion

The rapid development of single cell analysis tools (e.g., biophysical and chemical characterizations) can be seen based on the hundreds of review and technical papers currently published every year [109]. Clearly, the growing interest in this research field demonstrates its practical value from the viewpoint of proof of concept and applications. The traditional platform is a basic foundation to provide the most straight-forward mechanisms to analyze electrical properties of individual cell. However this approach suffers from low throughput, delicate protocol, requires an experienced operator and bulky experimental set-up [110]. For instance, in clinical application, high throughput devices are significantly required to test a large number of cells (e.g., blood) in order to obtain low numbers of meaningful data (e.g., CTC cells) [57]. Nevertheless, traditional platforms have provided fundamental insights to the microfluidic development. The microfuidics device is a promising technique to understand the cellular heterogeneity and overcome the limitation of traditional technique. For that reason, three microfluidics techniques (electrorotation, impedance flow cytometry and microelectrical impedance spectroscopy) have been developed to analyze and characterize the single cell's electrical properties. Among these techniques, microfluidic impedance flow cytometry (IFC) is a technique used widely in clinical diagnosis because of high throughput during count and differentiation of the WBCs. For example, parallel facing electrodes device achieved ~100 cells$\cdot s^{-1}$ and is capable of testing a large number of cells for obtain statistically meaningful data [81]. Microfluidic impedance flow cytometry has been demonstrated to distinguish various single cells (16 types of cell) based on the electrical properties conditions. Meanwhile, 13 types of cell were distinguished by electrorotation and microelectrical impedance spectroscopy.

Electrical measurements can also be incorporated with a cell sorting unit to collect cells having different physical properties for further biochemical assaying. AC dielectrophoretic (DEP) for sorting live cells from interfering particles of similar sizes by their polarizabilities under continuous flow was reported [111]. DEP forces induced by the AgPDMS electrodes were used to manipulate cells to move toward high or low electric field regions, depending on the relative polarizability between the cells and their suspending medium. Jun Yang *et al.* utilized magnetically activated cell sorting (MACS) for obtaining the subpopulations from human peripheral blood (B-lymphocytes and monocytes), thus performing the single cell electrical properties measurement by electrorotation techniques [48].

Microfluidic devices have demonstrated great potential in realizing electrical measurements on single cells at a higher testing speed and label free approach. Electrical measurements on single cells can be used to indicate possible diseases and it suitable for disease prescreening application. From prescreening processes, future examinations can be done to evaluate the disease condition. Table 4 shows a summary of comparisons between three microfluidic methods. The microfluidic techniques were discussed have some potential applications in biological and medical application [18,44,92,101,112].

Table 4. Comparisons between three microfluidic techniques.

Approaches Technique	Advantages	Disadvantages	Applications
Electrorotation	Capable to quantifying a cell's intrinsic electrical properties	Low throughput and limitation to low conductivity sucrose buffer solution	Monitor parasite; Cell separation
Impedance flow cytometry	High throughput	low specificity	Cell sorting and counting; Cell impedance variations; DNA hybridization detection
Microelectrical impedance spectroscopy	Characterizing ion channel activity	Low throughput and size-independent parameters	Cancerous stage screening; Toxin detection

5. Conclusions

The presented review of selected research works on single cell electrical properties provides information on technological development in single cell electrical characterization from traditional approaches to current microfluidic approaches. Microfluidics technology opens a new paradigm in cellular and microbiology research for early disease detection and provides critical information needed by research scientists and clinicians for improved clinical diagnosis and patient outcome. The recent excellent achievements in microfabrication techniques have enabled the rapid development of microfluidic technologies for further practical applications for the benefit of mankind. Furthermore, microfluidic technological progress has provided additional advantages such as reduced complexity of experiment handling, lower voltage on the electrodes, faster heat dissipation, small volume of reagents used, and *in situ* observation of the cell response [113].

Acknowledgments

This work was supported by grants from the Ministry of Education Malaysia (grant No. 4L640 and 4F351) and Universiti Teknologi Malaysia (grant No. 02G46, 03H82 and 03H80).

Author Contributions

Muhammad Asraf Mansor and Mohd Ridzuan Ahmad wrote and edited the review.

References

1. El-Ali, J.; Sorger, P.K.; Jensen, K.F. Cells on Chips. *Nature* **2006**, *442*, 403–411.
2. Suresh, S. Biomechanics and Biophysics of Cancer Cells. *Acta Biomater.* **2007**, *3*, 413–438.
3. Coley, H.M.; Labeed, F.H.; Thomas, H.; Hughes, M.P. Biophysical characterization of MDR breast cancer cell lines reveals the cytoplasm is critical in determining drug sensitivity. *Biochim. Biophys. Acta* **2007**, *1770*, 601–608.

4. Cho, Y.; Frazier, A.B.; Chen, Z.G.; Han, A. Whole-cell impedance analysis for highly and poorly metastatic cancer cells. *J. Microelectromech. Syst.* **2009**, *18*, 808–817.
5. Zhao, Y.; Chen, D.; Luo, Y.; Li, H.; Deng, B.; Huang, S.-B.; Chiu, T.-K.; Wu, M.-H.; Long, R.; Hu, H.; *et al.* A microfluidic system for cell type classification based on cellular size-independent electrical properties. *Lab Chip* **2013**, *13*, 2272–2277.
6. Byun, S.; Son, S.; Amodei, D.; Cermak, N.; Shaw, J.; Kang, J.H.; Hecht, V.C.; Winslow, M.M.; Jacks, T.; Mallick, P.; *et al.* Characterizing deformability and surface friction of cancer cells. *Proc. Natl. Acad. Sci. USA* **2013**, *110*, 7580–7585.
7. Zhao, Y.; Zhao, X.T.; Chen, D.Y.; Luo, Y.N.; Jiang, M.; Wei, C.; Long, R.; Yue, W.T.; Wang, J.B.; Chen, J. Tumor cell characterization and classification based on cellular specific membrane capacitance and cytoplasm conductivity. *Biosens. Bioelectron.* **2014**, *57*, 245–253.
8. Radke, S.M.; Alocilja, E.C. A high density microelectrode array biosensor for detection of *E. coli* O157:H7. *Biosens. Bioelectron.* **2005**, *20*, 1662–1667.
9. Bow, H.; Pivkin, I. V; Diez-Silva, M.; Goldfless, S.J.; Dao, M.; Niles, J.C.; Suresh, S.; Han, J. A microfabricated deformability-based flow cytometer with application to malaria. *Lab Chip* **2011**, *11*, 1065–1073.
10. Du, E.; Ha, S.; Diez-Silva, M.; Dao, M.; Suresh, S.; Chandrakasan, A.P. Electric impedance microflow cytometry for characterization of cell disease states. *Lab Chip* **2013**, *13*, 3903–3909.
11. Asphahani, F.; Zhang, M. Cellular impedance biosensors for drug screening and toxin detection. *Analyst* **2007**, *132*, 835–841.
12. Gabriel, C.; Gabriel, S.; Corthout, E. The dielectric properties of biological tissues: I. *Lit. Surv.* **1996**, *41*, 2231–2249.
13. Stoneman, M.R.; Kosempa, M.; Gregory, W.D.; Gregory, C.W.; Marx, J.J.; Mikkelson, W.; Tjoe, J.; Raicu, V. Correction of electrode polarization contributions to the dielectric properties of normal and cancerous breast tissues at audio/radiofrequencies. *Phys. Med. Biol.* **2007**, *52*, 6589–6604.
14. Yager, P.; Edwards, T.; Fu, E.; Helton, K.; Nelson, K.; Tam, M.R.; Weigl, B.H. Microfluidic diagnostic technologies for global public health. *Nature* **2006**, *442*, 412–418.
15. Whitesides, G.M. The origins and the future of microfluidics. *Nature* **2006**, *442*, 368–373.
16. Brehm-Stecher, B.F.; Johnson, E.A. Single-cell microbiology: tools, technologies, and applications. *Microbiol. Mol. Biol. Rev.* **2004**, *68*, 538–559.
17. Marcus, J.S.; Anderson, W.F.; Quake, S.R. Microfluidic single-cell mRNA isolation and analysis. *Anal. Chem.* **2006**, *78*, 3084–3089.
18. Han, A.; Yang, L.; Frazier, A.B. Quantification of the heterogeneity in breast cancer cell lines using whole-cell impedance spectroscopy. *Clin. Cancer Res.* **2007**, *13*, 139–143.
19. Kunzelmann, K. Ion Channels and Cancer. *J. Membr. Biol.* **2005**, *205*, 159–173.
20. Huang, S.-B.; Zhao, Y.; Chen, D.; Lee, H.-C.; Luo, Y.; Chiu, T.-K.; Wang, J.; Chen, J.; Wu, M.-H. A clogging-free microfluidic platform with an incorporated pneumatically driven membrane-based active valve enabling specific membrane capacitance and cytoplasm conductivity characterization of single cells. *Sens. Actuators B Chem.* **2014**, *190*, 928–936.
21. Jang, L.-S.; Wang, M.-H. Microfluidic device for cell capture and impedance measurement. *Biomed. Microdevices* **2007**, *9*, 737–743.

22. Abdolahad, M.; Sanaee, Z.; Janmaleki, M.; Mohajerzadeh, S.; Abdollahi, M.; Mehran, M. Vertically aligned multiwall-carbon nanotubes to preferentially entrap highly metastatic cancerous cells. *Carbon N. Y.* **2012**, *50*, 2010–2017.
23. Katsumoto, Y.; Tatsumi, K.; Doi, T.; Nakabe, K. Electrical classification of single red blood cell deformability in high-shear microchannel flows. *Int. J. Heat Fluid Flow* **2010**, *31*, 985–995.
24. Dondorp, A.M.; Kager, P.A.; Vreeken, J.; White, N.J. Abnormal blood flow and red blood cell deformability in severe malaria. *Parasitol. Today* **2000**, *16*, 228–232.
25. Zhao, Y.; Inayat, S.; Dikin, D.A.; Singer, J.H.; Ruoff, R.S.; Troy, J.B. Patch clamp technique: review of the current state of the art and potential contributions from nanoengineering. *Proc. Inst. Mech. Eng. Part N: J. Nanoeng. Nanosyst.* **2008**, *222*, 1–11.
26. Sakmann, B.; Neher, E. Patch clamp techniques for studying ionic channels in excitable membranes. *Annu. Rev. Physiol.* **1984**, *46*, 455–472.
27. Neher, E.; Sakmann, B. The patch clamp technique. *Sci. Am.* **1992**, *266*, 44–51.
28. Hamill, O.P.; Marty, A.; Neher, E.; Sakmann, B.; Sigworth, F.J. Improved Patch-clamp techniques for high-resolution current recording from cells and cell-free membrane patches. *Pflugers Arch. Eur. J. Physiol.* **1981**, *391*, 85–100.
29. Jain, G.; Muley, N.; Soni, P.; Singh, J.N.; Sharma, S.S. Patch clamp technique: conventional to automated. *Curr. Res. Inf. Pharm. Sci.* **2009**, *10*, 9–15.
30. DeFelice, L.J. *Electrical Properties of Cells: Patch Clamp for Biologists*; Springer Science & Business Media: New York, NY, USA, 1997.
31. Marty, A.; Neher, E. Tight-seal whole-cell recording. In *Single-Channel Recording*; Sakmann, B., Neher, E., Eds.; Springer US: Boston, MA, USA, 2009; pp. 31–52.
32. Ogden, D.; Stanfield, P. Patch Clamp techniques for single channel and whole-cell recording. In *Microelectrode Techniques*; The Company of Biologists Ltd.: Cambridge, UK, 1994; pp. 53–78.
33. Karmazínová, M.; Lacinová, L. Measurement of cellular excitability by whole cell patch clamp technique. *Physiol. Res.* **2010**, *59*, S1–S7.
34. Zhang, J.; Davidson, R.M.; Wei, M.D.; Loew, L.M. Membrane electric properties by combined patch clamp and fluorescence ratio imaging in single neurons. *Biophys. J.* **1998**, *74*, 48–53.
35. Willumsen, N.J.; Bech, M.; Olesen, S.-P.; Jensen, B.S.; Korsgaard, M.P.G.; Christophersen, P. High throughput electrophysiology: New perspectives for ion channel drug discovery. *Recept. Channels* **2003**, *9*, 3–12.
36. Bébarová, M. Advances in patch clamp technique: towards higher quality and quantity. *Gen. Physiol. Biophys.* **2012**, *31*, 131–140.
37. Ahmad, M.R.; Nakajima, M. Single cells electrical characterizations using nanoprobe via ESEM-nanomanipulator system. *IEEE Conf. Nanotechnol.* **2009**, *8*, 589–592.
38. Hafiz, A.; Sulaiman, M.; Ahmad, M.R. Rigid and conductive dual nanoprobe for single cell analysis. *J. Teknol.* **2014**, *8*, 107–113.
39. Esteban-Ferrer, D.; Edwards, M.A; Fumagalli, L.; Juárez, A.; Gomila, G. Electric polarization properties of single bacteria measured with electrostatic force microscopy. *ACS Nano* **2014**, *8*, 9843–9849.

40. Sulaiman, A.H.M.; Ahmad, M.R. Integrated dual nanoprobe-microfluidic system for single cell penetration. In Proceedings of the 2013 IEEE International Conference on Control System, Computing and Engineering (ICCSCE), Mindeb, 29 November–1 December 2013; pp. 568–572.
41. Unger, M.A.; Chou, H.P.; Thorsen, T.; Scherer, A.; Quake, S.R. Monolithic microfabricated valves and pumps by multilayer soft lithography. *Science* **2000**, *288*, 113–116.
42. Hansen, C.; Quake, S.R. Microfluidics in structural biology: Smaller, faster...better. *Curr. Opin. Struct. Biol.* **2003**, *13*, 538–544.
43. Reyes, D.R.; Iossifidis, D.; Auroux, P.-A.; Manz, A. Micro total analysis systems. 1. introduction, theory, and technology. *Anal. Chem.* **2002**, *74*, 2623–2636.
44. Javanmard, M.; Davis, R.W. A microfluidic platform for electrical detection of DNA hybridization. *Sens. Actuators B Chem.* **2011**, *154*, 22–27.
45. Yang, L.; Li, Y.; Griffis, C.L.; Johnson, M.G. Interdigitated microelectrode (IME) impedance sensor for the detection of viable salmonella typhimurium. *Biosens. Bioelectron.* **2004**, *19*, 1139–1147.
46. Valencia, P.M.; Farokhzad, O.C.; Karnik, R.; Langer, R. Microfluidic technologies for accelerating the clinical translation of nanoparticles. *Nat. Nanotechnol.* **2012**, *7*, 623–629.
47. Tian Wei-Cheng, E.F. *Microfluidics for Biological Applications*; Springer Science + Business Media, LLC: New York, NY, USA, 2008.
48. Yang, J.; Huang, Y.; Wang, X.; Wang, X.B.; Becker, F.F.; Gascoyne, P.R. Dielectric properties of human leukocyte subpopulations determined by electrorotation as a cell separation criterion. *Biophys. J.* **1999**, *76*, 3307–3314.
49. Cen, E.G.; Dalton, C.; Li, Y.; Adamia, S.; Pilarski, L.M.; Kaler, K.V.I.S. A Combined dielectrophoresis, traveling wave dielectrophoresis and electrorotation microchip for the manipulation and characterization of human malignant Cells. *J. Microbiol. Methods* **2004**, *58*, 387–401.
50. Arnold, W.M.; Zimmermann, U. Rotating-field-induced rotation and measurement of the membrane capacitance of single mesophyll cells of avena sativa sites. *Z. Naturforsch* **1982**, *37c*, 908–915.
51. Dalton, C.; Goater, A.D.; Burt, J. P.H.; Smith, H.V. Analysis of parasites by electrorotation. *J. Appl. Microbiol.* **2004**, *96*, 24–32.
52. Huang, J.P.; Yu, K.W. First-principles approach to electrorotation assay. *J. Phys. Condens. Matter* **2002**, *14*, 1213–1221.
53. Huang, J.; Yu, K.; Gu, G. Electrorotation of a pair of spherical particles. *Phys. Rev. E* **2002**, *65*, doi:org/10.1103/PhysRevE.65.021401.
54. Jones, T.B. Basic theory of dielectrophoresis and electrorotation. *IEEE Eng. Med. Biol. Mag.* **2003**, *22*, 33–42.
55. Morgan, H.; Green, N.G. *AC Electrokinetics: Colloids and Nanoparticles*; Microtechnologies and Microsystems Series; Research Studies Press: Hertfordshire, UK, 2003.
56. Han, S.-I.; Joo, Y.-D.; Han, K.-H. An electrorotation technique for measuring the dielectric properties of cells with simultaneous use of negative quadrupolar dielectrophoresis and electrorotation. *Analyst* **2013**, *138*, 1529–1537.

57. Zheng, Y.; Nguyen, J.; Wei, Y.; Sun, Y. Recent advances in microfluidic techniques for single-cell biophysical characterization. *Lab Chip* **2013**, *13*, 2464–2483.
58. Zhou, X.F.; Markx, G.H.; Pethig, R. Effect of biocide concentration on electrorotation spectra of yeast cells. *Biochim. Biophys. Acta* **1996**, *1281*, 60–64.
59. Hölzel, R. Non-invasive determination of bacterial single cell properties by electrorotation. *Biochim. Biophys. Acta* **1999**, *1450*, 53–60.
60. Dalton, C.; Goater, A.D.; Drysdale, J.; Pethig, R. Parasite viability by electrorotation. *Colloids Surfaces A Physicochem. Eng. Asp.* **2001**, *195*, 263–268.
61. Goater, A.D.; Burt, J.P. H.; Pethig, R. A combined travelling wave dielectrophoresis and electrorotation device: Applied to the concentration and viability determination of cryptosporidium. *J. Phys. D Appl. Phys.* **1997**, *30*, L65–L69.
62. De Gasperis, G.; Wang, X.; Yang, J.; Becker, F.F.; Gascoyne, P.R.C. Automated electrorotation: dielectric characterization of living cells by real-time motion estimation. *Meas. Sci. Technol.* **1998**, *9*, 518–529.
63. Cristofanilli, M.; de Gasperis, G.; Zhang, L.; Hung, M.C. Automated electrorotation to reveal dielectric variations related to HER-2/neu overexpression in MCF-7 sublines. *Clin. Cancer Res.* **2002**, *8*, 615–619.
64. Hughes, M.P. Computer-aided analysis of conditions for optimizing practical electrorotation. *Phys. Med. Biol.* **1998**, *43*, 3639–3648.
65. Wang, X.B.; Huang, Y.; Gascoyne, P.R.; Becker, F.F.; Hölzel, R.; Pethig, R. Changes in friend murine erythroleukaemia cell membranes during induced differentiation determined by electrorotation. *Biochim. Biophys. Acta* **1994**, *1193*, 330–344.
66. Becker, F.F.; Wang, X.B.; Huang, Y.; Pethig, R.; Vykoukal, J.; Gascoyne, P.R. Separation of human breast cancer cells from blood by differential dielectric affinity. *Proc. Natl. Acad. Sci. USA* **1995**, *92*, 860–864.
67. Hölzel, R. Electrorotation of single yeast cells at frequencies between 100 Hz and 1.6 GHz. *Biophys. J.* **1997**, *73*, 1103–1109.
68. Davey, H.M.; Kell, D.B. Flow cytometry and cell sorting of heterogeneous microbial populations: The Importance of single-cell analyses. *Microbiol. Rev.* **1996**, *60*, 641–696.
69. Shapiro, H.M. The evolution of cytometers. *Cytom. Part A* **2004**, *58*, 13–20.
70. Givan, A.L. *Flow Cytometry: First Principles*; John Wiley & Sons, Inc.: New York, NY, USA, 2001.
71. WH, C. High speed automatic blood cell counter and cell analyzer. *Proc. Natl. Electron. Conf.* **1956**, *12*, 1034–1040.
72. Sun, T.; Morgan, H. Single-cell microfluidic impedance cytometry: A review. *Microfluid. Nanofluid.* **2010**, *8*, 423–443.
73. Saleh, O.A.; Sohn, L.L. Quantitative sensing of nanoscale colloids using a microchip coulter counter. *Rev. Sci. Instrum.* **2001**, *72*, 4449.
74. Sohn, L.L.; Saleh, O.A.; Facer, G.R.; Beavis, A.J.; Allan, R.S.; Notterman, D.A. Capacitance cytometry: Measuring biological cells one by one. *Proc. Natl. Acad. Sci. USA* **2000**, *97*, 10687–10690.

75. Gawad, S.; Schild, L.; Renaud, P.H. Micromachined impedance spectroscopy flow cytometer for cell analysis and particle sizing. *Lab Chip* **2001**, *1*, 76–82.
76. Cheung, K.; Gawad, S.; Renaud, P. Impedance spectroscopy flow cytometry: On-chip label-free cell differentiation. *Cytom. Part A* **2005**, *65*, 124–132.
77. Sun, T.; Green, N.G.; Gawad, S.; Morgan, H. Analytical electric field and sensitivity analysis for two microfluidic impedance cytometer designs. *IET Nanobiotechnol.* **2007**, *1*, 69–79.
78. Schade-Kampmann, G.; Huwiler, A.; Hebeisen, M.; Hessler, T.; di Berardino, M. On-chip non-invasive and label-free cell discrimination by impedance spectroscopy. *Cell Prolif.* **2008**, *41*, 830–840.
79. Küttel, C.; Nascimento, E.; Demierre, N.; Silva, T.; Braschler, T.; Renaud, P.; Oliva, A.G. Label-free detection of babesia bovis infected red blood cells using impedance spectroscopy on a microfabricated flow cytometer. *Acta Trop.* **2007**, *102*, 63–68.
80. Haandbæk, N.; Bürgel, S.C.; Heer, F.; Hierlemann, A. Characterization of subcellular morphology of single yeast cells using high frequency microfluidic impedance cytometer. *Lab Chip* **2014**, *14*, 369–377.
81. Holmes, D.; Pettigrew, D.; Reccius, C.H.; Gwyer, J.D.; van Berkel, C.; Holloway, J.; Davies, D.E.; Morgan, H. Leukocyte analysis and differentiation using high speed microfluidic single cell impedance cytometry. *Lab Chip* **2009**, *9*, 2881–2889.
82. Haandbæk, N.; With, O.; Bürgel, S.C.; Heer, F.; Hierlemann, A. Resonance-enhanced microfluidic impedance cytometer for detection of single bacteria. *Lab Chip* **2014**, *14*, 3313–3324.
83. Van Berkel, C.; Gwyer, J.D.; Deane, S.; Green, N.G.; Green, N.; Holloway, J.; Hollis, V.; Morgan, H. Integrated systems for rapid point of care (PoC) blood cell analysis. *Lab Chip* **2011**, *11*, 1249–1255.
84. Han, X.; van Berkel, C.; Gwyer, J.; Capretto, L.; Morgan, H. Microfluidic lysis of human blood for leukocyte analysis using single cell impedance cytometry. *Anal. Chem.* **2012**, *84*, 1070–1075.
85. Evander, M.; Ricco, A.J.; Morser, J.; Kovacs, G.T.A; Leung, L.L.K.; Giovangrandi, L. Microfluidic impedance cytometer for platelet analysis. *Lab Chip* **2013**, *13*, 722–729.
86. Myers, F.B.; Zarins, C.K.; Abilez, O.J.; Lee, L.P. Label-free electrophysiological cytometry for stem cell-derived cardiomyocyte clusters. *Lab Chip* **2013**, *13*, 220–228.
87. Emaminejad, S.; Javanmard, M.; Dutton, R.W.; Davis, R.W. Microfluidic diagnostic tool for the developing world: Contactless impedance flow cytometry. *Lab Chip* **2012**, *12*, 4499–4507.
88. Sun, T.S.T.; Tsuda, S.; Zauner, K.-P.; Morgan, H. Single cell imaging using electrical impedance tomography. In Proceedings of the 4th IEEE International Conference on Nano/Micro Engineered and Molecular Systems, Shenzhen, China, 5–8 January 2009; pp. 858–863.
89. Chai, K.T.C.; Davies, J.H.; Cumming, D.R.S. Electrical impedance tomography for sensing with integrated microelectrodes on a CMOS microchip. *Sens. Actuators B Chem.* **2007**, *127*, 97–101.
90. Benazzi, G.; Holmes, D.; Sun, T.; Mowlem, M.C.; Morgan, H. Discrimination and analysis of phytoplankton using a microfluidic cytometer. *IET Nanobiotechnol.* **2007**, *1*, 94–101.
91. Katsumoto, Y.; Hayashi, Y.; Oshige, I.; Omori, S.; Kishii, N.; Yasuda, A.; Asami, K. Dielectric cytometry with three-dimensional cellular modeling. *Biophys. J.* **2008**, *95*, 3043–3047.

92. Cheung, K.C.; di Berardino, M.; Schade-Kampmann, G.; Hebeisen, M.; Pierzchalski, A.; Bocsi, J.; Mittag, A.; Tárnok, A. Microfluidic impedance-based flow cytometry. *Cytometry. A* **2010**, *77*, 648–666.
93. Bernabini, C.; Holmes, D.; Morgan, H. Micro-impedance cytometry for detection and analysis of micron-sized particles and bacteria. *Lab Chip* **2011**, *11*, 407–412.
94. Chen, J.; Zheng, Y.; Tan, Q.; Zhang, Y.L.; Li, J.; Geddie, W.R.; Jewett, M.A.S.; Sun, Y. A microfluidic device for simultaneous electrical and mechanical measurements on single cells. *Biomicrofluidics* **2011**, *5*, 14113.
95. Mernier, G.; Duqi, E.; Renaud, P. Characterization of a novel impedance cytometer design and its integration with lateral focusing by dielectrophoresis. *Lab Chip* **2012**, *12*, 4344–4349.
96. Zheng, Y.; Shojaei-Baghini, E.; Wang, C.; Sun, Y. Microfluidic characterization of specific membrane capacitance and cytoplasm conductivity of single cells. *Biosens. Bioelectron.* **2013**, *42*, 496–502.
97. Han, A.; Frazier, A.B. Ion channel characterization using single cell impedance spectroscopy. *Lab Chip*. **2006**, 6, 1412–1414.
98. Cho, S.; Thielecke, H. Micro Hole-based cell chip with impedance spectroscopy. *Biosens. Bioelectron.* **2007**, *22*, 1764–1768.
99. Kurz, C.M.; Büth, H.; Sossalla, A.; Vermeersch, V.; Toncheva, V.; Dubruel, P.; Schacht, E.; Thielecke, H. Chip-based impedance measurement on single cells for monitoring sub-toxic effects on cell membranes. *Biosens. Bioelectron.* **2011**, *26*, 3405–3412.
100. Mondal, D.; Roychaudhuri, C.; Das, L.; Chatterjee, J. Microtrap electrode devices for single cell trapping and impedance measurement. *Biomed. Microdevices* **2012**, *14*, 955–964.
101. Malleo, D.; Nevill, J.T.; Lee, L.P.; Morgan, H. Continuous differential impedance spectroscopy of single cells. *Microfluid. Nanofluid.* **2010**, *9*, 191–198.
102. Ameri, S.K.; Singh, P.K.; Dokmeci, M.R.; Khademhosseini, A.; Xu, Q.; Sonkusale, S.R. All electronic approach for high-throughput cell trapping and lysis with electrical impedance monitoring. *Biosens. Bioelectron.* **2014**, *54*, 462–467.
103. Sano, M.B.; Caldwell, J.L.; Davalos, R.V. Modeling and development of a low frequency contactless dielectrophoresis (cDEP) platform to sort cancer cells from dilute whole blood samples. *Biosens. Bioelectron.* **2011**, *30*, 13–20.
104. Tsai, S.-L.; Chiang, Y.; Wang, M.-H.; Chen, M.-K.; Jang, L.-S. Battery-powered portable instrument system for single-cell trapping, impedance measurements, and modeling analyses. *Electrophoresis* **2014**, *16*, 2392–2400.
105. Sin, M.L.Y.; Mach, K.E.; Wong, P.K.; Liao, J.C. Advances and challenges in biosensor-based diagnosis of infectious diseases. *Expert Rev. Mol. Diagn.* **2014**, *14*, 225–244.
106. Han, K.-H.; Han, A.; Frazier, A.B. Microsystems for isolation and electrophysiological analysis of breast cancer cells from blood. *Biosens. Bioelectron.* **2006**, *21*, 1907–1914.
107. Lan, K.C.; Jang, L.S. Integration of single-cell trapping and impedance measurement utilizing microwell electrodes. *Biosens. Bioelectron.* **2011**, *26*, 2025–2031.
108. Shah, P.; Zhu, X.; Chen, C.; Hu, Y.; Li, C.-Z. Lab-on-chip device for single cell trapping and analysis. *Biomed. Microdevices* **2014**, *16*, 35–41.

109. Klepárník, K.; Foret, F. Recent advances in the development of single cell analysis—A Review. *Anal. Chim. Acta* **2013**, *800*, 12–21.

110. Ahmad, I.; Ahmad, M. Trends in characterizing single cell's stiffness properties. *Micro Nano Syst. Lett.* **2014**, *2*, doi:10.1186/s40486-014-0008-5.

111. Lewpiriyawong, N.; Kandaswamy, K.; Yang, C.; Ivanov, V.; Stocker, R. Microfluidic characterization and continuous separation of cells and particles using conducting poly(dimethyl siloxane) electrode induced alternating current-dielectrophoresis. *Anal. Chem.* **2011**, *83*, 9579–9585.

112. Goater, A.D.; Pethig, R. Electrorotation and dielectrophoresis. *Parasitology* **1998**, *117*, S177–S189.

113. Čemažar, J.; Damijan Miklavčič, T.K. Microfluidic devices for manipulation, modification and characterization of biological cells in electric Fields—A Review. *J. Microelectron. Electron. Compon. Mater.* **2013**, *43*, 143–161.

7

Detecting Antigen-Specific T Cell Responses: From Bulk Populations to Single Cells

Chansavath Phetsouphanh [1,*], John James Zaunders [1,2] and Anthony Dominic Kelleher [1,2]

[1] Kirby Institute, University of New South Wales, 2031 Sydney, Australia;
E-Mails: jzaunders@kirby.unsw.edu.au (J.J.Z.); akelleher@kirby.unsw.edu.au (A.D.K.)

[2] Centre for Applied Medical Research, St. Vincent's Hospital, 2010 Sydney, Australia

* Author to whom correspondence should be addressed; E-Mail: c.phetsouphanh@gmail.com

Academic Editors: Fan-Gang Tseng and Tuhin Subhra Santra

Abstract: A new generation of sensitive T cell-based assays facilitates the direct quantitation and characterization of antigen-specific T cell responses. Single-cell analyses have focused on measuring the quality and breadth of a response. Accumulating data from these studies demonstrate that there is considerable, previously-unrecognized, heterogeneity. Standard assays, such as the ICS, are often insufficient for characterization of rare subsets of cells. Enhanced flow cytometry with imaging capabilities enables the determination of cell morphology, as well as the spatial localization of the protein molecules within a single cell. Advances in both microfluidics and digital PCR have improved the efficiency of single-cell sorting and allowed multiplexed gene detection at the single-cell level. Delving further into the transcriptome of single-cells using RNA-seq is likely to reveal the fine-specificity of cellular events such as alternative splicing (*i.e.*, splice variants) and allele-specific expression, and will also define the roles of new genes. Finally, detailed analysis of clonally related antigen-specific T cells using single-cell TCR RNA-seq will provide information on pathways of differentiation of memory T cells. With these state of the art technologies the transcriptomics and genomics of Ag-specific T cells can be more definitively elucidated.

Keywords: antigen-specific T cells; digital PCR; microfluidics; mingle-cell RNA-seq

---

1. Introduction

A wide variety of assays can be used to characterize T cell responses in order to determine the state and capability of the immune system. Such studies can reveal fundamental mechanisms underlying immunity aid the design of clinical diagnostics, help develop intervention therapies, and determine signatures of effective immune responses [1]. Primarily, research on antigen (Ag)-specific CD4+ T cells has been done using bulk-sorted populations by focusing on small sets of cells defined by selected markers that were hypothesized to identify homogenous sub-populations; for example, central *versus* effector memory cells defined by surface markers such as CD45RA/RO isoforms and CCR7 or CD62L expression [2–4]. However, accumulating data demonstrate that there is considerable heterogeneity within the Ag-specific population on the basis of genomic differences, cytokine secretion profiles, function, and trafficking markers [1,5–8].

Standard analytical technologies historically have measured the average response from highly heterogeneous populations. Common assays that detected cell proliferation, cytolytic activity, and cytokine expression have yielded valuable insights into disease pathogenesis and immunity to microbes such as viruses and tumour or self-antigens. However, these assays examined multiple parameters at the population level, where the implicit averaging of many measurements may mask the specific involvement of individual cells and the interactions that can occur between neighbouring cells. These technologies made it difficult to infer the characteristics of rare subsets of cells, such as Ag-specific T cell responses, without first purifying subsets of T cells. Even when purified, the T cell subsets were generally identified on the basis of a relatively small number of markers, compared to the much larger number of cell surface proteins expressed by T cells [9–11]. Single cell analyses are beginning to show that these approaches have underestimated heterogeneity.

Recently, single-cell analyses have focused on measuring the quality and breadth of a response. Variations in the expression of molecules between individual cells are thought to play an important role in functionally diversifying an immune response at the population level and also determining the diverse anatomical locations of individual cells. Advances in genome-wide quantitative analysis of single cells can provide an important vehicle that allows the investigator to make further insights into the variation between individual cells and to determine how these impact on the fine specificity of the nature and regulation of the immune response.

The challenge of understanding heterogeneity between cells, particularly tumour cells [12–14] has driven many of the major technological advances, resulting in the design of powerful instruments, protocols, and analysis protocols that enable the elucidation of DNA, RNA, and protein expression at the single-cell level. Flow cytometry has been widely adopted as the cornerstone of high-throughput analysis of specific protein expression and phosphorylation states of single cells within complex populations. Cell sorting has typically been used to purify up to six populations at a time from these mixtures of cells. The recent coupling of this technology with microfluidics and genome-wide deep sequencing at the single cell level has enabled further insights into cell biology. Single-cell genomics provides the basis for unbiased investigations into the molecular and functional consequences of cellular variability. In this review, the advantages and disadvantages of standard T cell response detection assays will be discussed. Newer technologies to more comprehensively define T cell responses at the single-cell level will be examined and the advances in single-cell genomics will be highlighted.

2. Standard Assays Are Insufficient for the Detection of T Cell Responses at the Single Cell Level

Measuring T lymphocyte proliferation after antigenic or mitogenic stimulation is an important parameter used in diagnosis of various immuno-deficiencies and in the monitoring a variety of immune responses. Measurement of the incorporation of tritiated thymidine [^{3}H] into lymphocyte DNA is a common approach used to determine the extent of antigen- or mitogen-driven cell proliferation [15]. Disadvantages of this assay include: The response of individual populations cells cannot be delineated without cell sorting; the inherent variability of the assay; the limitations and safety of handling radioactive material; and the labour-intensive nature of the protocols (*i.e.*, multiple wash steps involved, PBMC isolation, and inability to determine which cells are proliferating) [16].

More recent approaches that overcome some of these problems include the use of the cytoplasmic dyes, such as carboxyfluorescein succinimidyl ester (CFSE), to track lymphocyte proliferation [17]. CFSE covalently binds to amino groups on intracellular macromolecules that anchor the dye. CFSE is inherited equally by daughter cells after division, resulting in the halving of mean fluorescence with each generation. Disadvantages of this assay are: The time required for proliferation detection is usually 4–6 days; and high concentrations can cause cell toxicity and impair expression of activation markers [17,18]. However cell subpopulations can be tracked if markers are expressed stably. The use of multi-parameter flow cytometry allows other information about the proliferating cells and their phenotype to be accumulated. Further, the stability, or otherwise, of markers during the process of proliferation in response to antigen can be definitively determined.

ELISPOT (enzyme-linked immunospot) and intracellular cytokine staining (ICS) are widely used assays for the detection and enumeration of antigen-specific T cells. ELISPOT assays detect antigen-induced secretion of cytokines (usually IFN-γ) trapped by an immobilized antibody on a nitrocellulose membrane and visualized by an enzyme-coupled secondary antibody. ICS also detects cytokines produced by antigen-stimulated T cells, via the detection of the cytokines, such as IFN-γ, TNF, and IL-2, trapped within golgi/ER bodies in the cytoplasm through inactivation of granule secretion by brefeldin A or monensin, and visualized via fluorochrome-labelled monoclonal antibodies (mAb) after cell permeabilisation followed by flow cytometry.

The use of both assays is associated with implicit *a priori* assumptions regarding the importance of certain cytokines in the responses of interest (*i.e.*, IFN-γ) and do not take into account the other aspects of the T cell response; most often used to measure Th1 or Th2 responses, but can also be used to measure effector molecules such as Granzyme B. ELISPOT can only measure one or two parameters at a time and, without depletion or purification of cell subsets, the source of the cytokine cannot be determined. Multiple parameters in the ICS need to be optimized for each model system, such as stimulation, fixation/permeabilization, and appropriate controls need to be considered [19]. These assays are typically restricted to Th1 subset identification, as Th2 cytokine detection works poorly in these assays, and Th17 responses are usually only detectable when mitogens are used. This is due to a combination of factors including limitations in availability of mAb and the lower levels of cytokine production. Further, these assays do not allow live isolation of cells for downstream functional or molecular analysis and bulk populations with high cell numbers are required for each assay [20,21].

Detailed information of responding single cells in ELISPOT assays is limited to the level of markers used to purify cells prior to the assay, while ICS assays are limited by the number of additional fluorescence markers that can added to the assay. Each assay comes with issues of assay validation and quality control, with advantages and disadvantages depending on the nature of the responses detected. For example, standardization of assay requires the removal of differences in confounding variables when comparing experiments performed by the same or different users under the same conditions. Issues, such as choice of starting material, coating techniques, incubation and washing steps, and personal preference in reading spot development, need to be taken into account [22]. Additionally, currently-available flow cytometric analysis is generally limited to a maximum of 16 fluorochromes, but the use of up to 29 parameters has been recently reported [1,23,24]. The use of mAbs labeled with rare-earth isotopes, combining flow cytometric and mass cytometric technologies (CyTOF®), also greatly increases the number of parameters able to be defined on each cell [25,26]. A caveat with the CyTOF® platform is that it still lacks the sensitivity of standard fluorescences cytometry, up to three-fold less sensitive and requires more cells for staining. Nevertheless, the available number of parameters still does not approach the number of proteins that are variably expressed by T cells [27,28].

2.1. Using MHC-Multimers to Identify Antigen-Specific T Cells

Recombinant multimeric complexes of soluble recombinant MHC molecules often referred to as "tetramers" have emerged as a key tool for elucidation of the frequency of antigen-specific T cells *in vitro*, particularly in viral infections and post-vaccinations. Since 1996, there has been a revolution in the characterization of antigen-specific T cells, due to the development of reagents consisting of soluble multimerized MHC-I peptide complexes to detect epitope specific CD8+ T cells using flow cytometry through their increased avidity for TCR [29,30]. MHC-I tetramers combined with other staining techniques have been used to examine detailed information about antigen specific T cells, such as levels of activation, effector function, proliferation, and apoptosis [31], but importantly because they bind to the surface of T cells based only on their expression of TCR which recognize the incorporated epitope cells which are anergic, "stunned" or "exhausted" can be identified by this technology [32–36]. Further, because only surface staining is required, the cells can be sorted and used for functional assays and/or RNA profiling. This has led to the colossal expansion of information on antigen-specific CD8+ T cells and to a lesser extent antigen-specific CD4+ T cells, because of the more limited availability of MHC-II tetramers [37].

MHC-I-peptide complexes consist of an invariant light chain (β_2-microglobulin), a polymorphic heavy chain and a specific 8–10 amino acid peptide. The presence of a cognate peptide in the antigen-binding groove is essential for the formation of MHC-I molecules. This association is based on specific complimentary interactions between amino acid side chains at the anchor positions of the peptide and allele-specific pockets [38]. This helps ensure that the recombinant MHC-I heterodimers that refold are conformationally correct. Typically the formation of these soluble MHC-I multimers is dependent upon the specific biotinylation of a tail engineered to replace the transmembrane and cytoplasmic domains of the heavy chain. The biotinylated heterodimers are then bound to each of the four binding sites of fluorescently labeled streptavidin, forming a tetrameric complex. However, other strategies for multimeriziation have been employed to achieve the necessary increase in avidity between MHC-I and TCR to allow interactions stable enough for robust staining of T cells and subsequent identification by flow cytometry [39–42]. A possible drawback from this technique is that individual tetramers have to be designed for each epitope of interest, and then are only applicable to individuals carrying a particular class I allele. The shelf life of the constructs are variable, and positive and negative controls need to be carefully identified to ensure staining is specific, thus making the process very labour intensive, especially if multiple epitope-specific responses are to be studied simultaneously [37]. To overcome these limitations, dextramers (multimers with a dextran backbone) were developed, each dextramer molecule bears multiple fluorescein and streptavidin components. These dextramers were able to identify low frequency antigen-specific T cells and produced stronger signals than their tetramer counterpart [43].

In contrast to the relative ease of production MHC class I tetramers, the development of MHC class II tetramers has been more difficult. There are multiple reasons for this. Firstly, the peptide-binding grove depends on correct association of variable alpha and beta chains making the synthesis of some alleles inefficient. The characteristics determining binding of antigenic peptide to MHC-II are different to MHC-I in several ways. Secondly, the stability of the class II molecule is not dependent on the binding of peptide in the groove. The groove is open-ended at either end allowing longer peptides to bind and these may bind in different registers within the groove. Further, efficient loading of peptide occurs at acid pH, is dependent upon molecules such as HLA-DM and DO catalysing the efficient exchange of peptide for the CLIP peptide and the peptide receptive state of the class II molecule is rapidly lost. These conditions are difficult to reproduce *in vitro*. In addition, there is much greater promiscuity of peptide binding to class II than class I, which has resulted in much lower rates of definitive description of class II restricted epitopes [44,45]. Many methods have been explored in an attempt solve these problems. One of these methods is producing class II proteins by co-expressing the α and β subunits in mammalian or insect cells and relying on *in vivo*, rather than *in vitro*, re-folding, as is the case with proteins produced from *E. coli* [44]. Unfortunately, much of the material bound to the binding groove is not the epitope of interest and the exchange of peptide into the binding groove is often inefficient. To overcome this constructs expressing a fusion protein, consisting of peptide fused to MHC-II β N-terminus via a flexible linker region have been trialled. This covalently attached peptide has preferential access to the peptide-binding site, thus increasing the rate of the correct peptide occupying the binding groove [46,47]. Other modifications used to enhance the assembly of the subunits have included the use of leucine zippers and chimeric IgG Fc domains to promote assembly and stability of the heterodimer [48,49]. Combined these approaches have resulted in a slow, but steady, increase in the availability of the pool of reliable MHC-II multimers, though their availability is still far more limited than MHC-I equivalents. With these methods of MHC-II production, it is now theoretically possible to identify a subset of antigen-specific CD4+ T cells with multimeric complexes [37,50]. This will facilitate the isolation of single antigen-specific CD4+ T cells for downstream analysis [51,52]. MHC/peptide tetramers, termed streptamers, were developed for this purpose. Streptamers can be reversibly dissociated from binding to antigen-specific CD4+ T cells, this feature allows these cells to remain functionally active, contrary to conventional tetramers that impairs lytic function and proliferation of the bound cells [53].

2.2. Cell Surface Detection of Antigen-Specific T Cells

Apart from ICS and multimeric MHC-II complexes, other approaches have been reported to use flow cytometry to identify Ag-specific T cells, including cell surface trapping of cytokines using magnetic bead technology (Miltenyi cytokine-capture system®) [54], or the use of activation-induced trafficking of some intracellular markers to the cell surface such as CD40L expressed on CD4+ T cells [55,56]. However, both these assays are limited by their detection of only one effector molecule at a time and we know that many important responses, such as Tregs and CTL, are not detected by either of these methods.

2.3. CD25/OX40 Assay

CD4+ T cell responses are pivotal to the regulation of the immune system during viral infection. Due in great part to the difficulties associated with the synthesis and use of class II multimers and therefore an inability to identify and isolate epitope specific CD4+ T cells, our understanding of the molecular basis of CD4+ function has fallen behind that of CD8+ T cells. Thus an assay that allows live isolation of antigen specific T cells is necessary.

A recent flow cytometric assay, developed by Zaunders *et al.*, employs the co-expression of CD25 (IL-2Rα) and CD134/OX40 (a TNF receptor family member) to identify antigen-specific CD4+ T cells. The co-expression of these two molecules in CD4+ T cells is very low in peripheral blood. However, with antigenic stimulation *in vitro*, their expression becomes up regulated over time, peaking at approximately 44 h post-stimulation. Advantages of this assay are that it detects a more global population of antigen-specific cells that appear to include all Th1, Th2, Th17, Tfh, and Treg lineages and is, therefore, not biased by detection of a particular cytokine or effector molecule, resulting in higher levels of antigen-specific T cells than ICC and ELISPOT. Further, it allows live isolation of individual cells for downstream functional and molecular characterization [7,57–62].

3. Enhanced Flow Cytometry Techniques for Single-Cell Analysis

Flow cytometry has been extensively used to analyse protein expression within cells. The relatively recent development combining microscopic imaging and fluorescence-activated flow cytometry has allowed a more in-depth examination of single cells. Each cell can be examined for morphology, as well as the spatial localization of the protein molecules within the cell. Instruments, such as the Imagestream®, allow rapid acquisition and processing, enabling measurements of thousands of cells per second, which is an advantage over conventional microscopy platforms [63]. This combination technology has been useful for the determination of protein localization, cell morphology, observation of protein interactions during signalling cascades and cellular uptake of foreign particles [64,65]. Although highly useful for providing information on single cells, the measurements acquired are only obtained at a single snapshot in time.

Newer technologies that complement and augment the data generated from flow cytometry have been developed. These technologies can assess the functional and transcriptional dynamics of cells at the microliter to picolitre scale. The two types of techniques that have emerged as tools to detect immune responses are microfluidic systems and spatial arrays with nanolitre-scale wells. For a detailed review on nanolitre micro-well and micro-dense arrays please refer to a review by Lindstrom, *et al.* [66].

Micro-dense arrays contain up to 100,000 sub-nanolitre welled compartments that allow the isolation or distribution of single-cells and measurement of cellular function, protein secretion and mRNA in parallel. This platform provides the ability to examine cellular interaction between different cell types and provides highly-specific resolution not found when using bulk populations. These cells can be deposited into nanowells for co-culture by dispensing each cell onto the array sequentially and allowing cells to settle via gravity, these cells can then be used for further analyses [1]. Co-cultures of cells (e.g., one effector cell and one antigen-presenting cell) in individual wells can provide insight into intercellular signaling and interaction dynamics. This technology can be used to assess antigen-specific interactions between cognate T cells and their corresponding antigen presenting cell to detect activation, cytolysis and cytokine production [67]. This approach has demonstrated that HIV-specific CD8+ T cells responses initially involve either cytolysis or secretion of pro-inflammatory cytokines (*i.e.*, IFNγ) [68,69].

Nanolitre micro-well arrays enable the study of single-cell phenotyping of rare cells, such as antigen-specific T cells and B cells. Studies on these rare subsets have revealed structure-function relationships between molecular synapses and signalling cascades [70]. Co-culture studies together with barcoded antibody arrays enabled the examination of the influence of paracrine signalling molecules on tumour cell function and signalling networks via the multiplexed detection of both intracellular and secreted proteins/cytokines [71,72]. These nano-arrays have also been used to examine serial killing capacities of Natural Killer (NK) cells, whereby NK cells were able to kill MHC-I deficient tumour cells and also showed that simultaneous interaction with several target cells increases the cytotoxic responsiveness of NK cells [73]. This platform has been used to show to the extent of variability among homogeneous populations. Dura *et al.* investigated the heterogeneity of CD8 T cells upon antigen presentation and correlated this with early activation events. They discovered that the cells showed relatively homogeneous calcium mobilization with high antigen stimuli with uniform timings of activation. However, the response pattern became more heterogeneous with lower antigen concentrations. These cell populations could be grouped into distinct clusters. Measurement of early signalling events simultaneously revealed high heterogeneity in ERK phosphorylation in single-cells, despite uniform timing and stimulus strength [74]. This study demonstrates the potential of micro-scale tools to clarify the complex intercellular interactions initiating and regulating T cell activation through the measurement of multiple parameters over a substantial number of individual cells. However, there are some technical limitations in this system, which include the limited control on the fluidic microenvironment required to maintain the cells in culture, as well as the risk of cross-contamination between wells during the rinsing of the array chips [75].

4. Microfluidics and Digital PCR at the Single-Cell Level

Reverse transcription quantitative polymerase chain reaction (RT-PCR) has been extensively used to examine gene expression patterns in T cells and has been the basis for many systems biology approaches profiling cellular activity. It provides exceptional specificity and sensitivity and has been adapted for the measurement of gene expression in single cells [7,76]. The challenges of single-cell RT-PCR studies include the cumbersome and laborious steps in purifying mRNA from individual cells, and the difficulty in synthesizing and purifying cDNA from single cells [77]. These difficulties arise from the loss of starting material during cell isolation, lysis, and cDNA synthesis steps. The loss of material may be caused by mRNA degradation, adhesion to plastics, and inefficient reverse transcription [78,79]. Microfluidic devices have been developed as platforms to overcome these particular problems and to handle the low reaction volumes required for single-cell analyses, while allowing relatively high-throughput, whereby multiple samples can be run simultaneously on the same device under standardized conditions.

One of the commercially-available valve-based microfluidic qPCR systems that have been successfully used for single cell studies is the Dynamic Array™ (Fluidigm). This is a low-volume (nanolitre) system that allows high-throughput with assessment of up to 96 parameters in 96 cells per run. It allows low copy detection while being used for large-scale studies [76]. The concordance of copy number detection between microfluidics and digital PCR compared to conventional PCR has been demonstrated [80]. This technology offers a higher level of precision and can be used to measure rare transcripts [81], as well as small RNA species (*i.e.*, microRNA) [82,83].

Mingueneau *et al.* used this platform to examine early thymocyte differentiation of αβ T cells. Gene expression patterns during early T cell differentiation were measured. It was found that during transit through the CD4+CD8+ stage, these double-positive thymocytes showed a global repression of housekeeping genes, which is rare among other cells of the immune system and correlated highly with the expression of c-Myc. They also identified genetic signatures that distinguished cells destined for positive selection *versus* apoptosis [84]. Johnson *et al.* used the Fluidigm® platform to define transcriptional profiles of HIV-specific CD4+ T cells. In doing so they identified a distinct transcriptional signature of HIV-specific cytolytic CD4+ T cells compared to Th1 cells, and these signatures were similar to features found in HIV-specific CD8+ T cells. These cytolytic CD4+ T cells also showed comparable killing activity to their CD8 counterpart and worked co-operatively to destroy virally infected target cells [85].

These examples demonstrate the innovative uses of these technologies; however, only a selected number of genes can be examine within one cell at a given time. Although this technology expands on the fairly outdated capabilities of traditional PCR, the elucidation of the entire transcriptome would be too cumbersome for this platform. Separation of single cells using microfluidics and the extraction of total DNA or RNA for next-generation sequencing would solve this issue. An example of this is the Fluidigm® C1 single-cell isolation instrument which allows automated cell capture for RNA/DNA extraction and nucleic acid amplification that is highly useful for RNA-seq and downstream genomic analysis. Microtools such as these allow clinical samples consisting of very small cell numbers to be examined [86]. Examination of cells from cytobrushes and biopsies allows the comparison between cell phenotypes, cell-to-cell communication [87], and function at mucosal sites, which is important in many disease conditions, such as HIV-1 infection.

5. Single-Cell Genomics Analysis via Next-Generation Deep Sequencing

A new and important extension of bulk transcription technologies, single-cell RNA sequencing (RNA-seq) has emerged as a powerful tool for mRNA expression analysis and allows genome-wide transcriptomics to be explored [88]. The transcriptome encompasses an essential part of cell identity and function, as RNA is essential for regulation, house-keeping, effector and messenger roles. The state of a specific single-cell (e.g., antigen-specific T cells) can be assessed via the profiling of all coding and non-coding transcripts that eventually leads to genome-wide transcriptomics analysis. RNA-seq techniques involve the conversion of cellular RNA transcripts into cDNA and subsequent sequencing in parallel by using next-generation sequencing technologies [89]. This technology enables high-resolution analysis of single-cells, whereby important cellular events such as alternative splicing (*i.e.*, splice variants) and allele-specific expression can be studied, that will also aid the discovery of new genes [90].

Single-cell RNA sequencing has been to investigate the role of *de novo* hormone synthesis in T helper cells and its role in T cell homeostasis. Mahata *et al.* demonstrated that that Th2 cells produce the steroid, pregnenolone, which inhibits T helper cell proliferation and B cell class switching. They proposed that this lympho-steroid is produced in an intrinsic manner by Th2 cells during allergic immune responses to restore immune homeostasis [91]. Shalek *et al.* used single-cell transcriptomics to assess heterogeneity of mouse bone marrow-derived dendritic cells (BMDCs) responding to lipopolysaccharides. They found bimodal variance in the RNA abundance and splicing patterns within responding cells. The observed splicing patterns displayed high levels of heterogeneity between cells. They identified 137 highly variable, but co-regulated, antiviral response genes that may be propagated through an interferon feedback loop involving the transcription factors Stat2 and Irf7 [89]. These studies highlight the promise and power of single-cell RNA-seq. This technology can be used to uncover the functional diversity between cells, as well as, discovering new gene regulation circuits.

5.1. Detecting Clonally-Related Ag-Specific T Cells Using TCR RNA-seq

A single-cell method to assess T cell receptor beta chain (TRBV) and alpha chain (TRAV) sequence data of sorted CD4+ and CD8+ T cells has recently been published [92]. This method also describes a method allowing simultaneous measurement of mRNA transcripts for up to 34 CD4+ T cell lineage-defining transcription factors and cytokines. Three steps of nested PCR amplification from a single cell, including barcoding from each well of up to 20 separate 96-well plates, followed by pooling and deep sequencing, allows high throughput TCR sequencing and RNA profiling in parallel in 1000 s of single cells, which will dramatically increase our ability to consider relationships between TCR repertoire and T cell phenotypic or functional subsets in a range of immune responses. Previous approaches using cloning and Sanger sequencing have not been feasible for this number of cells. Also, the combination with other genomic data on the transcription factors and cytokines, using this approach is an extremely favourable combination of cost-effectiveness and high throughput, for the first time definitively matching information on clonality and genotype. Detailed analysis of clonally-related antigen-specific T cells will finally provide information on pathways of differentiation of memory T cells not possible by other currently available techniques.

6. Conclusions

The final goal of complete genomics of the full range of single antigen-specific T cells is getting closer, due to very substantial and rapid improvements in identification of antigen-specific T cells, particularly CD4+ T cells, combined with advances in single cell mRNA technology (Figure 1). Finally, it is hoped that such studies will discriminate favourable outcomes after vaccination, or after pathogenic infection and provide a road map for rationale, rather than empiric development of vaccines and immunotherapeutics for chronic infections and drug resistant cancers.

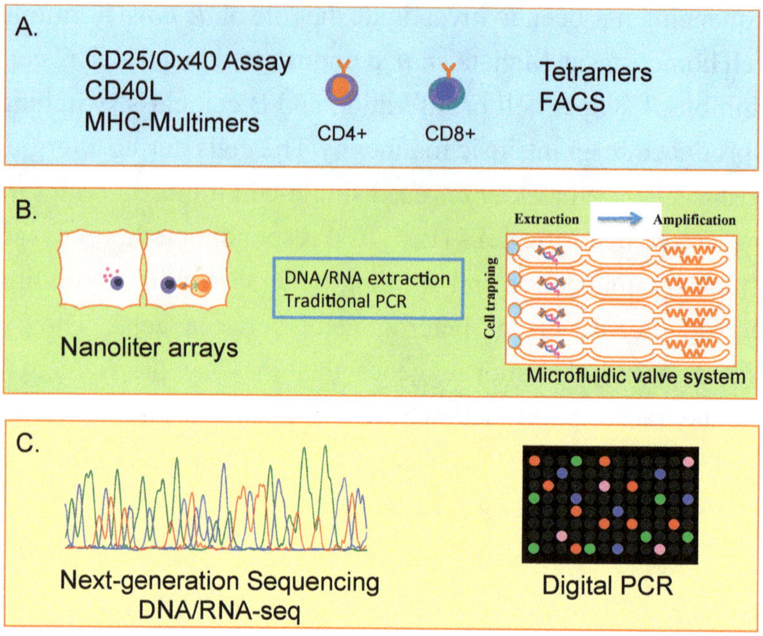

Figure 1. Analysis antigen-specific T cells at the single cell level pipeline. (**A**) Detection of antigen-specific CD4+ and CD8+ T cells using multiple assays and MHC-multimers; (**B**) Functional analysis of single Ag-specific cells via micro-well dense arrays and nanolitre arrays or single cell trapping using microfluidic valves for further molecular extraction/amplification; (**C**) Transcriptomic analysis using digital PCR and next-generation sequencing technologies (*i.e.*, RNA-seq) (coloured lines and dots represent fluorescence signals).

Acknowledgments

Authors would like to thank the NHMRC (National Health and Medical Research Council) (Canberra, Australia) for grant funding.

Author Contributions

Chansavath Phetsouphanh wrote manuscript; John James Zaunders and Anthony Dominic Kelleher edited and finessed manuscript for publication.

References

1. Chattopadhyay, P.K.; Gierahn, T.M.; Roederer, M.; Love, J.C. Single-cell technologies for monitoring immune systems. *Nat. Immunol.* **2014**, *15*, 128–135.
2. Sallusto, F.; Lanzavecchia, A. Heterogeneity of CD4+ memory T cells: Functional modules for tailored immunity. *Eur. J. Immunol.* **2009**, *39*, 2076–2082.
3. Sallusto, F.; Lenig, D.; Forster, R.; Lipp, M.; Lanzavecchia, A. Two subsets of memory T lymphocytes with distinct homing potentials and effector functions. *Nature* **1999**, *401*, 708–712.

4. Sallusto, F.; Geginat, J.; Lanzavecchia, A. Central memory and effector memory T cell subsets: Function, generation, and maintenance. *Annu. Rev. Immunol.* **2004**, *22*, 745–763.
5. Li, H.; Margolick, J.B.; Bream, J.H.; Nilles, T.L.; Langan, S.; Bui, H.T.; Sylwester, A.W.; Picker, L.J.; Leng, S.X. Heterogeneity of CD4+ and CD8+ T-cell responses to cytomegalovirus in HIV-infected and HIV-uninfected men who have sex with men. *J. Infect. Dis.* **2014**, *210*, 400–404.
6. Becattini, S.; Latorre, D.; Mele, F.; Foglierini, M.; de Gregorio, C.; Cassotta, A.; Fernandez, B.; Kelderman, S.; Schumacher, T.N.; Corti, D.; *et al.* T cell immunity. Functional heterogeneity of human memory $CD4^+$ T cell clones primed by pathogens or vaccines. *Science* **2015**, *347*, 400–406.
7. Phetsouphanh, C.; Xu, Y.; Amin, J.; Seddiki, N.; Procopio, F.; Sekaly, R.P.; Zaunders, J.J.; Kelleher, A.D. Characterization of transcription factor phenotypes within antigen-specific CD4+ T cells using qualitative multiplex single-cell RT-PCR. *PLoS ONE* **2013**, *8*, e74946.
8. Zaunders, J.; Jing, J.; Leipold, M.; Maecker, H.; Kelleher, A.D.; Koch, I. Computationally efficient multidimensional analysis of complex flow cytometry data using second order polynomial histograms. *Cytometry A* **2015**, doi:10.1002/cyto.a.22704.
9. Evans, E.J.; Hene, L.; Sparks, L.M.; Dong, T.; Retiere, C.; Fennelly, J.A.; Manso-Sancho, R.; Powell, J.; Braud, V.M.; Rowland-Jones, S.L.; *et al.* The T cell surface—how well do we know it? *Immunity* **2003**, *19*, 213–223.
10. McMichael, A.J.; Ogg, G.; Wilson, J.; Callan, M.; Hambleton, S.; Appay, V.; Kelleher, T.; Rowland-Jones, S. Memory CD8+ T cells in HIV infection. *Philos. Trans. R. Soc. Lond. B Biol. Sci.* **2000**, *355*, 363–367.
11. McMichael, A.J.; Callan, M.; Appay, V.; Hanke, T.; Ogg, G.; Rowland-Jones, S. The dynamics of the cellular immune response to HIV infection: Implications for vaccination. *Philos. Trans. R. Soc. Lond. B Biol. Sci.* **2000**, *355*, 1007–1011.
12. Chen, Y.C.; Allen, S.G.; Ingram, P.N.; Buckanovich, R.; Merajver, S.D.; Yoon, E. Single-cell Migration Chip for Chemotaxis-based Microfluidic Selection of Heterogeneous Cell Populations. *Sci. Rep.* **2015**, *5*, 9980.
13. Li, S.; Zhu, X.; Liu, B.; Wang, G.; Ao, P. Endogenous molecular network reveals two mechanisms of heterogeneity within gastric cancer. *Oncotarget* **2015**, *6*, 13607–13627.
14. Norton, K.A.; Popel, A.S.; Pandey, N.B. Heterogeneity of chemokine cell-surface receptor expression in triple-negative breast cancer. *Am. J. Cancer Res.* **2015**. *5*, 1295–1307.
15. Cavanagh, B.L.; Walker, T.; Norazit, A.; Meedeniya, A.C. Thymidine analogues for tracking DNA synthesis. *Molecules* **2011**, *16*, 7980–7993.
16. Last'ovicka, J.; Budinsky, V.; Spisek, R.; Bartunkova, J. Assessment of lymphocyte proliferation: CFSE kills dividing cells and modulates expression of activation markers. *Cell Immunol.* **2009**, *256*, 79–85.
17. Parish, C.R. Fluorescent dyes for lymphocyte migration and proliferation studies. *Immunol. Cell Biol.* **1999**, *77*, 499–508.
18. De Clerck, L.S.; Bridts, C.H.; Mertens, A.M.; Moens, M.M.; Stevens, W.J. Use of fluorescent dyes in the determination of adherence of human leucocytes to endothelial cells and the effect of fluorochromes on cellular function. *J. Immunol. Methods* **1994**, *172*, 115–124.

19. Karlsson, A.C.; Martin, J.N.; Younger, S.R.; Bredt, B.M.; Epling, L.; Ronquillo, R.; Varma, A.; Deeks, S.G.; McCune, J.M.; Nixon, D.F.; *et al.* Comparison of the ELISPOT and cytokine flow cytometry assays for the enumeration of antigen-specific T cells. *J. Immunol. Methods* **2003**, *283*, 141–153.
20. Pala, P.; Hussell, T.; Openshaw, P.J.M. Flow cytometric measurement of intracellular cytokines. *J. Immunol. Methods* **2000**, *243*, 107–124.
21. Streeck, H.; Frahm, N.; Walker, B.D. The role of IFN-gamma Elispot assay in HIV vaccine research. *Nat. Protoc.* **2009**, *4*, 461–469.
22. Janetzki, S.; Britten, C.M. The impact of harmonization on ELISPOT assay performance. *Methods Mol. Biol.* **2012**, *792*, 25–36.
23. Bendall, S.C.; Nolan, G.P.; Roederer, M.; Chattopadhyay, P.K. A deep profiler's guide to cytometry. *Trends Immunol.* **2012**, *33*, 323–332.
24. Chattopadhyay, P.K.; Roederer, M. A mine is a terrible thing to waste: High content, single cell technologies for comprehensive immune analysis. *Am. J. Transplant.* **2015**, *15*, 1155–1161.
25. Behbehani, G.K.; Bendall, S.C.; Clutter, M.R.; Fantl, W.J.; Nolan, G.P. Single-cell mass cytometry adapted to measurements of the cell cycle. *Cytometry A* **2012**, *81*, 552–566.
26. Nolan, G.P. Flow cytometry in the post fluorescence era. *Best Pract. Res. Clin. Haematol.* **2011**, *24*, 505–508.
27. Bendall, S.C.; Simonds, E.F.; Qiu, P.; Amir, E.-A.D.; Krutzik, P.O.; Finck, R.; Bruggner, R.V.; Melamed, R.; Trejo, A.; Ornatsky, O.I.; *et al.* Single-cell mass cytometry of differential immune and drug responses across a human hematopoietic continuum. *Science* **2011**, *332*, 687–696.
28. Blonder, J.; Issaq, H.J.; Veenstra, T.D. Proteomic biomarker discovery: It's more than just mass spectrometry. *Electrophoresis* **2011**, *32*, 1541–1548.
29. Altman, J.D.; Moss, P.A.; Goulder, P.J.; Barouch, D.H.; McHeyzer-Williams, M.G.; Bell, J.I.; McMichael, A.J.; Davis, M.M. Phenotypic analysis of antigen-specific T lymphocytes. *Science* **1996**, *274*, 94–96.
30. McHeyzer-Williams, M.G.; Altman, J.D.; Davis, M.M. Enumeration and characterization of memory cells in the TH compartment. *Immunol. Rev.* **1996**, *150*, 5–21.
31. Kurtulus, S.; Hildeman, D. Assessment of CD4(+) and CD8 (+) T cell responses using MHC class I and II tetramers. *Methods Mol. Biol.* **2013**, *979*, 71–79.
32. Yi, J.S.; Cox, M.A.; Zajac, A.J. T-cell exhaustion: Characteristics, causes and conversion. *Immunology* **2010**, *129*, 474–481.
33. Zajac, A.J.; Blattman, J.N.; Murali-Krishna, K.; Sourdive, D.J.; Suresh, M.; Altman, J.D.; Ahmed, R. Viral immune evasion due to persistence of activated T cells without effector function. *J. Exp. Med.* **1998**, *188*, 2205–2213.
34. Appay, V.; Rowland-Jones, S.L. The assessment of antigen-specific CD8+ T cells through the combination of MHC class I tetramer and intracellular staining. *J. Immunol. Methods* **2002**, *268*, 9–19.
35. Appay, V.; Almeida, J.R.; Sauce, D.; Autran, B.; Papagno, L. Accelerated immune senescence and HIV-1 infection. *Exp. Gerontol.* **2007**, *42*, 432–437.

36. Huygens, A.; Lecomte, S.; Tackoen, M.; Olislagers, V.; Delmarcelle, Y.; Burny, W.; van Rysselberge, M.; Liesnard, C.; Larsen, M.; Appay, V.; *et al.* Functional Exhaustion Limits CD4+ and CD8+ T-Cell Responses to Congenital Cytomegalovirus Infection. *J. Infect. Dis.* **2015**, *212*, 484–494.
37. Sims, S.; Willberg, C.; Klenerman, P. MHC-peptide tetramers for the analysis of antigen-specific T cells. *Expert Rev. Vaccines* **2010**, *9*, 765–774.
38. Leisner, C.; Loeth, N.; Lamberth, K.; Justesen, S.; Sylvester-Hvid, C.; Schmidt, E.G.; Claesson, M.; Buus, S.; Stryhn, A. One-pot, mix-and-read peptide-MHC tetramers. *PLoS ONE* **2008**, *3*, e1678.
39. Bakker, A.H.; Schumacher, T.N. MHC multimer technology: Current status and future prospects. *Curr. Opin. Immunol.* **2005**, *17*, 428–433.
40. Knabel, M.; Franz, T.J.; Schiemann, M.; Wulf, A.; Villmow, B.; Schmidt, B.; Bernhard, H.; Wagner, H.; Busch, D.H. Reversible MHC multimer staining for functional isolation of T-cell populations and effective adoptive transfer. *Nat. Med.* **2002**, *8*, 631–637.
41. Bakker, A.H.; Hoppes, R.; Linnemann, C.; Toebes, M.; Rodenko, B.; Berkers, C.R.; Hadrup, S.R.; van Esch, W.J.; Heemskerk, M.H.; Ovaa, H.; *et al.* Conditional MHC class I ligands and peptide exchange technology for the human MHC gene products HLA-A1, -A3, -A11, and -B7. *Proc. Natl. Acad. Sci. USA* **2008**, *105*, 3825–3830.
42. Hadrup, S.R.; Toebes, M.; Rodenko, B.; Bakker, A.H.; Egan, D.A.; Ovaa, H.; Schumacher, T.N. High-throughput T-cell epitope discovery through MHC peptide exchange. *Methods Mol. Biol.* **2009**, *524*, 383–405.
43. Batard, P.; Peterson, D.A.; Devevre, E.; Guillaume, P.; Cerottini, J.C.; Rimoldi, D.; Speiser, D.E.; Winther, L.; Romero, P. Dextramers: New generation of fluorescent MHC class I/peptide multimers for visualization of antigen-specific CD8+ T cells. *J. Immunol. Methods* **2006**, *310*, 136–148.
44. Cameron, T.O.; Norris, P.J.; Patel, A.; Moulon, C.; Rosenberg, E.S.; Mellins, E.D.; Wedderburn, L.R.; Stern, L.J. Labeling antigen-specific CD4(+) T cells with class II MHC oligomers. *J. Immunol. Methods* **2002**, *268*, 51–69.
45. Crawford, F.; Kozono, H.; White, J.; Marrack, P.; Kappler, J. Detection of antigen-specific T cells with multivalent soluble class II MHC covalent peptide complexes. *Immunity* **1998**, *8*, 675–682.
46. Kozono, H.; White, J.; Clements, J.; Marrack, P.; Kappler, J. Production of soluble MHC class II proteins with covalently bound single peptides. *Nature* **1994**, *369*, 151–154.
47. Cunliffe, S.L.; Wyer, J.R.; Sutton, J.K.; Lucas, M.; Harcourt, G.; Klenerman, P.; McMichael, A.J.; Kelleher, A.D. Optimization of peptide linker length in production of MHC class II/peptide tetrameric complexes increases yield and stability, and allows identification of antigen-specific CD4+T cells in peripheral blood mononuclear cells. *Eur. J. Immunol.* **2002**, *32*, 3366–3375.
48. Lebowitz, M.S.; O'Herrin, S.M.; Hamad, A.R.; Fahmy, T.; Marguet, D.; Barnes, N.C.; Pardoll, D.; Bieler, J.G.; Schneck, J.P. Soluble, high-affinity dimers of T-cell receptors and class II major histocompatibility complexes: Biochemical probes for analysis and modulation of immune responses. *Cell Immunol.* **1999**, *192*, 175–184.
49. Novak, E.J.; Liu, A.W.; Nepom, G.T.; Kwok, W.W. MHC class II tetramers identify peptide-specific human CD4(+) T cells proliferating in response to influenza A antigen. *J. Clin. Investig.* **1999**, *104*, R63–R67.

50. Kwok, W.W.; Gebe, J.A.; Liu, A.; Agar, S.; Ptacek, N.; Hammer, J.; Koelle, D.M.; Nepom, G.T. Rapid epitope identification from complex class-II-restricted T-cell antigens. *Trends Immunol.* **2001**, *22*, 583–588.
51. Long, H.M.; Chagoury, O.L.; Leese, A.M.; Ryan, G.B.; James, E.; Morton, L.T.; Abbott, R.J.; Sabbah, S.; Kwok, W.; Rickinson, A.B. MHC II tetramers visualize human CD4+ T cell responses to Epstein-Barr virus infection and demonstrate atypical kinetics of the nuclear antigen EBNA1 response. *J. Exp. Med.* **2013**, *210*, 933–949.
52. Petersen, S.H.; Odintsova, E.; Haigh, T.A.; Rickinson, A.B.; Taylor, G.S.; Berditchevski, F. The role of tetraspanin CD63 in antigen presentation via MHC class II. *Eur. J. Immunol.* **2011**, *41*, 2556–2561.
53. Neudorfer, J.; Schmidt, B.; Huster, K.M.; Anderl, F.; Schiemann, M.; Holzapfel, G.; Schmidt, T.; Germeroth, L.; Wagner, H.; Peschel, C.; *et al.* Reversible HLA multimers (Streptamers) for the isolation of human cytotoxic T lymphocytes functionally active against tumor- and virus-derived antigens. *J. Immunol. Methods* **2007**, *320*, 119–131.
54. Manz, R.; Assenmacher, M.; Pfluger, E.; Miltenyi, S.; Radbruch, A. Analysis and sorting of live cells according to secreted molecules, relocated to a cell-surface affinity matrix. *Proc. Natl. Acad. Sci. USA* **1995**, *92*, 1921–1925.
55. Chattopadhyay, P.K.; Yu, J.; Roederer, M. Live-cell assay to detect antigen-specific CD4+ T-cell responses by CD154 expression. *Nat. Protoc.* **2006**, *1*, 1–6.
56. Frentsch, M.; Arbach, O.; Kirchhoff, D.; Moewes, B.; Worm, M.; Rothe, M.; Scheffold, A.; Thiel, A. Direct access to CD4+ T cells specific for defined antigens according to CD154 expression. *Nat. Med.* **2005**, *11*, 1118–1124.
57. Hsu, D.C.; Kerr, S.J.; Iampornsin, T.; Pett, S.L.; Avihingsanon, A.; Thongpaeng, P.; Zaunders, J.J.; Ubolyam, S.; Ananworanich, J.; Kelleher, A.D.; *et al.* Restoration of CMV-specific-CD4 T cells with ART occurs early and is greater in those with more advanced immunodeficiency. *PLoS ONE* **2013**, *8*, e77479.
58. Zaunders, J.J.; Munier, M.L.; Seddiki, N.; Pett, S.; Ip, S.; Bailey, M.; Xu, Y.; Brown, K.; Dyer, W.B.; Kim, M.; *et al.* High levels of human antigen-specific CD4+ T cells in peripheral blood revealed by stimulated coexpression of CD25 and CD134 (OX40). *J. Immunol.* **2009**, *183*, 2827–2836.
59. Seddiki, N.; Cook, L.; Hsu, D.C.; Phetsouphanh, C.; Brown, K.; Xu, Y.; Kerr, S.J.; Cooper, D.A.; Munier, C.M.; Pett, S.; *et al.* Human antigen-specific CD4(+) CD25(+) CD134(+) CD39(+) T cells are enriched for regulatory T cells and comprise a substantial proportion of recall responses. *Eur. J. Immunol.* **2014**, *44*, 1644–1661.
60. Hsu, D.C.; Kerr, S.J.; Thongpaeng, P.; Iampornsin, T.; Pett, S.L.; Zaunders, J.J.; Avihingsanon, A.; Ubolyam, S.; Ananworanich, J.; Kelleher, A.D.; *et al.* Incomplete restoration of Mycobacterium tuberculosis-specific-CD4 T cell responses despite antiretroviral therapy. *J. Infect.* **2014**, *68*, 344–354.
61. Keoshkerian, E.; Helbig, K.; Beard, M.; Zaunders, J.; Seddiki, N.; Kelleher, A.; Hampartzoumian, T.; Zekry, A.; Lloyd, A.R. A novel assay for detection of hepatitis C virus-specific effector CD4(+) T cells via co-expression of CD25 and CD134. *J. Immunol. Methods* **2012**, *375*, 148–158.
62. Phetsouphanh, C.; Xu, Y.; Bailey, M.; Pett, S.; Zaunders, J.; Seddiki, N.; Kelleher, A.D. Ratios of effector to central memory antigen-specific CD4(+) T cells vary with antigen exposure in HIV+ patients. *Immunol. Cell Biol.* **2014**, *92*, 384–388.

63. Zuba-Surma, E.K.; Kucia, M.; Abdel-Latif, A.; Lillard, J.W.J.; Ratajczak, M.Z. The ImageStream System: A key step to a new era in imaging. *Folia. Histochem. Cytobiol.* **2007**, *45*, 279–290.
64. Beum, P.V.; Lindorfer, M.A.; Hall, B.E.; George, T.C.; Frost, K.; Morrissey, P.J.; Taylor, R.P. Quantitative analysis of protein co-localization on B cells opsonized with rituximab and complement using the ImageStream multispectral imaging flow cytometer. *J. Immunol. Methods* **2006**, *317*, 90–99.
65. George, T.C.; Basiji, D.A.; Hall, B.E.; Lynch, D.H.; Ortyn, W.E.; Perry, D.J.; Seo, M.J.; Zimmerman, C.A.; Morrissey, P.J. Distinguishing modes of cell death using the ImageStream multispectral imaging flow cytometer. *Cytometry A* **2004**, *59*, 237–245.
66. Lindstrom, S.; Andersson-Svahn, H. Overview of single-cell analyses: Microdevices and applications. *Lab Chip* **2010**, *10*, 3363–3372.
67. Torres, A.J.; Contento, R.L.; Gordo, S.; Wucherpfennig, K.W.; Love, J.C. Functional single-cell analysis of T-cell activation by supported lipid bilayer-tethered ligands on arrays of nanowells. *Lab Chip* **2013**, *13*, 90–99.
68. Varadarajan, N.; Julg, B.; Yamanaka, Y.J.; Chen, H.; Ogunniyi, A.O.; McAndrew, E.; Porter, L.C.; Piechocka-Trocha, A.; Hill, B.J.; Douek, D.C.; *et al.* A high-throughput single-cell analysis of human CD8(+) T cell functions reveals discordance for cytokine secretion and cytolysis. *J. Clin. Investig.* **2011**, *121*, 4322–4331.
69. Varadarajan, N.; Kwon, D.S.; Law, K.M.; Ogunniyi, A.O.; Anahtar, M.N.; Richter, J.M.; Walker, B.D.; Love, J.C. Rapid, efficient functional characterization and recovery of HIV-specific human CD8+ T cells using microengraving. *Proc. Natl. Acad. Sci. USA* **2012**, *109*, 3885–3890.
70. Dura, B.; Voldman, J. Spatially and temporally controlled immune cell interactions using microscale tools. *Curr. Opin. Immunol.* **2015**, *35*, 23–29.
71. Elitas, M.; Brower, K.; Lu, Y.; Chen, J.J.; Fan, R. A microchip platform for interrogating tumor-macrophage paracrine signaling at the single-cell level. *Lab Chip* **2014**, *14*, 3582–3588.
72. Kravchenko-Balasha, N.; Wang, J.; Remacle, F.; Levine, R.D.; Heath, J.R. Glioblastoma cellular architectures are predicted through the characterization of two-cell interactions. *Proc. Natl. Acad. Sci. USA* **2014**, *111*, 6521–6526.
73. Vanherberghen, B.; Olofsson, P.E.; Forslund, E.; Sternberg-Simon, M.; Khorshidi, M.A.; Pacouret, S.; Guldevall, K.; Enqvist, M.; Malmberg, K.J.; Mehr, R.; *et al.* Classification of human natural killer cells based on migration behavior and cytotoxic response. *Blood* **2013**, *121*, 1326–1334.
74. Dura, B.; Dougan, S.K.; Barisa, M.; Hoehl, M.M.; Lo, C.T.; Ploegh, H.L.; Voldman, J. Profiling lymphocyte interactions at the single-cell level by microfluidic cell pairing. *Nat. Commun.* **2015**, *6*, 5940.
75. Lindstrom, S.; Mori, K.; Ohashi, T.; Andersson-Svahn, H. A microwell array device with integrated microfluidic components for enhanced single-cell analysis. *Electrophoresis* **2009**, *30*, 4166–4171.
76. Lecault, V.; White, A.K.; Singhal, A.; Hansen, C.L. Microfluidic single cell analysis: From promise to practice. *Curr. Opin. Chem. Biol.* **2012**, *16*, 381–390.
77. White, A.K.; VanInsberghe, M.; Petriv, O.I.; Hamidi, M.; Sikorski, D.; Marra, M.A.; Piret, J.; Aparicio, S.; Hansen, C.L. High-throughput microfluidic single-cell RT-qPCR. *Proc. Natl. Acad. Sci. USA* **2011**, *108*, 13999–14004.

78. Marcus, J.S.; Anderson, W.F.; Quake, S.R. Microfluidic single-cell mRNA isolation and analysis. *Anal. Chem.* **2006**, *78*, 3084–3089.
79. Marcus, J.S.; Anderson, W.F.; Quake, S.R. Parallel picoliter rt-PCR assays using microfluidics. *Anal. Chem.* **2006**, *78*, 956–958.
80. Teles, J.; Enver, T.; Pina, C. Single-cell PCR profiling of gene expression in hematopoiesis. *Methods Mol. Biol.* **2014**, *1185*, 21–42.
81. Whale, A.S.; Huggett, J.F.; Cowen, S.; Speirs, V.; Shaw, J.; Ellison, S.; Foy, C.A.; Scott, D.J. Comparison of microfluidic digital PCR and conventional quantitative PCR for measuring copy number variation. *Nucleic Acids Res.* **2012**, *40*, e82.
82. Moltzahn, F.; Hunkapiller, N.; Mir, A.A.; Imbar, T.; Blelloch, R. High throughput microRNA profiling: Optimized multiplex qRT-PCR at nanoliter scale on the fluidigm dynamic arrayTM IFCs. *J. Vis. Exp.* **2011**, *54*, 2552.
83. Jang, J.S.; Simon, V.A.; Feddersen, R.M.; Rakhshan, F.; Schultz, D.A.; Zschunke, M.A.; Lingle, W.L.; Kolbert, C.P.; Jen, J. Quantitative miRNA expression analysis using fluidigm microfluidics dynamic arrays. *BMC Genom.* **2011**, *12*, 144.
84. Mingueneau, M.; Kreslavsky, T.; Gray, D.; Heng, T.; Cruse, R.; Ericson, J.; Bendall, S.; Spitzer, M.H.; Nolan, G.P.; Kobayashi, K.; *et al.* The transcriptional landscape of alphabeta T cell differentiation. *Nat. Immunol.* **2013**, *14*, 619–632.
85. Johnson, S.; Eller, M.; Teigler, J.E.; Maloveste, S.M.; Schultz, B.T.; Soghoian, D.Z.; Lu, R.; Oster, A.F.; Chenine, A.L.; Alter, G.; *et al.* Cooperativity of HIV-specific cytolytic CD4+ T cells and CD8+ T cells in control of HIV viremia. *J. Virol.* **2015**, *89*, 7494–7505.
86. Yin, H.; Marshall, D. Microfluidics for single cell analysis. *Curr. Opin. Biotechnol.* **2012**, *23*, 110–119.
87. Shalek, A.K.; Satija, R.; Shuga, J.; Trombetta, J.J.; Gennert, D.; Lu, D.; Chen, P.; Gertner, R.S.; Gaublomme, J.T.; Yosef, N.; *et al.* Single-cell RNA-seq reveals dynamic paracrine control of cellular variation. *Nature* **2014**, *510*, 363–369.
88. Saliba, A.E.; Westermann, A.J.; Gorski, S.A.; Vogel, J. Single-cell RNA-seq: Advances and future challenges. *Nucleic Acids Res.* **2014**, *42*, 8845–8860.
89. Shalek, A.K.; Satija, R.; Adiconis, X.; Gertner, R.S.; Gaublomme, J.T.; Raychowdhury, R.; Schwartz, S.; Yosef, N.; Malboeuf, C.; Lu, D.; *et al.* Single-cell transcriptomics reveals bimodality in expression and splicing in immune cells. *Nature* **2013**, *498*, 236–240.
90. Quake, T.K.S.R. Single-cell genomics. *Nat. Methods* **2011**, *8*, 311–314.
91. Mahata, B.; Zhang, X.; Kolodziejczyk, A.A.; Proserpio, V.; Haim-Vilmovsky, L.; Taylor, A.E.; Hebenstreit, D.; Dingler, F.A.; Moignard, V.; Gottgens, B.; *et al.* Single-cell RNA sequencing reveals T helper cells synthesizing steroids de novo to contribute to immune homeostasis. *Cell Rep.* **2014**, *7*, 1130–1142.
92. Han, A.; Glanville, J.; Hansmann, L.; Davis, M.M. Linking T-cell receptor sequence to functional phenotype at the single-cell level. *Nat. Biotechnol.* **2014**, *32*, 684–692.

8

Technologies for Single-Cell Isolation

Andre Gross [1,2,†], Jonas Schoendube [1,2,†], Stefan Zimmermann [1,†], Maximilian Steeb [1], Roland Zengerle [1,3,4] and Peter Koltay [1,2,*]

1 Laboratory for MEMS Applications, IMTEK–Department of Microsystems Engineering, University of Freiburg, Georges-Koehler-Allee 103, Freiburg 79110, Germany; E-Mails: andre.gross@imtek.uni-freiburg.de (A.G.); jonas.schoendube@imtek.uni-freiburg.de (J.S.); stefan.zimmermann@imtek.uni-freiburg.de (S.Z.); maximilian.steeb@outlook.de (M.S.); Roland.Zengerle@imtek.uni-freiburg.de (R.Z.)

2 Cytena GmbH, Georges-Koehler-Allee 103, Freiburg 79110, Germany

3 Hahn-Schickard, Georges-Koehler-Allee 103, Freiburg 79110, Germany

4 BIOSS–Centre for Biological Signalling Studies, University of Freiburg, Freiburg 79110, Germany

† These authors contributed equally to this work.

* Author to whom correspondence should be addressed; E-Mail: Peter.Koltay@imtek.uni-freiburg.de

Academic Editor: Fan-Gang Tseng

Abstract: The handling of single cells is of great importance in applications such as cell line development or single-cell analysis, e.g., for cancer research or for emerging diagnostic methods. This review provides an overview of technologies that are currently used or in development to isolate single cells for subsequent single-cell analysis. Data from a dedicated online market survey conducted to identify the most relevant technologies, presented here for the first time, shows that FACS (fluorescence activated cell sorting) respectively Flow cytometry (33% usage), laser microdissection (17%), manual cell picking (17%), random seeding/dilution (15%), and microfluidics/lab-on-a-chip devices (12%) are currently the most frequently used technologies. These most prominent technologies are described in detail and key performance factors are discussed. The survey data indicates a further increasing interest in single-cell isolation tools for the coming years. Additionally, a worldwide patent search was performed to screen for emerging technologies that might become relevant in the future. In total 179 patents were found, out of which 25 were evaluated by screening the title and abstract to be relevant to the field.

Keywords: single-cell analysis; single-cell handling; single-cell separation; single-cell technologies; flow cytometry; laser microdissection; limiting dilution; microfluidics

1. Introduction

With regards to heterogeneous cell populations, such as those found in many tumors, the generation and analysis of single cells has an increasing impact on various fields of life sciences and biomedical research [1]. The analysis of heterogeneous cell populations in bulk is only able to provide averaged data about the population, by which important information about a small but potentially relevant subpopulation is possibly lost in the background. Cancer development is based on a complex interrelation of mutations, selection, and clonal expansion resulting in a mosaic out of different subclones within a single tumor [2]. If rare subclones, which lead to only subtle genomic signals, can be detected at all from studying bulk populations, it takes a large sequencing and computational effort [3]. In contrast, the analysis of single cells, representing such a subpopulation, can provide very detailed information—information which may be used for therapeutic decisions in an increasingly personalized medicine. A further need for the analysis of single-cells relates to very rare cells like circulating tumor cells, which are surrounded by billions of normal blood cells and have an increasing clinical impact as a so-called liquid biopsy [4]. However, at present, the isolation and separation of single cells is still a technically challenging task. Main challenges are the yield and quality or in other words the integrity and purity of the cells as well as the throughput and the sensitivity of single cell isolation methods. Today, a large variety of technologies for single-cell separation, isolation, and sorting are already available that are applied according to the scientific objective. These technologies can be briefly classified according to their:

- Level of automation, distinguishing manual methods for cell separation like microscope-assisted picking from automated devices such as fluorescence activated cell sorters (FACS).
- Ability to isolate specific/individual cells, distinguishing statistical methods (*i.e.*, the separation of cells according to a certain statistical probability) from a controlled cell separation (*i.e.*, a cell is specifically selected and confirmed to be single).
- Compatibility with certain application requirements, distinguishing technologies mainly applied for production of monoclonal cell cultures (derived from single cells) from technologies preferably used for single-cell genome/proteome analysis.

In this review, we present the current single-cell isolation technologies in consideration of their compatibility with requirements for downstream life-science applications. In order to identify the most relevant technologies to be presented, a market study of single-cell technologies was conducted by the authors by means of an online survey. After briefly describing the methodology of the market study, the five most-used technologies are reviewed in the following in detail. In addition to these most widely adopted technologies, further emerging technologies were identified through a patent search at the European Patent Office (EPO) for worldwide patents about single-cell separation technologies. Many of these patented technologies are probably not commercialized so far, but nevertheless have the potential to enable advances in the field of single-cell research and might become more relevant and well known in the future. Although this review does not claim to be exhaustive, it might turn out be a helpful guide through the heterogeneous field of state-of-the-art single-cell handling technologies.

2. Market Study of Single-Cell Technologies

As part of this work, a market survey about the German market for single-cell technologies has been conducted in the summer of 2014. More than 3000 contacts of potential survey participants have been manually selected as statistical population as described in the following. The criterion for (academic) research contacts to be considered for participation was to be active in the field of cell biology. For company contacts, the criterion was to be listed in the publicly available database for biotechnological companies "Biotechnology Database", provided by BIOCOM AG, Berlin, Germany [5].

The data for the marked study was generated by an online questionnaire in June and July 2014. Invitations were sent by e-mail to the previously identified group of contacts. In total 210 participants from German universities, research institutes, and industry have responded to the invitation, out of which 102 have completely filled out the questionnaire. Of all participants 17% are affiliated to universities, 16% to university hospitals, 11% to non-university research organizations, 5% to commercial companies and 51% did not specify their affiliation. Most participants categorized themselves as belonging to the fields of research in medicine (46%) or biology (42%). Figure 1 shows the distribution of the participants in their general and specific fields of research. *Basic research* appears to be slightly more widespread than *applied research*, which can potentially be attributed to the immaturity of the young field of single-cell research. With regards to the more specific research fields, *immunology* and *oncology* were the areas, where most participants worked.

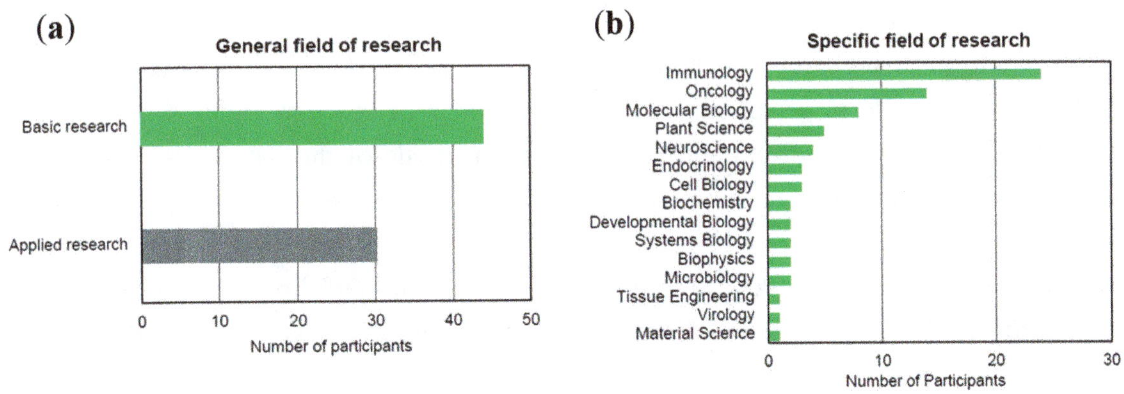

Figure 1. (**a**) Distribution of participants in their general fields of research. 44 participants stated *basic research* and 30 *applied research* as their general field; and (**b**) Distribution of participants in their specific fields of research. *Immunology* and *oncology* were most frequently named.

According to the survey participants, the most commonly used technologies for handling of single cells today are FACS respectively flow cytometry (33%), manual cell picking (17%), laser microdissection (also 17%), random seeding/limiting dilution (15%) and microfluidics/lab-on-a-chip devices (12%). Technologies like optical tweezers and others were mentioned less often (in total 6%). Figure 2a shows the most commonly used technologies in Germany in 2014. The five most prominent technologies will be in the focus of Section 3 of this review. Comparable results were obtained by a worldwide market study performed also in 2014 by HTStec (Cambridge, UK) [6], in collaboration with the authors of this paper. This study ranks the same technologies as the top five most extensively used amongst researchers worldwide (Figure 2b).

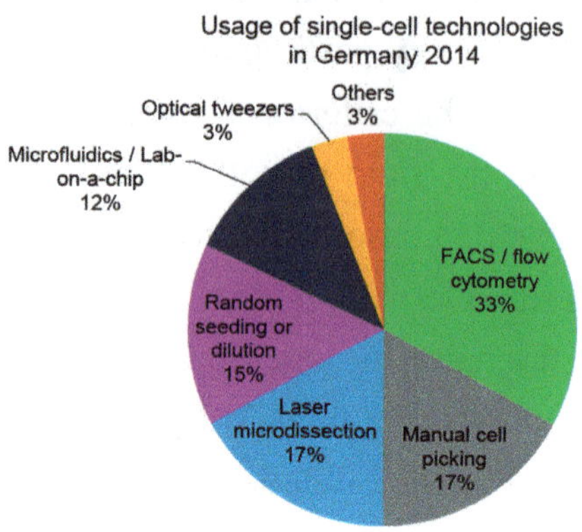

(a)

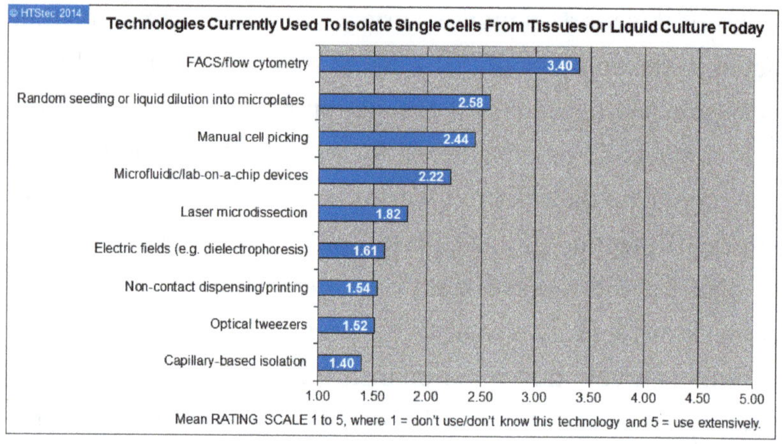

(b)

Figure 2. (**a**) The usage of technologies for handling single-cells in Germany in 2014. This data was derived as part of this work by a survey amongst 210 participants from German universities, research institutes and industry; and (**b**) Extensiveness of use of different single-cell technologies (data from "Single Cell Technologies Trends 2014"[6], reproduced with permission from HTStec Limited, Single Cell Technologies Trends 2014, HTStec 2014

A further finding of the survey was, that on average approximately 14 single-cell experiments are performed by the respondents per month, which corresponds to 164 experiments per year. The most frequently given answer was 1–5 experiments per month (mode of the data set). This indicates, that single-cell separation and handling is not a routine procedure yet, but performed regularly by those active in the field.

Finally, the participants were asked to rank the importance of the following criteria for selection of a specific instrument for single-cell isolation: *acquisition costs*, *maintenance & running costs*, *number of cells needed* (minimum to operate the device), *cell viability* (after isolation), *single-cell yield*, *compatibility with existing workflows*, *throughput* (in terms of single cells per second), and *space needed in the laboratory* (for the instrument). It turned out, that all of these criteria are considered to be important (*i.e.*, ranking larger 2.5 out of 5). The lowest ranking had *space needed in the laboratory* (2.75 of 5) and the highest ranking had *cell viability* and *single-cell yield* (4.12 of 5). Certainly, the relative importance of these criteria depend on the specific application, but it is for example noteworthy, that *cell viability* is ranked in average higher than *throughput* (3.52 of 5).

3. Single-Cell Isolation Technologies

Based on the market survey above, the methods and technologies presented hereafter are the most widespread technologies used for single-cell handling. In general, the applied methods strongly depend on the nature and origin of the sample and the processing or analysis to be performed on the cells once being isolated. To illustrate the diversity of sample nature, separation technology, and target applications Figure 3 shows schematics of the working principle of the five methods to be considered in detail in the following.

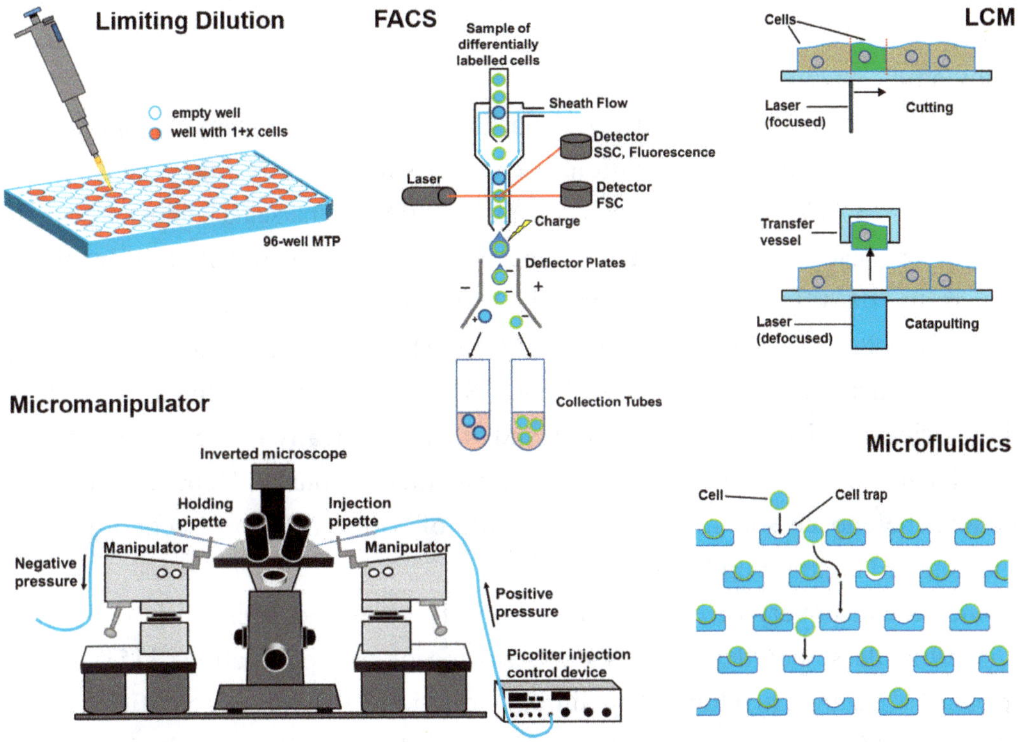

Figure 3. Schematic overview of single-cell separation technologies discussed in the following. The five technologies were identified through market studies as the most commonly used technologies for the handling of single cells (*cf.* (compare to) Figure 1).

3.1. Flow Cytometry

Thanks to the early pioneers of flow cytometry, since the 1970s researchers have access to ever more powerful flow cytometry instruments. Amongst others, patents and methods developed by Andrew Moldavan 1934 [7], Frank T. Gucker 1947 [8], Wallace H. Coulter 1953 [9], Mark Fulwyler 1965 [10,11], and Wolfgang Dittrich and Wolfgang Göhde 1968 [12] paved the way for the success of commercial flow cytometry [13].

Amongst the various types of flow cytometers, mainly Fluorescence Activated Cell Sorting (FACS) systems provide the ability to isolate single cells, thus they are focus of this section. FACS systems employ laser excitation and offer various analysis options. Cellular properties like relative size and granularity can be extracted as forward scatter (FSC) and side scatter (SSC), respectively. In addition a huge palette of functional properties can be measured by fluorescent staining. In FACS systems, cell suspensions are pressure driven through a flow cell. There they are lined up by a sheath flow liquid exploiting the effect of so called hydrodynamic focusing (see Figure 3). Upon such an arrangement, the cell stream rapidly passes by a laser beam to provide optical excitation and then optical detectors are used downstream to capture cell specific signals. The signals depend on the cells' respective physical, chemical, or optical properties–often enhanced by synthetic markers such as fluorescent dyes. Apart from size analysis and counting, the bypassing cells can also be sorted. After analysis, the cells are suspended in a closed system of small channels, the cell stream is forced through a small nozzle (typically 60–100 μm orifice diameter) and thereby a liquid jet is formed. By targeted vibrational actuation (e.g., by ultrasound) this jet breaks apart into a continuous stream of free flying droplets some of which carry cells. Using electrically charged plates for deflection of droplets containing cells of interest, these droplets can be guided to a collector vessel (typically a tube or micro well plate).

Popular systems like the FACS-Aria™ III (Becton, Dickinson and Company, Franklin Lakes, NJ, USA) provide up to six different colored excitation lasers and simultaneous fluorescent read-out in up to 18 color channels (FACS-Aria III brochure, BD, 2015, http://static.bdbiosciences.com/documents/BD_FACSAria_III_brochure.pdf). The system is able to generate up to 100,000 droplets per second and analyze up to 70,000 events per second (FACSAria III technical data sheet, BD, 2015, http://static.bdbiosciences.com/documents/ BD_FACSAria_III_tech_specs.pdf). Similar systems are available from Beckman Coulter (Brea, CA, USA), Sony Biotechnology Inc. (San Jose, CA, USA.), Bio-Rad Laboratories Inc. (Hercules, CA, USA), and others. Typically FACS systems provide different sort modes specialized either on high throughput or enrichment or purity. Depending on the application, type of cells, and the chosen sort mode the actual rate of sorted cells per second can strongly differ between some hundred up to several thousand cells.

Not only since the discovery of hybridoma cells by Koehler and Milstein [14] FACS has become an accepted, worldwide standard in analysis and sorting of cell populations [15], especially due to the known potential hazards in cloning by limiting dilution [16]. Prominent fields of research and application for the FACS technology are for example: DNA content analysis, immunophenotyping, quantification of soluble molecules [17], cell cycle analysis, hematopoietic stem cells, apoptosis, quantification of subpopulations [18], microbial analysis [19], and cancer diagnostics [20,21]. The sample range covers nearly every cell type from blood, bone marrow, tumor, plants, protoplasts, yeast, to bacteria and even viruses.

3.2. Laser Capture Microdissection

Laser capture microdissection (LCM) is an advanced technique to isolate individual cells or cell compartments from mostly solid tissue samples [22,23]. Figure 4 illustrates three different variations of the working principle. A tissue section is observed through a microscope and the target cell or compartment is visually identified. The operator marks the section to be cut off on the display by drawing a line around it. Along this trajectory the laser cuts the tissue and the isolated cell (or compartment) is–if required–extracted. While the cutting procedure using the laser is usually the same, there are several methods to extract the dissected tissue:

Contact-based extraction is done by laser cutting followed by extraction via adhesion, employing adhesive tube caps or heat-absorbing transfer foils, locally made adhesive by infrared (IR) lasers (*cf.* in Figure 4a).

Contact-free gravity-assisted microdissection (GAM) is featuring an inversely mounted substrate placed over a collector tube. Once being cut out by the laser, the target cell (or compartment) falls down into the collector (*cf.* in Figure 4b, Leica LMD7000).

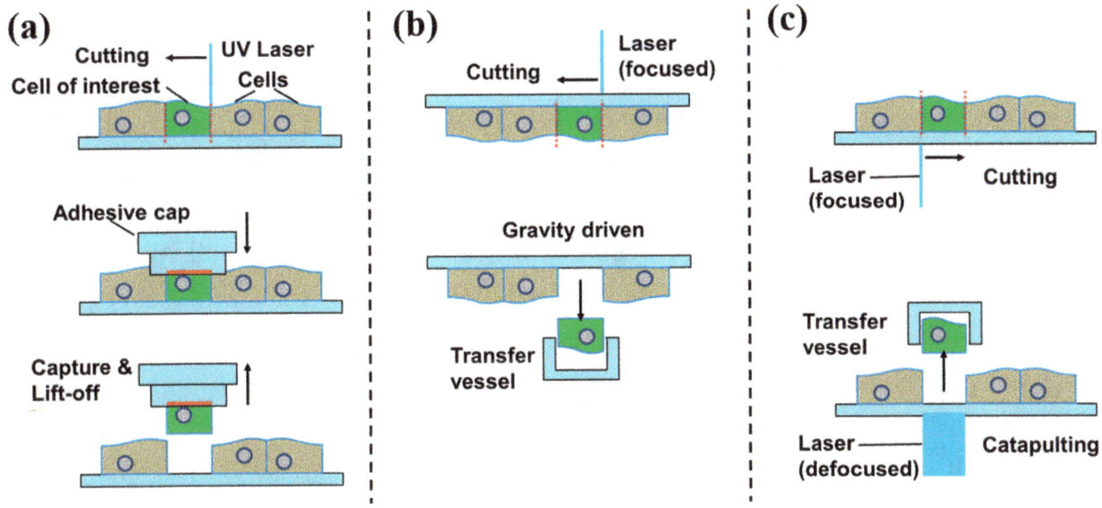

Figure 4. Schematic view on laser capture microdissection (LCM) methods. (**a**) Contact-based via adhesive tapes; (**b**) Cutting with a focused laser followed by capture with a vessel. Cut-out section extracted by gravity; and (**c**) Cutting with a focused laser followed by pressure catapulting with a defocused laser pulse.

Contact-free laser pressure catapulting (LPC) uses a short defocused laser pulse to ignite a local plasma below the previously cut cell (or compartment) [24,25]. The plasma impulse catapults the cell (or compartment) vertically against gravity into a nearby collector container (*cf.* in Figure 4c, Zeiss PALM MicroBeam LCM).

Samples are typically provided fixed in formalin, embedded in paraffin, or cryo-fixed [26]. Some LCM systems even allow for dissection of living tissue, enabling the extraction of live cells for culture or analysis (Leica LMD7000 with Live Cell Cutting (LCC)).

Analysis of solid tissue is of great interest when investigating heterogeneous tissue sections regarding their cellular structure as well as physiological and pathological processes [27]. In solid tumor research linking the molecular information of individual cells to their specific location in the tissue has become an important research field. Particularly, the access to cells *in situ* is of interest [28]. In combination with immune histological staining, LCM is a powerful tool for solid sample analysis on the single-cell level [29]. In the past years, various applications in single-cell analysis based on LCM extracted cells have been published: Single-cell RT-PCR [30], short tandem repeat analysis (STR) analysis in forensics [31], Western blot and mass spectrophotometry [32].

3.3. Limiting Dilution

Today many laboratories and companies use hand-pipettes or pipetting robots to isolate individual cells through dilution of the cell suspension. Due to the statistical distribution of the cells in the suspension, the number of cells in a highly diluted sample can be as low as one single cell per aliquot, when the suspension is split into small volumes (aliquots). This process is termed limiting dilution and is well known for decades for the production of monoclonal cell cultures [33–36]. Besides antibody production (as done by hybridomas), other applications such as cell-based assays, *etc.* also require cell populations grown from a single-cell.

Such seeding of cells in low concentration is indeed simple to carry out with standard pipetting tools, but it is not very efficient since the probability of achieving a single-cell in an aliquot is of statistical nature. The probability to obtain a certain number of cells per aliquot (*i.e.*, 0, 1, 2, *etc.*) is described by Poisson's distribution [37]. In order to achieve a sufficiently high probability for the appearance of single cells while at the same time minimizing the probability for multiple cells, the sample has to be strongly diluted. Typically a density of less than one cell per aliquot is applied (e.g., cell concentration smaller than 1 cell/10 μL = 100 cells/mL, if the aliquot volume is 10 μL). Often cited protocols e.g., recommend 0.5 to 0.9 cells per aliquot [16,35,38]. Table 1 presents the respective distribution of cells per well and the ideally achievable number of empty wells, single cells, and multiple cells according to Poisson's distribution. Obviously, on average only about one third of the prepared wells in a cell culture plate will contain a single-cell. Which of the wells indeed contain single cells has to be confirmed after seeding the cells in a separate process (e.g., by microscopy) due to the statistical nature of the separation method.

Table 1. Statistical probability for the number of cells per aliquot according to Poisson's distribution for cell concentrations of 0.5 and 0.9 cells per aliquot.

0.5 Cells/Aliquot		**0.9 Cells/Aliquot**	
Cell Number/Well	**Probability**	**Cell Number/Well**	**Probability**
0	61%	0	41%
1	30%	1	37%
2	8%	2	16%
3	1%	3	5%
4	0%	4	1%

3.4. Manual Cell Picking

Micromanipulators for manual cell picking typically consist of an inverted microscope combined with micro-pipettes movable through motorized mechanical stages. Micropipettes are made of ultrathin glass capillaries coupled to an aspiration and dispensation unit. The cell sample is typically provided as suspension in a dish or well-plate. Via microscope observation the operator selects a specific cell, moves the micro-pipette in close proximity and aspirates the cell by applying suction to the micropipette. The aspirated liquid volume including the selected cell can be transferred to a collection vessel (e.g., a well of a well-plate), where it is released by dispensation. This process is commonly performed manually.

Micromanipulators enable the controlled separation of selected, living cells from suspension and even allow for isolation of prokaryotic cells [39]. The fields of application range from bacterial analysis [40] to reproductive medicine [41] and forensics [42].

3.5. Microfluidics

Vast numbers of microfluidic or lab-on-a-chip devices have been proposed for single-cell analysis and handling in the past years (for reviews see [43] or [44]). In this section, we focus on methods that enable isolation of single cells for further downstream analysis or culture. General cell separation techniques that offer no control at the single-cell level [45] are not within the scope of this section. Though many different microfluidic devices for single-cell separation and handling have been published in the literature, most of these devices use at least one of the three following microfluidic principles to isolate single cells:

- Droplet-in-oil-based isolation as for example published in [46] (Figure 5a);
- Pneumatic membrane valving as for example published in [47] (Figure 5b);
- Hydrodynamic cell traps as for example published in [48] (Figure 5c).

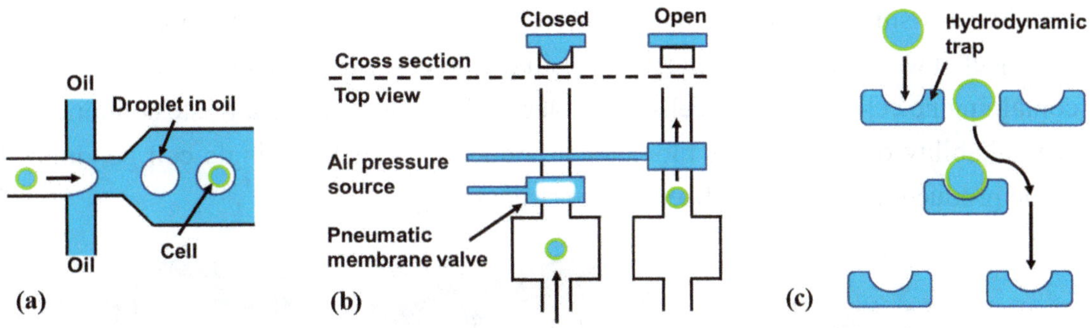

Figure 5. Schematic overview of different microfluidic methods for single-cell isolation. (**a**) An aqueous stream of cells is broken up into individual droplets-in-oil containing random distribution of cells; (**b**) Pneumatic membrane valves use air pressure to close a microfluidic channel by membrane deflection. This stops the flow and can trap a cell; and (**c**) Hydrodynamic traps are passive elements that only fit single cells and hold them at one position.

Droplet-based microfluidics uses channels filled with oil to hold separated aqueous droplets (similar to an emulsion). Within these droplets, single cells can be contained and thus be isolated (Figure 5a). Droplet-based microfluidic concepts can separate single cells either randomly according to Poisson's distribution [46] (similar to the limiting dilution method discussed above) or with even higher yields of over 80% [49]. The biggest advantage of droplets-in-oil-based cell separation and sorting technologies in general is the tremendous throughput of up to several thousand single-cells per second [49].

Pneumatic membrane valves use pressurized air to deflect an elastomer membrane. This membrane deflection closes a microfluidic channel below (Figure 5b). This allows for digitally opening or closing channels in a microfluidic network. Valve-based approaches need a cell detection unit or an operator to isolate cells individually. Typically, these systems are limited in throughput, compared to the droplets-in-oil-technology described before.

Hydrodynamic traps are passive structures in a microfluidic channel that allow only one cell to enter the "trap" (Figure 5). Typically, double occupation is minimized by adjusting the trap size to the average cell size in a given sample. Such systems can operate on a large number of cells in parallel by using a large number of traps [48]. The commercial system C1 from Fluidigm Corp. for example, allows for isolation and subsequent genetic analysis, of up to 96 individual cells in parallel. Hydrodynamic trapping can even be integrated into handheld pipettes to enable manual single-cell pipetting [50] without the need of micromanipulation under a microscope.

Furthermore, approaches to miniaturize flow cytometers by use of microfluidic technologies have been proposed [51]. One of the goals of this field of research is to bring the advantages of flow cytometers (see Section 3.1) such as cell sorting and counting to small and affordable devices, which can potentially be portable. Microfluidic lab-on-a-chip technologies for single-cell applications demonstrate exciting opportunities, but are mostly specifically designed to serve a particular application and therefore exhibit only little flexibility regarding upstream sample preparation and downstream analysis methods. To increase the flexibility and enable easier interfacing with other upstream and downstream methods, the authors of this review have proposed previously a flexible single-cell isolation system (Figure 6), based on inkjet-like single-cell printing [52,53]. This so called single-cell printer (SCP) uses an imaging system and automated object recognition algorithms to detect cells in a microfluidic dispenser chip that can produce droplets similar to an inkjet printer. Cells are classified in the nozzle of the chip and subsequently ejected within a microdroplet (60 μm in diameter) to be deposited onto various substrates. Droplets containing no cell or multiple cells are deflected in flight towards a waste container by vacuum suction. The suitability of the SCP for biomedical applications such as single-cell genomics [54] and clonal cell line production has already been demonstrated [53].

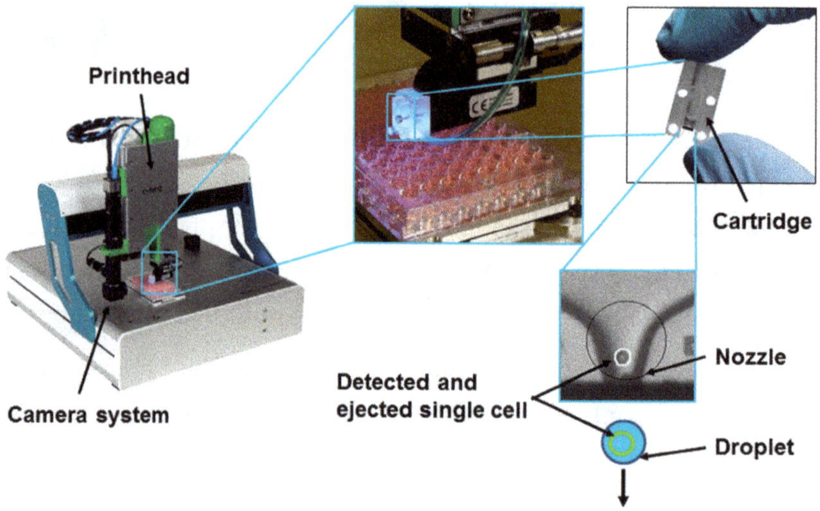

Figure 6. Single-cell printer (SCP) for single-cell isolation. A microfluidic dispenser chip integrated in a polymer cartridge is filled with cell suspension. An automated object recognition algorithm detects cells in the dispenser nozzle prior to the dispensation. This allows for ejection of droplets containing one single-cell only and their deposition in direction of the arrow on various substrates, such as micro-well plates.

4. Patent Search for Single-Cell Separation Technologies

In order to identify emerging single-cell separation technologies currently not well known by the scientific community, a patent search has been performed. The reasoning behind this complementary approach is that a market analysis like described above can only be expected to reveal technologies that are sufficiently well known within the user community, while novel technologies that are probably in a pre-commercialization phase are not necessarily known by this group of persons. Still, such emerging technologies could become relevant or even displace existing technologies in the future. In order to screen for novel technologies, we performed a patent search at the European Patent Office (EPO) in the worldwide database for patents. In detail, we were applying the "Smart search" on the EPO homepage (http://www.epo.org/searching/free/espacenet.html) using combinations of search termini based on Boolean operations: (txt = "single cell" and (txt = isolation OR txt = separation)) NOT (txt = fuel OR txt = solar). This search led to 179 results, which can be found as a complete list in the supplement (Table S1). The titles and abstracts of the 179 results were carefully reviewed in consideration of their relevance as actual isolation technology for single biological cells. Conversely, the patents considered to be particularly relevant should not represent:

- Only an analysis method of single cells.
- A common method using the term "single cell suspension" without addressing specifically a method for single-cell isolation.
- A cell separation method, which is already established and only part of a patented workflow.
- Other, in this context, irrelevant methods by using the term "cell" in a non-biological context such as for a battery or a chamber in a technical device. Therefore, the terms "fuel" and "solar" were already excluded from the original search from the beginning (see above).

According to these criteria 25 patents were selected from the list of search results based on the information provided in the title and abstract of these patents. The resulting list of relevant patents is presented in Table 2.

Table 2. Patents of single-cell isolation technologies identified in the worldwide database of the European Patent Office (EPO).

Title	EPO Publication Number	Reference
Methods for multiplex analytical measurements in single cells of solid tissues	AU2013315409 (A1)	[55]
Single-cell isolation screen adapted with pipettor tip	CN104195036 (A)	[56]
An integrated microfluidic device for single-cell isolation, cell lysis and nucleic acid extraction *	CA2817775 (A1)	[57]
System and method for capturing and analyzing cells *	US2014349867 (A1)	[58]
Single-cell automatic analysis device based on dual-optical-path micro-fluidic chip *	CN203929785 (U)	[59]
Microfluidic devices and methods for cell sorting, cell culture and cells based diagnostics and therapeutics *	US2014248621 (A1)	[60]
High-throughput single-cell imaging, sorting, and isolation *	US8934700 (B2); US2014247971 (A1)	[61]
Automatic single cell analysis method based on microfluidic system *	CN103926190 (A)	[62]
Apparatus for single cell separation and position fixing *	US2013129578 (A1); US8475730 (B2)	[63]
Method and apparatus for single cell isolation and analysis	US2012315639 (A1)	[64]
Apparatus for magnetic separation of cells	US2012045828 (A1)	[65]
Method and apparatus for the discretization and manipulation of sample volumes *	CN102187216 (A)	[66]
Plate for separating single cell	JP2011152108 (A); JP5622189 (B2)	[67]
Array apparatus for separation of single cell *	KR20110037345 (A); KR101252829 (B1)	[68]
Device and method for continuously analyzing single-cell contents by miniflow control chip at high speed *	CN101923053 (A); CN101923053 (B)	[69]
Complete set of equipment for single cell gel electrophoresis test	CN201662556 (U)	[70]
Single cell analysis of membrane molecules *	US2009173631 (A1)	[71]

Table 2. *Cont.*

Title	Publication Number	Reference
Single-cell inclusion analytical method based on micro-fluidic chip *	CN101393124 (A)	[72]
Analytical system based on porous material for highly parallel single cell detection *	US2008020453 (A1)	[73]
Single cell isolation apparatus and method of use	US6538810 (B1)	[74]
Cell isolation and screening device and method of using same *	WO03011451 (A1)	[75]
Cell transfer mechanism and cell fusion apparatus *	JPH0731457 (A)	[76]
Device for automatically testing single cell dielectric spectrum based on composite dielectrophoresis	CN201075104 (Y)	[77]
High-pass cell separation device and use method therefor *	CN1962845 (A)	[78]
Cell inclusions analysis method based on microfluid chip *	CN1734265 (A)	[79]

*, indicate patents which can be assigned to the field of microfluidics.

Considering that isolation of biological cells means handling of particles in liquid typically in the range of 10–100 μm in diameter, the majority of these patents (17 out of 25) can be assigned to the field of microfluidics, as indicated by the appearance of "microfluidic" or terms inherent to microfluidics in the title, abstract and/or original document of the respective patents. Since microfluidics has been described in section 3.5 in general terms, the authors refrain from a full discussion of these technologies. The remaining rather "non-microfluidic" technologies utilize a targeted lysis within a tissue [55], a separation sieve for a conventional pipette tip [56], antibody-conjugated magnetic beads [64,65], adjustable permeable wells on a plate [67], electrophoresis [70], tailor-made wells on a membrane [74], and dielectrophoresis [77] for the isolation of single cells.

5. Future Potential of Single-Cell Technologies

There are currently several indicators that suggest that the field of single-cell analysis is going to grow further in the coming years. One of them is the recent acknowledgments of the field by the journal *Nature Methods.* The increased interest is supported by *Nature Methods* selecting "single-cell methods" as *method to watch 2011* [80] and "single-cell sequencing" as *method of the year 2013* [81]. The demand for single-cell handling technologies is expected to grow simultaneously with the field of single-cell analysis.

Figure 7 shows the expected increase in importance of single-cell technologies as seen by the participants of the market study, described in Section 2. In the online survey participants were asked to rank the importance of single-cell research from *not important* to *very important* to their present as well as for their future work. For the present year (2014) the answer with the highest score (mode of the data set) was *fairly important*. For 2017 the expectation with the highest score was *very important* (the highest possible score). These data were acquired amongst German researchers. A similar trend is expected to hold for worldwide researchers, which is in accordance to market data published [6].

Furthermore, participants, who did not have any experience with single-cell technologies in 2014, were asked about their future plans. Out of these, 79% planned to work with single-cell technologies within the next five years. 48% even planned to start working with single-cell technologies in 2015.

In recent years, the field of single-cell analysis has seen a few commercial products hit the market and more are expected to come. Applications currently envisioned or implemented include next generation sequencing (NGS) of single cells [82], isolation of circulating tumor cells for diagnostic purposes [83], or single-cell proteomics [84].

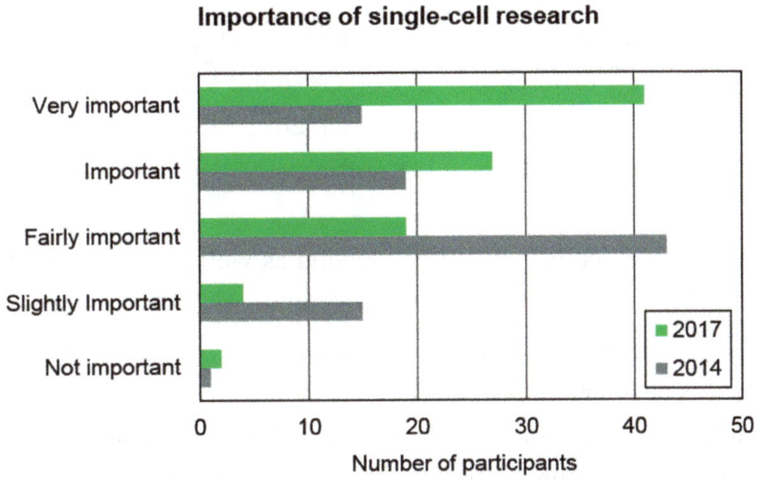

Figure 7. Importance of single-cell analysis to German cell researchers in 2014 and estimated for 2017. This data was derived by a survey amongst 210 participants from German universities, research institutes and industry. A strong growth of interest is expected over the next years.

6. Discussion

The requirements for technologies to separate and isolate single cells from samples of different nature are as heterogeneous as the purpose for which the cells are used downstream of the separation and isolation process. Depending on the specific requirements, some of the mentioned technologies might be more suitable than others to enable a specific single cell application. In the following, strengths as well as shortcomings of the reviewed commercial technologies are listed and compared to these requirements qualitatively. However, quantitative performance parameters for the various technologies cannot be determined, since this would require focusing on a specific application, which is out of scope of this article. The actual result of a single-cell isolation process—especially with respect to efficiency and cell viability—is depending on many factors like cell type, sample preparation, device calibration, sorting mode, substrate, and many more factors that are hard to quantify in general (e.g., operator skill for manual methods). Thus, the assessment of the different technologies has to remain more general and instead general properties of the technologies like throughput or cell viability, *etc.* are discussed.

6.1. General Aspects

Defining general requirements in the field of single-cell handling and analysis is not trivial due to the heterogeneity of the applications mentioned above. The following non-exhaustive list seeks to cover general requirements that many applications have in common.

Sample nature and origin defines to a large extent which technologies can be used at all. The information to be determined from the cells of interest could largely differ: in solid samples often tissue architecture or cell-cell interactions are of interest, while in cell suspensions heterogeneity studies of the cell population are mostly the primary objective.

Cell integrity is often required throughout the cell isolation process. Especially when the genome or proteome is the target of the analysis, cell integrity should be kept prior to lysis to avoid early degradation of DNA/RNA.

Cell viability is required when isolating single-cells for the purpose of production of monoclonal cell cultures or for studying stem cell differentiation. Cells respond to stress factors like mechanical forces, radiation, chemical changes in their environment, *etc.* which can lead to differentiation, reduced viability or even apoptosis. Technologies should provide sufficiently "gentle" extraction and handling when operating on living cells.

Throughput in terms of single cells isolated per second as well as the targeted total number of single cells is another important factor. Especially when large populations with low abundance of target cells are given (e.g., for CTC (Circulating tumor cells) applications), high throughput or high parallelization is mandatory and manual procedures are prohibitive.

Rare samples with a low amount of cells require technologies capable of dealing with small sample volumes and providing low dead volumes. In this context the separation yield (see below) is often a key issue as well in this context to prevent the loss of cells of interest.

Purity of the isolated single cells is crucial when analyzing cellular DNA and RNA. Isolating the cell of interest and while excluding any other contamination from the liquid suspension (e.g., cell fragments, free-DNA, *etc.*) is of highest importance. Transfer volume (the droplet or pipetting volume the cell is typically enclosed in during transfer to the target) and cross-contamination come into play here as well, as they determine the amount of contaminants in the aliquot containing the single cell.

Efficiency concerning yield of the single-cell isolation process can be of importance when using homogeneous samples containing a large number of cells (e.g., for cloning). However, it's a lot more important when performing single-cell analysis on a rare cell sample or when using complex and costly reagents. Besides the costs per cell that come into play from the economical point of view (e.g., for amplification and library preparation for next-generation sequencing [85]), the analysis of individually selected cells rather than randomly seeded ones or the analysis of a complete population (100% analysis) can impose strict requirements on the separation efficiency. Preventing aliquots that are empty or occupied with multiple cells can be mandatory in certain applications while in others it might be more important that no cell must be lost in the isolation process.

Besides these most relevant general criteria for the assessment of different technologies each application might impose additional specific requirements that have to be carefully considered. For applications with living cells, for example, often carry over-free and sterile operation conditions are requested which calls for technologies relying on disposable components (e.g., microfluidic chips).

Obviously, each single cell separation technology exhibits specific features with respect to the above mentioned aspects that have to be matched to the application under consideration. Usually, this problem is addressed from the side of the application (e.g., single-cell analysis, monoclonal cell cultures, *etc.*) for which the requirements are usually well known. In the following the view point should be shifted towards considering the specific features of the previously discussed technologies.

6.2. Flow Cytometry

FACS systems provide high throughput in terms of single-cell analysis and sorting. Paired with high flexibility in terms of cell type, standardized substrates, and sorting modes FACS is a powerful tool. Moreover its suitability for rare cell sorting (subpopulations < 1%) has increasing diagnostic prospect when analyzing heterogeneous cell samples. Some of the FACS systems are able to deposit single cells in micro-well plates with high purity and yield within the time frame of minutes to enable further downstream analysis such as NGS (Next generation sequencing). The popularity and wide-spread use of FACS systems makes them accessible to a broad range of users.

Nevertheless, for certain applications FACS systems are still limited to some extent. Cells must be in suspension meaning tissues need to be dissociated resulting in loss of cellular functions and cell-cell interactions as well as tissue architecture [86]. Subpopulations with similar marker expression are difficult to differentiate and overlap of emission spectra between fluorochromes may lead to an increasing noise level making low-intensity samples unavailable for detection. Further, FACS sorting may have non-negligible effects on cell viability, which was demonstrated for Chinese Hamster Ovary (CHO) cells and for a human monocytic cell line (THP1) by trypan blue exclusion and necrosis/apoptosis assays [87].

Minimal sample volume for FACS systems is in the range of several hundreds of microliters to milliliters. This is due to the fact, that typically long sections of tubing cause high dead volumes, preventing the use of rare samples, especially when the entire sample needs to be analyzed. And finally, sterile operation is difficult to achieve by FACS systems in general due to the complex system consisting of non-disposable components. However, such limitations are not always present or significant, but strongly dependent on instrument, process parameters (e.g., speed, laser type, *etc.*), cell type, and application.

6.3. Laser Capture Microdissection

Whenever single cells need to be isolated from solid samples (e.g., tissue, biopsies) LCM systems are the commonly applied tool of choice. For such applications a single cell's location within the tissue architecture is essential to know. LCM systems enable separation and isolation of individual cells for downstream analysis. Based on optical microscopes coupled with a coaxial cutting laser and computer assisted control, such systems are relatively easy to handle. The sharply focused pulse lasers cut with sub-micrometer precision and without introducing deleterious heat to the tissue. Another benefit might be that operators actually decide on every cell to select and isolate rather than leaving the decision to automated systems or statistical distributions.

Although modern LCM systems provide a higher level of user-friendliness and automation, the selection and isolation process remains operator based and therefore strongly limits the throughput. The integrity of extracted cells is important for reliable downstream analysis of biomolecules such as DNA, RNA, and proteins [88]. Depending on the quality of fixation [32] and cell extraction method used (adhesive tape, gravity, catapulting) single-cell integrity might be compromised [89]. It might remain unclear, if the cell was actually transferred and/or if any contaminants (e.g., fragments of adjacent cells) were transferred to the substrate along with the cell of interest, especially when using contact-based cell extraction (adhesive cap) [90].

6.4. Micromanipulator

Micromanipulators are microscope assisted picking tools allowing for targeted isolation of individual single cells from suspensions, which is a feature not shared by many other technologies. The operator selects the cells to be isolated and performs the aspiration, transfer, and dispensation. Similar to LCM systems, the targeted isolation of a specific cell under microscope vision is one of the key benefits of this technology.

Although being a very flexible technology in terms of cell types and substrates, the serial and manual process of obtaining single cells limits the overall throughput. Furthermore, with the majority of the systems it is not possible to observe and control the correct transfer of the single-cell to its target location. Once the micropipette leaves the microscope's optical focus plane the transfer volume containing the cell is unobservable as is the actual transition to the target. To actually confirm if a single-cell has successfully been transferred additional observation of the target is required. Recent approaches target improvement on this towards fully automated isolation and placement of single cells assisted by video systems and image processing algorithms [91].

6.5. Limiting Dilution

In many pharmaceutical companies, fully automated pipetting robots perform limiting dilution in great numbers at considerable throughput. The process is simple, reproducible and to a certain extent cost-efficient since the degree of automation is very high. However, due to the statistical nature of the process and the lack of control over an individual cell, often further technologies are required downstream to prove the presence of single cells in a specific well, such as automatic microscopic imaging systems. However, limiting dilution remains a simple, gentle, and relatively cost efficient process to obtain single cells with reasonable throughput, but lacking the controlled isolation and sorting as well as proof of single-cell presence. Combined with upstream sorting or enrichment techniques it can constitute an appropriate tool to easily separate viable single cells for downstream analysis.

6.6. Microfluidics

Microfluidics plays an increasing role in establishing entire workflows for single-cell separation, isolation, and analysis. Although the number of commercially available systems is still low, this field of research itself is highly dynamic. Microfluidic systems can be operated with very low volumes regarding cell sample as well as reagents, which is advantageous for rare cell applications as well as from an economical point of view. Developed as disposables and once produced with mass-fabrication techniques those systems could provide an attractive alternative, especially with regard to clinical applications: when using disposables, cross-contamination between subsequent samples is not an issue.

Recent hurdles preventing a broader market-entry of microfluidic technologies might be the low degree of flexibility offered by a specific microfluidic chip. Microfluidic systems–unlike the more established technologies discussed before – are often restricted to one single application (e.g., genomic single-cell analysis). Regarding the complexity and variety of single-cell applications throughout the field, and in view of continuously newly established analysis methods, this likely is one of the greatest drawbacks of microfluidic technologies for research applications. Leading scientists and microfluidic companies often regret the absence of a "killer application" supporting their technology (This is true for the entire microfluidic field, not only for those devices targeting single-cell applications.) to enter the markets [92–94]. Still, when a sufficiently standardized workflow of isolation, sorting, and analysis can be established by a microfluidic approach, there are probably few other technologies that can provide similar performance in that specific case.

6.7. Patents and Emerging Technologies

The patent search performed for this review revealed that 68% (17 out of 25) of the related patents cover microfluidic devices. This patent trend has been present since the middle of the 90s [95] and emphasizes the strong driving forces generated by research and emerging industries. Besides microfluidics few patents deal with alternative technologies e.g., separation by dielectrophoresis or size/affinity based filtering. Partially such alternative technologies are also combined with microfluidic approaches at chip level. Whether or not such special technologies will become relevant in the future depends to a large degree on the applications that can be enabled by them. As outlined before, a strong match between application requirements and technology features can render any technology relevant in the context of a well matching application.

7. Conclusions

The review focused on the most frequently used single-cell isolation technologies as derived from a market survey amongst German scientists. It can be concluded that there is no universal technology suitable for all single-cell applications. However, the vast majority of applications can be addressed by at least one of the technologies available today.

Regarding performance FACS systems have the main benefit of high throughput and sorting capability but are cost-intensive and potentially harmful to some cells. LCM is ideal for isolation of single cells from solid tissue and quite unique in this regard. Micromanipulator assisted cell picking is a manual process and therefore slow, but provides maximum control over individual cells. Limiting dilution relies on statistical distribution, is simple to implement and can be automated. However, the presence of single cells often needs to be verified subsequently. Apart from these established technologies, microfluidic technologies allow for integration of entire application specific workflows and only require small amounts of sample and reagents. Thus, microfluidic technologies have potential but are still lacking broad commercial presence, probably due to a lower degree of flexibility. To summarize the topic, Table 3 provides an overview about the key features of the presented technologies.

The worldwide patent search carried out with the objective to identify emerging technologies, resulted in 179 patents out of which 25 have been identified to be particularly relevant to the field. 17 out of those 25 are related to the field of microfluidics, again emphasizing the importance of this evolving field. From the market study it can also be concluded, that the field of single-cell analysis can be expected to grow significantly in the following years and the need for single-cell isolation technologies is likely to increase simultaneously. Presumably, established commercial technologies as well as novel microfluidic devices will contribute equally to advances in this field in the coming years.

Table 3. Selected features of discussed technologies. Rating based on the authors' personal experience and knowledge.

Technology	Automation Level	Throughput	Impact on Cell Integrity	Control over Individual Cell [1]	Compatibility with Established Workflows [2]
Fluorescence-Activated Cell Sorting (FACS)	Automatic	High	Often impairing	Yes	High
Limiting dilution	Manual or automatic	Moderate	Gentle	No	High
Micromanipulation	Manual	Low	Moderate	Yes	Moderate
Laser-capture microdissection	Manual	Low	Often impairing	Yes	Low
Microfluidics (Lab-on-a-Chip)	Automatic	Low to high	Diverse	Typically not	Low
Microfluidics (inkjet-like printing)	Automatic	Moderate	Gentle	Yes	High

[1], Possibility for active selection of single cells before their isolation in contrast to random distribution of individual cells; [2], Compatibility with commercially available substrates such as microtiter plates, tubes, slides, *etc.*

Acknowledgments

We gratefully acknowledge financial support by EU FP7 project SICTEC (GA 606167) and national BMWi/ESF funded project SCOD (FKZ 03EFBBW057). Further we would like to thank the University of Mannheim, Germany for their support in conducting the German market study and HTStec, UK for cooperation regarding their international market study.

Author Contributions

The authors contributed as follows: Andre Gross, writing chapters 3.1–3.4, 6, Figure 4; Jonas Schoendube, writing chapters 2, 3.5, 5, Figures 1,2,5,6,7; Stefan Zimmermann, patent search, writing chapters 1, 4, Figure 3; Maximilian Steeb executed the survey for the German market study; Peter Koltay, writing chapter 7, editing and proof-reading; Roland Zengerle, design of the study, editing and proof-reading. This review presents only major technologies and techniques and does not claim to be comprehensive and complete in all fields. Technology ratings like in Table 3 are personal opinions of the authors based on their experience in the field.

References

1. Blainey, P.C.; Quake, S.R. Dissecting genomic diversity, one cell at a time. *Nat. Methods.* **2014**, *11*, 19–21.
2. Van Loo, P.; Voet, T. Single cell analysis of cancer genomes. *Curr. Opin. Genet. Dev.* **2014**, *24*, 82–91.
3. Ding, L.; Wendl, M.C.; McMichael, J.F.; Raphael, B.J. Expanding the computational toolbox for mining cancer genomes. *Nat. Rev. Genet.* **2014**, *15*, 556–570.
4. Ilie, M.; Hofman, V.; Long, E.; Bordone, O.; Selva, E.; Washetine, K.; Marquette, C.H.; Hofman, P. Current challenges for detection of circulating tumor cells and cell-free circulating nucleic acids, and their characterization in non-small cell lung carcinoma patients. What is the best blood substrate for personalized medicine? *Ann. Transl. Med.* **2014**, *2*, 107.

5. BIOCOM AG. *Biotechnology Database*. Available online: http://www.biotechnologie.de/BIO/Navigation/EN/Databases/biotechnology-db.html (accessed on 30 June 2015).
6. HTStec. *Single Cell Technologies Trends 2014*; Available online: http://selectbiosciences.com/MarketReportsID.aspx?reportID=83 (accessed on 30 June 2015).
7. Moldavan, A. *Photo-Electric Technique for the Counting of Microscopial Cells*; Science: New York, NY, USA, 1934; Volume 80, pp. 188–189.
8. Gucker, F.T.; O'Konski, C.T. A photoelectronic counter for colloidal particles. *J. Am. Chem. Soc.* **1947**, *69*, 2422–2431.
9. Coulter, W.H. Means for Counting Particles Suspended in a Fluid. U.S. Patent 2,656,508 A, 20 October 1953.
10. Fulwyler, M.J. Electronic separation of biological cells by volume. *Science* **1965**, *150*, 910–911.
11. Fulwyler, M.J. Particle Separator. U.S. Patent 3,380,584 A, 30 April 1968.
12. Dittrich, W.; Goehde, W. Automatisches Mess- und Zaehlgeraet fuer die Teilchen einer Dispersion. D.E. Patent 1,815,352 A1, 14 January 1971.
13. Shapiro, H.M. *Practical Flow Cytometry*, 4th ed.; Wiley-Liss: New York, NY, USA, 2003.
14. Kohler, G.; Milstein, C. Continuous cultures of fused cells secreting antibody of predefined specificity. *Nature* **1975**, *256*, 495–497.
15. Herzenberg, L.A.; Parks, D.; Sahaf, B.; Perez, O.; Roederer, M.; Herzenberg, L.A. The history and future of the fluorescence activated cell sorter and flow cytometry: A view from Stanford. *Clin. Chem.* **2002**, *48*, 1819–1827.
16. Underwood, A.P.; Bean, P.A. Hazards of the limiting-dilution method of cloning hybridomas. *J. Immunol. Methods* **1988**, *107*, 119–128.
17. Brown, M.; Wittwer, C. Flow cytometry: Principles and clinical applications in hematology. *Clin. Chem.* **2000**, *46*, 1221–1229.
18. Valet, G. Past and present concepts in flow cytometry: A European perspective. *J. Biol. Regul. Homeost. Agents* **2003**, *17*, 213–222.
19. Davey, H.M.; Kell, D.B. Flow cytometry and cell sorting of heterogeneous microbial populations: The importance of single-cell analyses. *Microbiol. Rev.* **1996**, *60*, 641–696.
20. Lacombe, F.; Belloc, F. Flow cytometry study of cell cycle, apoptosis and drug resistance in acute leukemia. *Hematol. Cell Ther.* **1996**, *38*, 495–504.
21. McCoy, J.P.; Carey, J.L. Recent advances in flow cytometric techniques for cancer detection and prognosis. *Immunol. Ser.* **1990**, *53*, 171–187.
22. Emmert-Buck, M.R.; Bonner, R.F.; Smith, P.D.; Chuaqui, R.F.; Zhuang, Z.; Goldstein, S.R.; Weiss, R.A.; Liotta, L.A. Laser Capture microdissection. *Science* **1996**, *274*, 998–1001.
23. Espina, V.; Heiby, M.; Pierobon, M.; Liotta, L.A. Laser capture microdissection technology. *Expert Rev. Mol. Diagn.* **2007**, *7*, 647–657.
24. Vogel, A.; Noack, J.; Hüttman, G.; Paltauf, G. Mechanisms of femtosecond laser nanosurgery of cells and tissues. *Appl. Phys. B* **2005**, *81*, 1015–1047.
25. Vogel, A.; Lorenz, K.; Horneffer, V.; Hüttmann, G.; von Smolinski, D.; Gebert, A. Mechanisms of laser-induced dissection and transport of histologic specimens. *Biophys. J.* **2007**, *93*, 4481–4500.

26. Esposito, G. Complementary techniques: Laser capture microdissection–increasing specificity of gene expression profiling of cancer specimens. *Adv. Exp. Med. Boil.* **2007**, *593*, 54–65.
27. Fink, L.; Kwapiszewska, G.; Wilhelm, J.; Bohle, R.M. Laser-microdissection for cell type- and compartment-specific analyses on genomic and proteomic level. *Exp. Toxicol. Pathol.* **2006**, *5*, 25–29.
28. Fink, L.; Bohle, R.M. Laser microdissection and RNA analysis. *Methods Mol. Biol.* **2005**, *293*, 167–185.
29. Nakamura, N.; Ruebel, K.; Jin, L.; Qian, X.; Zhang, H.; Lloyd, R.V. Laser capture microdissection for analysis of single cells. *Methods Mol. Med.* **2007**, *132*, 11–18.
30. Keays, K.M.; Owens, G.P.; Ritchie, A.M.; Gilden, D.H.; Burgoon, M.P. Laser capture microdissection and single-cell RT-PCR without RNA purification. *J. Immunol. Methods* **2005**, *302*, 90–98.
31. Vandewoestyne, M.; Deforce D. Laser capture microdissection in forensic research: A review. *Int. J. Leg. Med.* **2010**, *124*, 513–521.
32. DeCarlo, K.; Emley, A.; Dadzie, O.E.; Mahalingam, M. Laser capture microdissection: Methods and applications. *Methods Mol. Biol.* **2011**, *755*, 1–15.
33. Lefkovits, I.; Pernis, B. *Immunological Methods*; Elsevier Science: Burlington, MA, USA, 1979.
34. Goding, J.W. Antibody production by hybridomas. *J. Immunol. Methods* **1980**, 285–308.
35. Fuller, S.A.; Takahashi, M.; Hurrell, J.G. Cloning of hybridoma cell lines by limiting dilution. *Curr. Protoc. Mol. Biol.* **2001**, doi:10.1002/0471142727.mb1108s01.
36. Yokoyama, W.M.; Christensen, M.; Dos Santos, G.; Miller, D.; Ho, J.; Wu, T.; Dziegelewski, M.; Neethling, F.A. Production of monoclonal antibodies. *Curr. Protoc. Immunol.* **2013**, doi:10.1002/0471142735.im0205s102.
37. Staszewski, R. Cloning by limiting dilution: An improved estimate that an interesting culture is monoclonal. *Yale J. Biol. Med.* **1984**, *57*, 865–868.
38. Smith, R. Cell technology for cell products. Proceedings of the 19th ESACT Meeting, Harrogate, UK, 5–8 June 2005; Springer: Dordrecht, The Netherlands, 2007.
39. Fröhlich, J.; König, H. New techniques for isolation of single prokaryotic cells. *FEMS Microbiol. Rev.* **2000**, *24*, 567–572.
40. Brehm-stecher, B.F.; Johnson, E.A. Single-cell microbiology: Tools, technologies, and applications. *Microbiol. Mol. Biol. Rev.* **2004**, *68*, 538–559.
41. Wright, G.; Tucker, M.J.; Morton, P.C.; Sweitzer-Yoder, C.L.; Smith, S.E. Micromanipulation in assisted reproduction: A review of current technology. *Curr. Opin. Obstet. Gynecol.* **1998**, *10*, 221–226.
42. Li, C.-X.; Wang, G.-Q.; Li, W.-S.; Huang, J.-P.; Ji, A.-Q.; Hu, L. New cell separation technique for the isolation and analysis of cells from biological mixtures in forensic caseworks. *Croat. Med. J.* **2011**, *52*, 293–298.
43. Sims, C.E.; Allbritton, N.L. Analysis of single mammalian cells on-chip. *Lab Chip* **2007**, *7*, 423–440.
44. Lecault, V.; White, A.K.; Singhal, A.; Hansen, C.L. Microfluidic single cell analysis: From promise to practice. *Curr. Opin. Chem. Biol.* **2012**, *16*, 381–390.
45. Gossett, D.R.; Weaver, W.M.; Mach, A.J.; Hur, S.C.; Tse, H.T.K.; Lee, W.; Amini, H.; di Carlo, D. Label-free cell separation and sorting in microfluidic systems. *Anal. Bioanal. Chem.* **2010**, *397*, 3249–3267.

46. Brouzes, E.; Medkova, M.; Savenelli, N.; Marran, D.; Twardowski, M.; Hutchison, J.B.; Rothberg, J.M.; Link, D.R.; Perrimon, N.; Samuels, M.L. Droplet microfluidic technology for single-cell high-throughput screening. *Proc. Natl. Acad. Sci. USA* **2009**, *106*, 14195–14200.
47. Gomez-Sjoberg, R.; Leyrat, A.A.; Pirone, D.M.; Chen, C.S.; Quake, S.R. Versatile, fully automated, microfluidic cell culture system. *Anal.Chem.* **2007**, *79*, 8557–8563.
48. Di Carlo, D.; Wu, L.Y.; Lee, L.P. Dynamic single cell culture array. *Lab Chip* **2006**, *6*, 1445–1449.
49. Edd, J.F.; di Carlo, D.; Humphry, K.J.; KÇôster, S.; Irimia, D.; Weitz, D.; Toner, M. Controlled encapsulation of single-cells into monodisperse picolitre drops. *Lab Chip* **2008**, *8*, 1262–1264.
50. Zhang, K.; Han, X.; Li, Y.; Li, S.Y.; Zu, Y.; Wang, Z.; Qin, L. Hand-held and integrated single-cell pipettes. *J. Am. Chem. Soc.* **2014**, *136*, 10858–10861.
51. Ateya, D.A.; Erickson, J.S.; Howell, P.B.; Hilliard, L.R.; Golden, J.P.; Ligler, F.S. The good, the bad, and the tiny: A review of microflow cytometry. *Anal. Bioanal. Chem* **2008**, *391*, 1485–1498.
52. Yusof, A.; Keegan, H.; Spillane, C.D.; Sheils, O.M.; Martin, C.M.; O'Leary, J.J.; Zengerle, R.; Koltay, P. Inkjet-like printing of single-cells. *Lab Chip* **2011**, *11*, 2447–2454.
53. Gross, A.; Schondube, J.; Niekrawitz, S.; Streule, W.; Riegger, L.; Zengerle, R.; Koltay, P. Single-cell printer: Automated, on demand, and label free. *J. Lab. Autom.* **2013**, *18*, 504–518.
54. Stumpf, F.; Schoendube, J.; Gross, A.; Rath, C.; Niekrawietz, S.; Koltay, P.; Roth, G. Single-cell PCR of genomic DNA enabled by automated single-cell printing for cell isolation. *Biosens. Bioelectron.* **2015**, *69*, 301–306.
55. Finski, A.; Macbeath, G. Methods for Multiplex Analytical Measurements in Single Cells of Solid Tissues. Patent AU2013315409 (A1), 2 April 2015.
56. Liu, D. Single-Cell Isolation Screen Adapted with Pipettor Tip. Patent CN104195036 (A), 10 December 2014.
57. Yao, C.D. An Integrated Microfluidic Device for Single-Cell Isolation, Cell Lysis and Nucleic Acid Extraction. Patent CA2817775 (A1), 29 November 2014.
58. Handique, K.; Gogoi, P.; Javdani, S.S.; Zhou, Y. System and Method for Capturing and Analyzing Cells. US2014349867 (A1), 27 November 2014.
59. Liu, Y.; Li, J.; Chen, Y.; Jian, D.; Gao, J. Single-Cell Automatic Analysis Device Based on Dual-Optical-Path Micro-Fluidic Chip. Patent CN203929785 (U), 5 November 2014.
60. Collins, J. Microfluidic Devices and Methods for Cell Sorting, Cell Culture and Cells Based Diagnostics and Therapeutics. Patent US2014248621 (A1), 4 September 2014.
61. Bharadwaj, R.; Fathollahi, B. High-Throughput Single-Cell Imaging, Sorting, and Isolation. Patent US8934700 (B2), US2014247971 (A1), 4 September 2014.
62. Liu, Y.; Ban Qing, G.J. Automatic Single Cell Analysis Method Based on Microfluidic System. Patent CN103926190 (A), 16 July 2014.
63. Jeong, O.C. Apparatus for Single Cell Separation and Position Fixing. Patent US2013129578 (A1), US8475730 (B2), 23 May 2013.
64. Deng, G.Y.; Zhang, J.; Tian, F. Method and Apparatus for Single Cell Isolation and Analysis. Patent US2012315639 (A1), 13 December 2012.
65. Davis, R.W.; Jeffrey, S.S.; Mindrinos, M.N.; Pease, R.F.; Powell, A.A.; Talasaz, A.H. Apparatus for Magnetic Separation of Cells. Patent US2012045828 (A1), 23 February 2012.

66. Chiu, D.T.; Cohen, D.E.; Jeffries, G.D.M. Method and Apparatus for the Discretization and Manipulation of Sample Volumes. Patent CN102187216 (A), 14 September 2011.
67. Matsutani, A.; Takada, A. Plate for Separating Single Cell. Patent JP2011152108 (A), JP5622189 (B2), 11 August 2011.
68. Park, J.W.; Jung, M.Y.; Park, S.H. Array Apparatus for Separation of Single Cell. Patent KR20110037345 (A), KR101252829 (B1), 13 April 2011.
69. Yin, X.F.; Xu, C.X.; Liu, J.H. Device and Method for Continuously Analyzing Single-Cell Contents by Miniflow Control Chip at High Speed. Patent CN101923053 (A), CN101923053 (B), 22 December 2010.
70. Zhang, C.K.; Shi, J.Y.; Zeng, Y.C.; Shen, C.F.; Chen, Y,X.; Chen, L. Complete Set of Equipment for Single Cell Gel Electrophoresis Test. Patent CN201662556 (U), 1 December 2010.
71. Boone, T.; Singh, S. Single Cell Analysis of Membrane Molecules. Patent US2009173631 (A1), 9 July 2009.
72. Wu, D.P.; Yu, L.F.; Lin, B.C.; Qin, J.H. Single-Cell Inclusion Analytical Method Based on Micro-Fluidic Chip. Patent CN101393124 (A), 25 March 2009.
73. Ehben, T.; Zilch, C. Analytical System Based on Porous Material for Highly Parallel Single Cell Detection. Patent US2008020453 (A1), 24 January 2008.
74. Karanfilov, C. Single Cell Isolation Apparatus and Method of Use. Patent US6538810 (B1), 25 March 2003.
75. Wang, E.; Kim, E.; Campbell, S.; Kirk, G.L.; Casagrande, R. Cell Isolation and Screening Device and Method of Using Same. Patent WO03011451 (A1), 13 February 2003.
76. Suzuki, H.; Yasunaka, T. Cell Transfer Mechanism and Cell Fusion Apparatus. Patent JPH0731457 (A), 3 February 1995.
77. Ni, Z.H.; Song, C.F.; Yi, H.; Zhu, S.C. Device for Automatically Testing Single Cell Dielectric Spectrum Based on Composite Dielectrophoresis. Patent CN201075104 (Y), 18 June 2008.
78. Shen, B.Y. High-Pass Cell Separation Device and Use Method Therefor. Patent CN1962845 (A), 16 May 2007.
79. Lin, B.Y. Cell Inclusions Analysis Method Based on Microfluid Chip. Patent CN1734265 (A), 15 February 2006.
80. De Souza, N. Single-cell methods. *Nat. Methods* **2011**, *9*, 35.
81. Editorial. Method of the Year 2013. *Nat. Methods* **2013**, *11*, 1.
82. Shapiro, E.; Biezuner, T.; Linnarsson, S. Single-cell sequencing-based technologies will revolutionize whole-organism science. *Nat. Rev. Genet.* **2013**, *14*, 618–630.
83. Yu, M.; Stott, S.; Toner, M.; Maheswaran, S.; Haber, D.A. Circulating tumor cells: Approaches to isolation and characterization. *J. Cell Biol.* **2011**, *192*, 373–382.
84. Wu, M.; Singh, A.K. Single-cell protein analysis. *Curr. Opin. Biotechnol.* **2012**, *23*, 83–88.
85. Mardis, E.R. The impact of next-generation sequencing technology on genetics. *Trends Genet.* **2008**, *24*, 133–141.

86. Jahan-Tigh, R.R.; Ryan, C.; Obermoser, G.; Schwarzenberger, K. Flow cytometry. *J. Investig. Dermatol.* **2012**, *132*, doi:10.1038/jid.2012.282.
87. Mollet, M.; Godoy-Silva, R.; Berdugo, C.; Chalmers, J.J. Computer simulations of the energy dissipation rate in a fluorescence-activated cell sorter: Implications to cells. *Biotechnol. Bioeng.* **2008**, *100*, 260–272.
88. Bevilacqua, C.; Makhzami, S.; Helbling, J.-C.; Defrenaix, P.; Martin, P. Maintaining RNA integrity in a homogeneous population of mammary epithelial cells isolated by Laser Capture Microdissection. *BMC Cell Biol.* **2010**, *11*, 95.
89. Liu A. Laser capture microdissection in the tissue biorepository. *J. Biomol. Tech.* **2010**, *21*, 120–125.
90. Fend, F. Laser capture microdissection in pathology. *J. Clin. Pathol.* **2000**, *53*, 666–672.
91. Lu, Z.; Moraes, C.; Zhao, Y.; You, L.D.; Simmons, C.A.; Sun, Y. A micromanipulation system for single cell deposition. In Proceeding of 2010 IEEE International Conference on Robotics and Automation (ICRA 2010), Anchorage, Alaska, 3–8 May 2010; IEEE: New York, NY, USA, May 2010; pp. 494–499.
92. Blow, N. Microfluidics: In search of a killer application. *Nat. Methods* **2007**, *4*, 665–668.
93. Whitesides, G.M. The origins and the future of microfluidics. *Nature* **2006**, *442*, 368–373.
94. Becker, H. Hype, hope and hubris: The quest for the killer application in microfluidics. *Lab Chip* **2009**, *9*, 2119–2122.
95. Haber, C. Microfluidics in commercial applications; an industry perspective. *Lab Chip* **2006**, *6*, 1118–1121.

9

A Review of Cell Adhesion Studies for Biomedical and Biological Applications

Amelia Ahmad Khalili [1] **and Mohd Ridzuan Ahmad** [1,2,*]

[1] Department of Control and Mechatronic Engineering, Faculty of Electrical Engineering, Universiti Teknologi Malaysia, Johor 81310, Malaysia; E-Mail: amelia.ahmadkhalili@gmail.com

[2] Institute of Ibnu Sina, Universiti Teknologi Malaysia, Johor 81310, Malaysia

* Author to whom correspondence should be addressed; E-Mail: ridzuan@fke.utm.my

Academic Editor: Fan-Gang Tseng

Abstract: Cell adhesion is essential in cell communication and regulation, and is of fundamental importance in the development and maintenance of tissues. The mechanical interactions between a cell and its extracellular matrix (ECM) can influence and control cell behavior and function. The essential function of cell adhesion has created tremendous interests in developing methods for measuring and studying cell adhesion properties. The study of cell adhesion could be categorized into cell adhesion attachment and detachment events. The study of cell adhesion has been widely explored via both events for many important purposes in cellular biology, biomedical, and engineering fields. Cell adhesion attachment and detachment events could be further grouped into the cell population and single cell approach. Various techniques to measure cell adhesion have been applied to many fields of study in order to gain understanding of cell signaling pathways, biomaterial studies for implantable sensors, artificial bone and tooth replacement, the development of tissue-on-a-chip and organ-on-a-chip in tissue engineering, the effects of biochemical treatments and environmental stimuli to the cell adhesion, the potential of drug treatments, cancer metastasis study, and the determination of the adhesion properties of normal and cancerous cells. This review discussed the overview of the available methods to study cell adhesion through attachment and detachment events.

Keywords: cell adhesion; attachment event; detachment event; adhesion strength

1. Introduction

Adhesion plays an integral role in cell communication and regulation, and is of fundamental importance in the development and maintenance of tissues. Cell adhesion is the ability of a single cell to stick to another cell or an extracellular matrix (ECM). It is important to understand how cells interact and coordinate their behavior in multicellular organisms. In vitro, most mammalian cells are anchorage-dependent and attach firmly to the substrate [1]. According to the "cell adhesion model", the more a cell sticks the more it shows the greater number of chemical bonds it has on its surface [2,3].

Cell adhesion is involved in stimulating signals that regulate cell differentiation, cell cycle, cell migration, and cell survival [4]. The affinity of cells to substrate is a crucial consideration in biomaterial design and development. Cell adhesion is also essential in cell communication and regulation, and becomes of fundamental importance in the development and maintenance of tissues. Changes in cell adhesion can be the defining event in a wide range of diseases including arthritis [5,6], cancer [4,7,8], osteoporosis [9,10], and atherosclerosis [11,12]. Cell adhesiveness is generally reduced in human cancers. Reduced intercellular adhesiveness allows cancer cells to disobey the social order, resulting in destruction of histological structure, which is the morphological hallmark of malignant tumors [8]. Tumor cells are characterized by changes in adhesivity to ECM, which may be related to the invasive and metastatic potential. Alterations in cell-matrix and cell-cell interactions are cell type- and oncogene-specific. For example, while the transfection of rodent fibroblast cells with *Src* and *Ras* oncogenes reduces the adhesiveness to fibronectin (Fn) by impairing $\alpha5\beta1$ integrins, the activation of oncogene *ErbB2* in breast cancer up-regulates $\alpha5\beta1$ integrin and enhances adhesion [13,14]. The adhesion of highly invasive cancer cells altered the biomechanics of endothelial cells [15]. Mierke [15] reported that *MDA-MB-231* cells' attachment may lower the endothelial cells' stiffness by breaking down the cells' barrier function through remodelling of the actin cytoskeleton.

Different requirements for cell adhesion are needed for various types of applications, and are dependent on the cell's specific applications [16]. Various techniques to analyze cell adhesion have been applied to understand different fields of study including biomaterial studies [17], the effects of biochemical treatments and environmental stimuli to the cell culture [18], and determination of adhesion properties of normal and cancerous cells [19]. Biomaterials designed in biomedical engineering that have to interact with blood, like those in artificial heart valves or blood vessels, are required not to be adherent to cells or plasma proteins to avoid thrombosis and embolism. On the other hand, materials used in scaffolds for tissue generation are needed to act as substrate to promote the cells' adhesion, subsequent proliferation, and biosynthesis [16]. Adhesion between cells allows blood clot formations that may lead to heart failure by restricting the blood supply to the heart muscles [16].

1.1. Focal Adhesion

Cells transmit extracellular or intracellular forces through localized sites at which they are adhered to other cells or an extracellular matrix. The adhesion sites are formed by transmembrane proteins called integrins to anchor the cell to a matrix or adhesion molecules to other cells [20]. Both the integrins and adhesion molecules are attached to the tensile members of the cytoskeleton, the actin filaments, through the focal adhesion (FA) complex (Figure 1), a highly organized cluster of molecules [21]. The cytoskeletal structure holds the nucleus and maintains the shape of the cell [22–24]. As a pathway for force transmission to the cytoskeleton, integrins play an important role in mechanotransduction through FA proteins connecting the integrin domains to the actin filaments to form the adhesion complex [24]. Upon binding, integrins cluster into FA complexes that transmit adhesive and traction forces [25,26]. The FA formation is important in cell signaling to direct cell migration [27], proliferation, and differentiation [28,29] for tissue organization, maintenance, and repair [28].

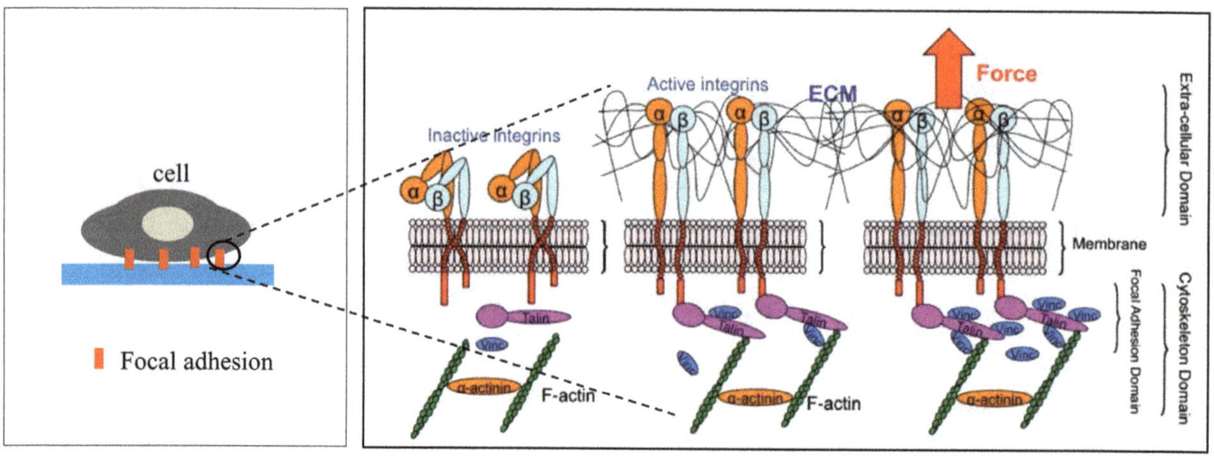

Figure 1. Schematic representation of activated integrin and formation of ECM-integrin-cytoskeleton linkages in the focal adhesion site upon application of an external tensile load. Reproduced "in part" from [24] with permission of The Royal Society of Chemistry.

1.2. Phases of Cell Adhesion and Spreading

1.2.1. Passive *in Vitro* Cell Adhesion

Passive *in vitro* cell adhesion is the cell adhesion process in a static medium culture, e.g., culture flasks, petri dishes. During static *in vitro* cell-matrix attachment and spreading, cells undergo morphologic alterations driven by passive deformation and active reorganization of the cytoskeleton. Integrin receptors and heterodimeric transmembrane proteins play a central role in cell adhesion and spreading. Specific integrin binding provides not only a mechanical linkage between the intracellular actin cytoskeleton and ECM, but also the bidirectional transmembrane signaling pathways [29–33]. Integrins recognize soluble ligands and insoluble ECM proteins and their interaction regulates cell responses such as cytoskeleton formation. The binding of integrins with their ECM proteins activates the Rho *GTPase* family (including *Rho*, *Rac*, and *Cdc42*), which is involved in cell spreading and migration [34,35], and *Rho* controls stress fiber formation and the assembly of focal adhesions [34].

The process of static *in vitro* cell adhesion is characterized by three stages (Table 1): attachment of the cell body to its substrate (initial stage), flattening and spreading of the cell body, and the organization of the actin skeleton with the formation of focal adhesion between the cell and its substrate [35]. Cell spreading appears to be accompanied by the organization of actin into microfilament bundles. The strength of adhesion becomes stronger with the length of time a cell is allowed to adhere to a substrate or another cell. The initial adhesive interaction between the cells and the substrate are driven by the specific integrin-mediated adhesion and starts with the binding of single receptor-ligand pairs [36]. This will initiate the subsequent receptor-ligand bonds and quickly enhance in number, thus increasing the total adhesion strength [37]. The adhesion properties of cells could be determined by studying various cell-substrate contact times [38].

Following initial attachment, cells continue flattening and spreading on the substrate, resulting in the decrement of cell height (the cell flattens) and increment of contact area (Phase I) [16]. Next, the cell spreads beyond the projected area of the unspread spherical cell (Phase II) [36]. The spreading process is the combination of continuing adhesion with the reorganization and distribution of the actin skeleton around the cell's body edge [16]. Cells will reach their maximum spread area through expansion and adhesion strength will become stronger (Phase III).

Table 1. Evaluation of passive *in vitro* cell adhesion intervention and stages [36].

Cell Adhesion Phases	**Phase I**	**Phase II**	**Phase III**
Schematic diagram of cell adhesion	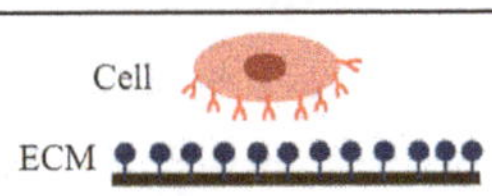	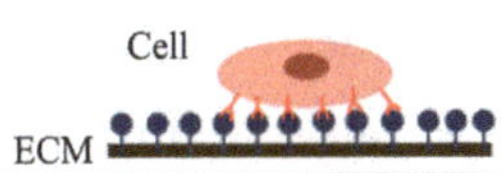	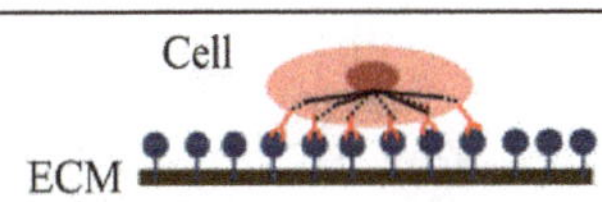
Schematic diagram of the transformation of cell shape	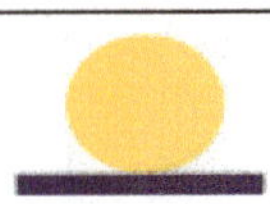 Initial attachment	 Flattening	 Fully spreading and structural organization
Cell adhesion intervension	Electrostatic interaction	Integrin bonding	Focal adhesion
Adhesion stages	Sedimentation	Cell attachment	Cell spreading and stable adhesion

1.2.2. Dynamic *in Vivo* Cell Adhesion

The adhesion of cells to the extracellular matrix *in vivo* under blood flow is a dynamic process. Cells will undergo dynamic adhesion alterations during their morphogenesis, tissue remodeling, and other responses to environmental cues [39]. *In vivo* dynamic cell adhesion is mediated through molecular bonding between cell-surface receptors and their ligands or counter-receptors on the other cell surfaces in the extracellular matrix. Shear flow is a crucial factor to initiate cell adhesion as it mediates the activation of β-integrin via E-selectin signaling [40]. The adhesive bond is defined as the sum of non-covalent interactions, e.g., hydrogen bonds, electrostatic interactions, van der Waals forces, dipole-dipole interactions between two macro molecules [39]. Leukocytes or hematopoietic progenitors and tumor cells are the cells involved in dynamic vascular cell adhesion *in vivo*.

Leukocytes and hematopoietic progenitors are the essential cells in the human immune response system that will migrate from one site to another to provide the effector function. Tumor cell interactions with endothelium and the subendothelial matrix constitutes a crucial factor in determining the metastatic potential of the cells and organ preferences of cancer metastasis [41,42].

The cell adhesion cascade and signaling events *in vivo* involve three basic steps: selectin-mediated rolling, chemokine-triggered activation, and integrin-dependent arrest [43]. Initial recognition of dynamic cell adhesion *in vivo* involves the "docking" phase, which occurs between the rolling of cells to endothelial cells and to cell arrest (Figure 2), mediated by a weak and transient adhesion mechanism involving carbohydrate-carbohydrate and/or carbohydrate-protein interactions. The molecules involved at this stage are cell-surface conjugates, selectins, chemokines, or immunoglobulins (Igs) [42]. The cell adhesion cascade begins as a cell tethers to roll on the vessel's wall. Molecular bonding between the adhesion molecules must form rapidly for cells to tether, and the bonding must break rapidly for cells to roll [39]. The rolling cells transduce signals from adhesion receptors and chemokine receptors that cause cells to roll slower and then to arrest, a prerequisite for emigration through the vasculature into underlying tissue [39].

Subsequently, cells will established stable bonds with endothelial cells during the activation-dependent "locking" phase, mediated largely by integrins and modulated by a host of bioactive mediators resulting from the activated cells [42]. Integrin-mediated adhesion is characterized by at least two events: arrest from rolling, which is mediated by increased cell avidity to endothelium, and a post-binding phase of adhesion stabilization [43]. In the "locking" phase, cell adhesion strengthening and spreading happens similarly to the static *in vitro* adhesion followed by intravascular crawling and transmigration (paracellular or transcellular) (Figure 2) [43]. This then permits the adhered cells to emigrate out of the vasculature. The intraluminal crawling facilitates the cell adhesion to emigration [44]. The post-adhesion events strengthen cell attachment to the endothelium, and into the molecules that are involved in cells' transendothelial migration [45,46].

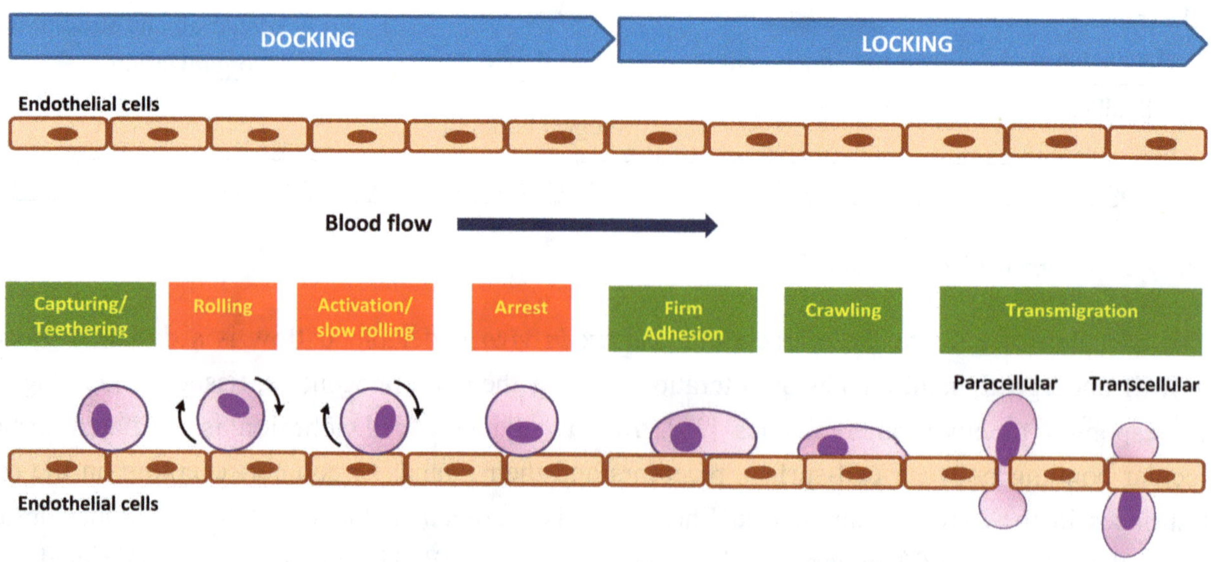

Figure 2. Dynamic *in vivo* cell adhesion cascade with "docking" and "locking" phases. The basic cascade steps are labeled in red boxes and steps recovered later are labeled in green boxes.

2. Types of Adhesion Studies

The mechanical interactions between a cell and its ECM can influence and control cell behavior and function. The essential function of cell mechanobiology and its progressively important role in physiology and disease have created tremendous interests in developing methods for measuring the mechanical properties of cells. In general, cell adhesion studies can be categorized into cell attachment and detachment events. Numerous techniques have been developed to analyze cell adhesion events through the study of single cells as well as the populations of cells. Cell adhesion attachment events are focusing on the cell attachment mechanism to the substrate, while the detachment events involve the application of load to detach the adhered cells on the substrate (Figures 3 and 4).

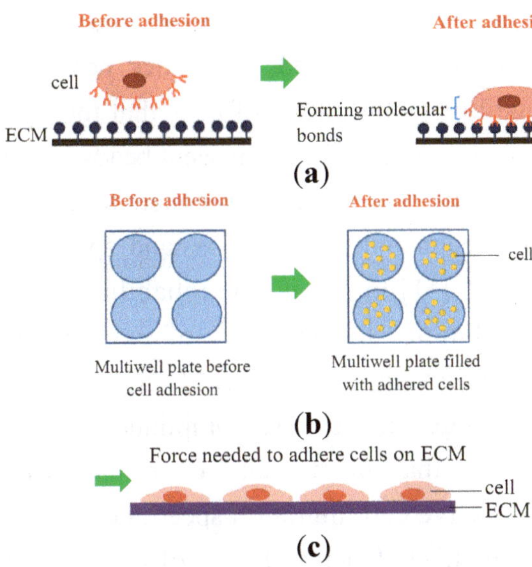

Figure 3. Schematic diagram of cell adhesion attachment events for (**a**) single cell studies via the formation of molecular bonds; (**b**) cell population studies via static adhesion (e.g., wash assay technique); and (**c**) cell population studies via dynamic adhesion (e.g., microfluidic technique).

2.1. Cell Adhesion Attachment Events

Cell attachment studies cover the analysis from the formation of a molecular bond between the cell's surface receptors and the complementary ligands (on the ECM's surface) to the observation of a population of cells' responses through the cells' behavior and changes of morphology during the attachment events. In cell migration, the cell adhesion plays a pivotal role in the driving force production [27]. The adhesion events could be grouped into single cell and cell population analysis (Figure 5). For the single cell study, the experiments were performed to analyze the interaction forces between the individual cell and its substrate (Figure 3a), observing the individual cell's morphology changes, studying the cell's migration, and measuring the cell's traction forces using polyacrylamide (PA) gel-based traction force microscopy (PA-TFM), micropattering technique, and three-dimensional traction force quantification (3D-TFM). Population studies involve the analysis of attachment events for a group of cells. It is important in analyzing the adhesion behavior of cells toward treatments or different physiological conditions (e.g., wash assay and microfluidic techniques), for the understanding the cell adhesion kinetics (e.g., resonance frequency technique), in determining the biocompatibility of biomaterials for tissue engineering, cancer metastasis studies, and also the potential of drug treatments (e.g., microfluidic techniques).

2.1.1. Techniques to Study Cell Attachment Events

Attachment Events: Single Cell Approach

Polyacrylamide-Traction Force Microscopy (PA-TFM). Numerous techniques have been developed to understand cell adhesion by characterizing single cells during their attachment events. PA-TFM is one of the widely used techniques to study single cells' traction force, the force exerted by cells through contact to the substrate surface. Cells will be cultured on the polyacrylamide gel functionalized with the cells' adhesive ligands and fluorescent beads embedded near the gel surface [47]. Upon the adhesion attachment events, cells will generate traction forces that move the beads and the movement will be quantified by tracking the displacement of the fluorescent beads [47–49]. Reinhart-King *et al.* [50] reported cell spreading increased with the increasing concentration of arginine-glycine-aspartic acid (RGD)-peptide by measuring the magnitude, direction, and spatial location of mechanical forces exerted by endothelial cells. Sabass *et al.* [51] improved the reliability and spatial resolution of traction force microscopy to 1 um by combining the advances of experimental computational methods. An epithelial wound-healing assay was developed by Ng *et al.* [52] to study the migration of individual *MCF10A* on PA substrates with a range of substrate compliances. Their findings showed that the wound could initiate a wave of motion that directs cells' coordination towards the wound edge and substrate stiffness influenced the collective cell migration speed, persistence, and directionality as well as the coordination of cell movements [52]. Traction forces of human metastatic breast, prostate, and lung cancer cell lines were found to be higher than non-metastatic counterparts, suggesting the cellular contractile force involve in metastasis and the physiological environment might regulate cellular force generation [48]. Wen *et al.* [53] reported stem cell differentiation was regulated by the stiffness of planar matrices independently of protein tethering and porosity. Beside single cell analysis, there have also been some adhesion studies done on the population of cells. Endothelial cells were reported to exert greater traction forces compared to single cells once in contact with the adjacent cell, thus suggesting an increase in cellular contractility with contact [54].

Micropatterning. Micropatterning (also known as microfabricated elastomeric post array (mPADs) or micropillar) is a method that provides a micrometer scale: a soft, three-dimensional complex and dynamic microenvironment for both single cell studies and also for the multi-cellular arrangements in populations of cells. It relies on basic elastic beam theory, which makes force quantification easier and more reliable, as there is only one traction force field for each micropost/micropillar displacement map [55]. Cell micropatterning comprises the fabrication and use of a culture substrate with microscopic features that impose a defined cell adhesion pattern. It is an efficient method to investigate the sensitivity and response of a cell to specific microenvironmental cues [56]. At the basic level, micropatterning approaches involve controlling cellular attachment, shape, and spreading as a function of the engineered spatial properties of the cultured surface [57]. Micropattering could be used to study cell adhesion for both the single cell level and also for the population of cells. Tan *et al.* [26] found that cell morphology regulates the magnitude of traction force generated by cells. These findings demonstrate a coordination of biochemical and mechanical signals to regulate cell adhesion and mechanics, which introduce the use of arrays of mechanically isolated sensors to manipulate and measure the mechanical interactions of cells [26]. Mandal *et al.* [58] introduced the micropatterned surfaces combined with the thermo-responsive poly(*N*-isopropyla-crylamide) (PNIPAM) as an actuator which induces cell detachment when the temperature is reduced below 32 °C. It has been reported that the micropatterning technique is able to independently tune the biochemical, mechanical, and spatial/topography properties of biomaterials that could provide the opportunity to control cell fate for tissue engineering and regenerative medicine applications [59]. Laminar atheroprotective fluid shear stress has been found to induce increments in traction force generation and endothelial cell alignment, which are associated with inflammation and atherosclerosis progression [60].

Three-Dimensional Traction Force Quantification (3D-TFM). The ability to grow cells within ECM gels (3D culture) is a major advantage to understand *in vivo* cellular cell behaviors, ranging from differentiated function to maintenance of stem cell functions [61,62]. The 3D-TFM technique uses 3D matrixes such as agarose, collagen, hyaluronic acid, fibrin, or matrigel for the cell culture. Individual cells are grown inside the gel matrix embedded with fluorescent beads surrounding the cell. Bead dispersion in the 3D gel will be observed to estimate the cellular contractility of the cell during migration [63–67]. In contrast to 2D migration, cell migration through a dense 3D network of extracellular matrix proteins is possible only when the cell generates sufficient tractions to overcome the steric hindrance of the surroundings [68]. Koch *et al.* [69] used the method to develop the 3D traction map of several tumor cell lines (*MDA-MB-231* breast carcinoma, *A-125* lung carcinoma) and found that the directionality is important for cancer cell invasion rather than the magnitude of traction, and the invasive cells elongated with spindle-like morphology as opposed to the more spherical shape of non-invasive cells [69]. The disruptive effect of the nocodazole drug on the neuronal processes has been analyzed using matrigel-embeded microbeads and neuron cells [65]. Kutys and Yamada [70] managed to explored pathways controlling the migration involving the GEF/GAP interaction of *βPix* with *srGAP1* that is critical for maintaining suppressive crosstalk between *Cdc42* and *RhoA* during 3D collagen migration. Fraley *et al.* [66] reported that instead of forming aggregates in the 3D matrix, the focal adhesion proteins diffused and distributed throughout the cytoplasm and were responsible in modulating cell motility.

Attachment Events: Population Approach

Wash Assay. In the population cell adhesion studies, the process of cell attachment can be divided into two types; static culture and dynamic culture, depending on the cell adherence mechanism during the cell culturing. Static culture is the stagnant condition of the cell culture medium during the incubation for cell adhesion, which is applicable for the culturing of cells inside microwell plates (Figure 3b), petri dishes, culture flasks, and cell cultures on the ECM-coated cantilever inside the chamber. The static culture was used in the wash assay and resonance frequency techniques. In the wash assay, cells were cultured in 96 multiwell plates for the cell attachment events, followed by cell washing before adhesion analysis (e.g., cell count, protein/DNA count, or antibody binding) was carried out [71–74]. Wash assays provide basic qualitative adhesion data by determining the fraction of cells which remain adhered after one or more washings [75]. Cells that remain adhered to the substrate after washing will be analyzed for further quantitative analysis such as cell count, quantification of DNA content, protein count, or antibody binding. Treatment of *D. mucronata* crude extract to the cancerous *wehi-164* cells significantly modulated their attachment and spreading behavior to the fibronectin-coated multiwell plates [74]. Chen *et al.* [73] extracted the adhered HeLa cells from the collagen-coated multiwell for further enrichment process and analysis. Park *et al.* [71] developed adhesion-based assay for high-throughput screening for the discovery of small-molecule regulators of the integrin CD11b/CD18 to further understand the mechanism of integrin activation and binding.

Resonance Frequency. The integration of advanced microelectronics technology with signaling processing and biological sensing interfaces has grown widely to develop biosensor devices. Piezoelectric sensor is the acoustic sensor which is able to detect label-free and selective biological events in real time. Quartz crystal microbalance (QCM) is one of the widely used piezoelectric acoustic wave resonator [76] biosensors for the study of cell adhesion and cell spreading. It is comprised of a thickness shear mode resonator made from a thin (AT-cut) quartz crystal sandwiched between two metal electrodes [77]. The sensor is coated with ECM before cells are mounted on it and placed in the chamber for cell adhesion to occur. Whole cells will act as sensing elements in the cell-based biosensors, as cells continuously react to the environment. The piezoelectric resonators will perform shear oscillations parallel to the sensor faces [78] upon cell adhesion activity, which will propagate through the sensor in a direction perpendicular to its surface. During attachment and cell spreading on the sensor surface, changes in resonant frequency could be detected upon the interactions between the cell membrane and the substrate, and upon the changes in the fractional surface coverage by the cells [77–85]. The resonance frequency of the sensor alters when a foreign mass attaches to the sensor's active surface and the frequency shifts will represent the mass of the absorbed material [77]. The adhesion process and the molecular interactions will produce signals representing cell adhesion kinetics determined by the sensor [36,83,86,87]. This technique has been found able to monitor cell attachment and spreading of animal cells on a particular surface quantitatively and in real time.

Zhu *et al.* [87] found the mechanisms governing elasticity and adhesion are coupled and affected differently during aging. Heitmann and Wegener [78] reported that the resonance frequency technique not only can be used to monitor cell adhesion but can also present as an actuator to perturb cell-substrate contact sites without causing damage to the cell when using the driving voltage for normal monitoring amplitude. The propagation properties of the acoustic wave in materials vary according to the different properties of the materials, and their energy is dissipated in the presence of fluids or viscoelastic materials [76,88]. The variation of the sensor resonance frequency (Δf) and acoustic wave energy dissipation can be used to gain direct measurements of the physical properties of the layers in contact with the chip [76,88,89]. The obvious difference in total frequency shift is caused by the different numbers of attached cells, since the geometries and the fundamental resonance frequency are similar. Besenbacher *et al.* [84,85] analyzed cell adhesion and spreading of *MC3T3-E1* and *NIH3T3* cells on different precoated biocompatible surfaces. Braunhut *et al.* [90–93] have developed and optimized the performance of the QCM biosensor and carried out a study on the effects of various types of anti-tumor drugs (e.g., nocodazole, taxol, taxane) on the adhesion of different types of cancer cells. In the advancement of resonance frequency technology, the sensor has been shown capable of acting as both sensor and actuator. It can be used as a sensor for cell adhesion and as an actuator to induce oscillations on the growth surface [78]. Recently, the usage of resonance frequency technique in cell adhesion has emerged to provide a platform for small sample sizes of cells in a dynamic fluid condition. Resonance frequency technique is used as an actuator to provide the acoustic wave needed for the dynamic fluid condition in the device and as sensor to measure cell adhesion. Warrick *et al.* [94] have developed a high-content adhesion assay to overcome the limitation of the available methods to perform analysis on small animal biopsies and rare cell isolations. Hartmann *et al.* [95] produced a new tool for the dynamic analysis of cell adhesion that provides small cells' sample volume, a short measuring time, and flexibility for different types of substrates that are suitable for studying the implant material compatibility.

Microfluidics. In contrast to the static culture, the dynamic culture applies fluid movement during the cell culturing and adhesion process. Low fluid shear flow is needed to help the cell attachment process as it mimics the blood flow in the human body. Cells are continuously exposed to hemodynamic forces generated by blood flow in most biological systems. The balance between the adhesive forces generated by the interactions of membrane-bound receptors and their ligands with the dispersive hydrodynamic forces determines cell adhesion [96]. Cell adhesion attachment events dynamic culture can be observed using microfluidic technique. The advantages of microfluidic systems (fluid manipulation and control, low fluid intake and miniaturization) encouraged their use in dynamic culturing for cell adhesion studies. This technique was used to study the ability of cells to adhere and to observe cell spreading, tracking, and migration inside the channel under the influence of fluid flow. Rupprecht *et al.* [97] reported that cell shape, movements, and the rate of cell division were found to be similar in petri dishes and microfluidic channels. Alapan *et al.* [98] applied the microfluidic technique to analyze the adhesion properties and deformability of human red blood cells in flow using blood samples from 12 different subjects. From the study of cell adhesion properties and dynamics, the technology has grown and been upgraded into the development of tissue-on-a-chip and, later, organ-on-a-chip for biomedical and pathological studies. Cell monolayers were grown in the microfluidic channel and made to mimic the human vascular system to be used for important bioengineering and biomedical analysis. A multi-step microfluidic device has been developed by Chaw *et al.* (2007) to analyze the deformation and biological and migratory capability of various tumor cell lines (*HepG2*, *HeLa*, and *MDA-MB 435S*) to the lining of *HMEC* cells inside the channels [99]. Nalayanda *et al.* [100] have built a series of bio-mimetic devices for the model of alveolar-capillary membranes while Song *et al.* [101] have developed a cancer cell metastasis model using microfluidic technology. Fu *et al.* [102] studied tumor cell adhesion to the endothelium cell layer under shear flow by combining micro-particle imaging velocimetry (μPIV) technique with flow chamber assay to understand the interactions between leukocytes and tumor cells near the endothelium wall region. A microfluidic model for organ-specific extravasation of circulating tumor cells (CTCs) has been build by Riahi *et al.* [103] to demonstrate the extravasation of *MDA-MB-231* cancer cells by analyzing the cells' adhesion capability to the endothelial monolayer inside the channel.

2.2. Cell Adhesion Detachment Events

Cell adhesion detachment studies involve load application to the adhered cells on the ECM to free the cells from their cell-matrix bonding (Figure 4). The applied force that produces cell detachment is quantified as the cell's adhesion strength. Many types of assays have been developed to measure cells' adhesion strengths and can be divided into single cell and population cell studies (Figure 4b,c). Measuring cells' adhesion strengths has become an emerging interest in various areas of study, including biomaterial compatibility in the human body, characterizing different stages of cancer cells, drug treatments for diseases, and the discovery of biomarkers for early disease diagnosis. Cells will be cultured and allowed to adhere on the ECM-coated matrix followed by cell detachment processes and adhesion strength measurement. For the single cell approach, the detachment process is focused on an individual cell and the measured value represents the adhesion strength of a single cell. Single cell detachment techniques were carried out either to fully detach a single cell from its substrate (whole cell detachment) for obtaining a single cell's adhesion strength or to focus on the load needed for breaking the molecular bonds to further understand cell adhesion kinetics. Techniques used for whole cell detachment are the cytodetachment and micropipette aspiration techniques, while to perform molecular bond breakage, the single cell force spectroscopy (SCFS) technique was used (Figure 4). Cell detachment events for the cell population approach were carried out by applying load at the population on adhered cells. Following the detachment process, the fraction of cells remaining on the substrate is quantified after varying loads of global force or stress have been applied. The force or stress value at which 50% of the cells detach is determined as the population adhesion strength. The adhesion measurement techniques for the cell population approach can be divided into four categories depending on the loading method applied to detach cells: centrifugation assay, spinning disk, flow chamber, and microfluidics (Figure 5).

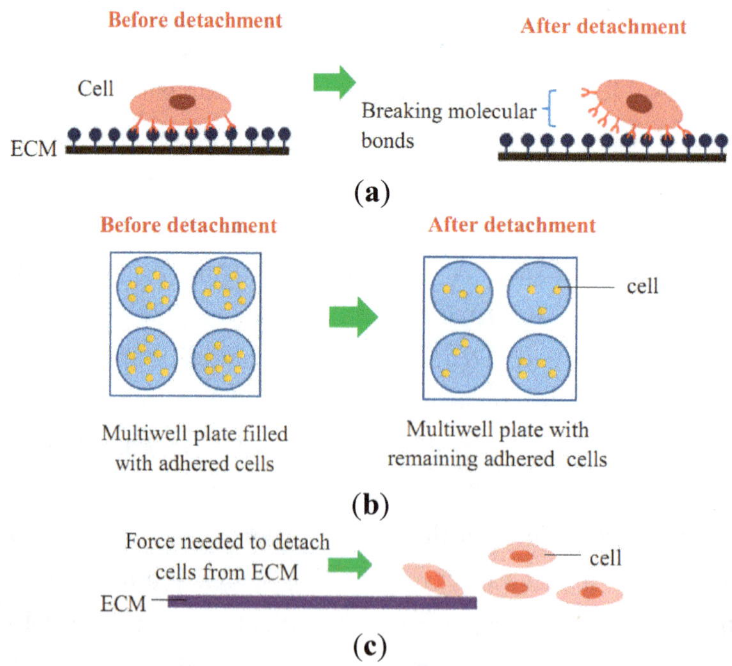

Figure 4. Schematic diagram of cell adhesion detachment events for (**a**) single cell studies via the breakage of molecular bonds (e.g., SCFS, micropipette aspiration, and optical tweezer techniques); (**b**) cell population studies via static adhesion (e.g., centrifugation technique); and (**c**) cell population studies via dynamic adhesion (e.g., spinning disk, flow chamber, and microfluidic techniques).

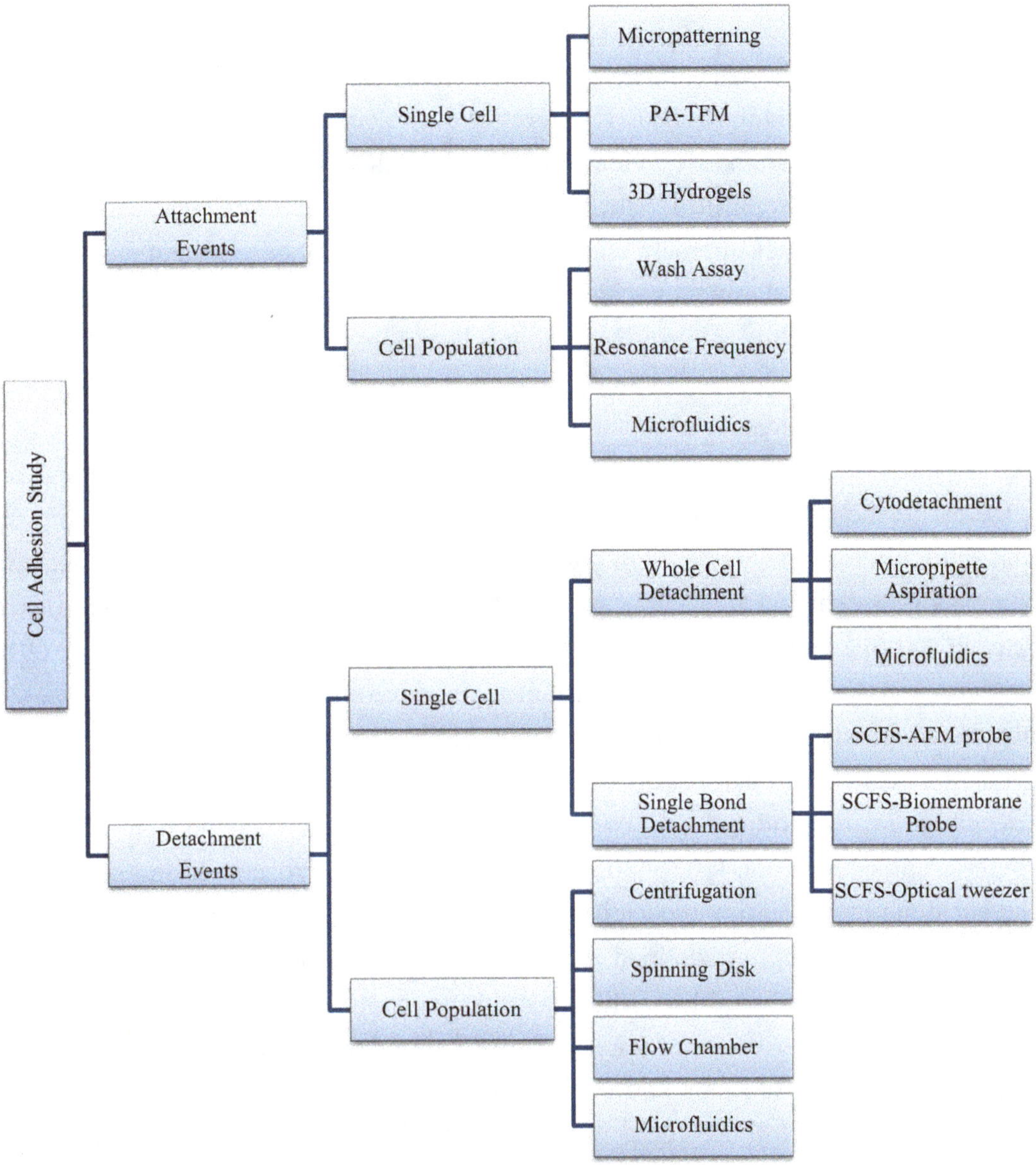

Figure 5. Summary of techniques involved in cell adhesion studies, categorized by the adhesion attachment events and detachment events.

2.2.1. Techniques to Study Cell Detachment Events

Single Cell Approach

Cytodetachment. Cytodetachment technique uses an atomic force microscopy (AFM) probe to physically detach individual cells in an open medium environment such as petri dish [75]. Cells have to be attached to a functionalized matrix before the single cell-probe alignment and probe translation application are carried out to detach the cell. Force is quantified by measuring the elastic deflection of the probe used to detach the cell and then divided by the cell area to calculate the average shear stress for each cell [75]. Yamamoto *et al.* [104] studied the force needed to detach murine fibroblast *L929* on four different materials by using cytodetachment technique. Using an image analysis system and fiber optic sensor, the apparent cell adhesive area was measured and the adhesive strength and detachment surface energy were calculated by dividing them by the cell adhesive area [104]. Their findings showed that cells on collagen-coated polystyrene produce the highest adhesion strength and cell detachment surface energy, and the adhesive properties of the cells between both polystyrene and glass are almost the same [104]. Human cervical carcinoma cells (*NHIK 3025*) were found to attached stonger and faster to the hydrophilic substrate using technique 1 at 37 °C when compared to 23 °C. There are multiple adhesion measurement studies combining the cytodetachment method with the laser tweezer work station [105] and optical tweezer technique to further understand the temporal effects of cell adhesion by analyzing the molecular binding between the cell and the substrate [16,105]. The study of bovine articular chrondocyte cell adhesion with different ECM-coated substrate was carried out by using a combined cantilever glass beam with a carbon filament as a cytodetacher. Huang *et al.* [16] used rabbit articular chrondocyte cells to study the mechanical adhesiveness of the cells using the cytodetachment method. They examined membrane cell tether formation using optical tweezers and cytoskeleton change through fluorescent staining. Findings showed that chrondocytes exhibited increasing mechanical adhesiveness and tether formation force with the increment of seeding time. Yang *et al.* [106] reported that the adhesion force of human fetal osteoblast (*hFOB*) cells was influenced by the cell's shape grown on the Ca-P grooved micropattern surface.

Micropipette Aspiration Technique. Micropipette aspiration is a widely used technique for measuring the mechanical properties of single cells. For single cell adhesion measurement, this technique detaches an immobilized cell by applying suction force to a portion of the cell surface employed by micropipette suction [107] under observation via a microscope. The force will release the cell from the substrate by increasing the aspiration pressure or by translating the pipette away from the substrate [75]. Adhesion strength is defined as the minimum force needed to detach a single cell from its substrate. Micropipette aspiration technique is able to measure various mechanical properties of cells, such as the membrane stiffness of chondrocytes and endothelial cells [108], the cortical tension of neutrophils [108], and the adhesion strength of human umbilical vein endothelial cells on different substrates [109]. Gao *et al.* [109] used cells in the phase they called the "adhered cells in round shape" phase to eliminate the influences of floating cells, weakly adhered cells, and spread cells and obtained high reproducibility and sensitivity of the measurement on different substrates. They reported that the sensitivity and accuracy of their technique could reach 8×10^{-12} N [109]. Micropipette aspiration technique was found able to identify comparable differences between the adhesion strength of normal and cancerous cells. Palmer *et al.* [18] have developed a single cell measuring apparatus using micropipette aspiration (SCAMA) that is able to measure and differentiate the adhesion strength of normal and cancerous prostate and breast cells. They found the measurement made on analogous human prostate cancer and normal cells showed a comparable three-fold difference in adhesiveness [18].

Single Cell Force Spectroscopy (SCFS). Force spectroscopy measurement methods were developed to measure the strength of cell adhesion down to single cell level. Commonly, the methods will use a microscope to observe the cell while force is applied to detach the cell using a nano/micromanipulator or micropipette. The imaging mode is used to study the structures and mechanics of isolated biomolecules [110–112], components of the cell nucleus [113,114], and subcellular cytoskeletal structures [110,115], while force mode is used to determine the mechanical properties of various cell types. The methods differ in the type of manipulation applied to the cell and the type of force measurements made. Examples of single cell force spectroscopy techniques used to measure cell detachment are the AFM probe techniques, biomembrane probe (BFP), and optical tweezer methods. Among these methods, AFM-based techniques are widely used to study various types of cell adhesion, effects of surface treatments and environmental conditions, and biomaterials compatibility.

AFM Probe Force Measurement. AFM probe force measurement is widely used to measure the stiffness [116–119] and adhesion strength of individual cells against mechanical force [16,104,105,120] due to its versatility and precision. By immobilizing individual cells to an AFM cantilever, the living cell will be converted into a probe for the measurement of adhesion strength between cell-cell or cell-matrix adhesions [38]. This probe is attached to a cell or ECM-coated substrate (cell adhesion occurs) and the cantilever is withdrawn at a constant speed to detach the cell from its binding place. Cantilever deflection is recorded as force-distance curve and the highest force recorded represents the cell's adhesion strength [38]. When the probe encounters the single cell surface, various forces between the cantilever probe and cell lead to a deflection of the cantilever according to Hooke's law [120]. Adhesion strength of a single cell could be monitored as a function of adhesion time and environmental conditions by using AFM probe force measurement [75]. Various aspects of adhesion could be studied using AFM probe SCFS without restriction on the type of cell adhesion molecules and cell types used. Puench *et al.* [121] reported that the technique is able to quantitatively determine the adhesion of primary grastulating cells and provide insight into the role of *Wnt11* signaling in modulating cell adhesion at single cell level. Weder *et al.* [122] studied the adhesion of human osteosarcoma cell lines (*Saos-2*) during different phases of the cell cycle and found that the cells are loosely attached to the substrate during the cells' round up (*M* phase) compared to during the interphase. Hoffmann *et al.* [123] determined the influence of the activating *NK* cell receptor 2B4 on the early adhesion processes of *NK* cells using AFM-probe SCFS. In addition, AFM is flexible and can be integrated with the standard modern inverted and transmission optical microscope. Lee *et al.* [105] developed a cell-detachment apparatus to measure the adhesion force of single cells integrated with a laser tweezer work station for cell manipulation observation. They studied the effect of experimental medium on the cell adhesion force of *NIH/3T3* fibroblast cells and found the cell adhesion strength increased with culturing time, and growth factor was found to enhance the adhesion strength between the cell and substrate. The study was continued with the combination of the optical trapping technique to further study the single cell adhesion properties of *MCDK* cells in different phases of adhesion [119]. Their findings showed that focal adhesion kinase (FAK) plays a role in enhancing the binding and spreading of *MDCK* cells through all the different phases of cell adhesion. Beaussart *et al.* [124] studied the adhesion forces of medically important microbes using AFM-SCFS and showed that procedures are applicable to pathogens such as *Staphylococcus epidermidis* and *Candida albicans*.

Biomembrane Force Probe (BFP). BFP is a sensitive technique that allows the quantification of single molecular bonds. It is a versatile tool that can be used in a wide range of forces (0.1 pN to 1 nN) and loading rates (1–10^6 pN/s) [125]. This technique uses a force transducer made of a biotinylated erythrocyte (such as biotinylated red blood cell (bRBC)) maintained by a glass micropipette. A streptavidin-coated glass microbead was attached to the bRBC (with known stiffness) and tuned by the controlled aspiration pressure applied by the holding micropipette. The assembly formed by the bRBC and the microbead constitutes a powerful nanodynamometer and is the force transducer (probe) used in the BFP [125]. The probe is brought into contact with the targeted cell and adhesion (bond formation) will occur between the probe and cell, followed by the detachment process where targeted cells are pulled away from the probe using a piezoelectric manipulator. Evans *et al.* [126,127] developed a transducer capable of measuring force ranging from 10^{-2} to 10^{3} picoNewton (pN) for probing molecular adhesions and structures of a living cell interface [126] to improve their previous probing method [127]. The BFP technique was used to quantify the human neutrophil (PMN) membrane unbinding forces and the kinetics rate for the membrane unbinding was found to increase as an exponential function of the pulling force [128]. Gourier *et al.* [125] proved the capability of the technique to quantify the local changes of gamate (oocyte and spermatozoan) membrane adhesion and probe the mechanical behavior of the oocyte membrane at a micrometer scale.

Optical Tweezers (OT). OT uses a highly focused laser beam to trap and manipulate microscopic, neutral objects such as small dielectric spherical particles that experience two kinds of forces: namely, the scattering force and the gradient force [129,130]. The technique is able to measure forces ranging from sub-pN up to several hundreds of pN with good precision (<1 pN) and is applicable for the study of interfacial interactions and non-covalent bonds, e.g., receptor-ligand bonds [129]. Single cell adhesion studies have been explored using OT [131–136] involving various cell types and purposes. Askenasy and Farkas [137] used OT for studying the cellular adhesion of hematopoietic stem cells (HSC) to the bone marrow stroma, and a forward scatter analysis (FORSA) has been integrated with the OT to investigate the binding force associated with cell-cell interactions and molecular interactions [132]. Thoumine *et al.* [131] were able to produce information on the receptor-ligand adhesion kinetics of fibroblast cells to fibronectin by coupling the experimental results with a probabilistic model of receptor-ligand kinetics. They gained information on the number, strength, and reaction rates of the bonds. Optical tweezers have also been applied to study the adhesion of *Saccharomyces cerevisiae* cells [134,135,138] and to map the adhesion force during the formation and maturation of cell adhesion sites of mouse embryonic fibroblasts [136].

Detachment Events: Population Approach

Centrifugation Assay. Centrifugation assay is one of the frequently used techniques to measure cells adhesion strength due to their simplicity and the wide availability of equipment in most laboratories. Cells will be seeded in a multiwell plate and undergo treatments by culturing (cell culturing is similar to wash assay) before being spun for the cell detachment process. During the spinning, cells will experience a body force acting in the direction normal to the bottom of the plate that pulls them away from the surface [75]. To assess adhesion strength, the number of cells before and after application of load in the centrifuge is quantified. The fraction of cells that remains adhered after centrifugation can be determined by measuring the amount of radiation emitted from radio-labeled cells [139,140], quantifying the amount of cells or cellular genetic material [139], or by using automated fluorescence analysis [140,141]. In many cases, the assay is used to assess the relative effect of treatments such as ECM protein type and concentration or the inhibition of a specific cellular function.

This method was used by Channavajjala *et al.* [141] to understand the importance of cell attachment to HIV-1 Tat protein and their finding shows that Tat protein mediates a significant but weak attachment of *HT 1080* cells compared to the cells binding to ECM proteins. García *et al.* [142–145] used centrifugation assay to study cell adhesion and integrin binding and the effects of surface functionality of self-assembled. Harbers *et al.* [146,147] demonstrate the importance of flanking amino acids in the developing ligands with tuneable activity and the relative adhesion strength of each ligand by high-throughput assays for rapidly testing receptor-ligand engagement. High-throughput capabilities of the centrifugation method have been applied by Reyes and Garcia [144], who analyzed the adhesion capabilities of different cells (*HT-1080*, *NIH3T3* and *MC3T3-E1*) on different concentrations of collagen or fibronectin. Koo *et al.* [140] used the method to study the effect of different ligand density and clustering to show that biophysical cues such as ligand spatial arrangement and ECM rigidity are central to the governance of cell responses to the external environment.

Spinning Disk. The spinning disk technique utilizes shear stress generated from a rotating disk device. Cells are first seeded on circular glass coverslipsor on the surface of a disk (typical diameter 10–50 mm). These disks are later fixed onto a rotating device, that is placed inside a chamber filled with buffer solution [75,148]. The rotating device has a rotating range from 500 to 3000 rpm. The adherent fractions of cells are generally quantified using microscopy and by counting the number of cells before and after spinning using either a manual procedure [149] or automated image processing software [150]. The spinning disk technique has been used to investigate various types of cell-substrate interactions for wide range of applications such as quantifying the adhesion strength of an osteoblast-like cell on bioactive glass [150,151], human bone marrow cells on hydroxyapatite [152,153], and *MC3T3-E1* cells on RGD peptides on self-assembled monolayers [154]. Lee *et al.* [154] used this method to quantify the nonspecific and specific contributions of bone cells on immobilized RGD peptides and quantitatively demonstrated both the possibilities and limitations of enhancing the osteogenic response of RGD-immobilized biomaterials by a change in peptide. The method has been used to study the role of focal adhesion kinase, which is essential for the focal adhesion function and the cell's adhesion strength [154,155]. Boettiger *et al.* observed the effect of transformed oncogene *v-src* on the adhesion strength of chick embryo fibroblasts [156] and human osteosarcoma cells [157] to understand its effect on integrin function. Lynch *et al.* [158] investigated the nature of signaling mechanisms that regulate integrin function in a steady-state adhesion and during cell motion using cells exposed to the insulin-like growth factor I (*IGF-I*). The effect of surface charges on different substrates has also been studied using *HT-1080* cells following fibronectin coating [159]. Reutelingsperger *et al.* [148] have investigated the effects of differential shear stresses on cell-cell and cell-matrix interactions in a monolayer of endothelial cells. García *et al.* [160] were able to measure the short- and long-term adhesion strength of *IMR-90* human fibroblasts adhered to fibronectin-coated glass using the spinning disk technique.

Flow Chamber. There are two types of flow chambers used for adhesion strength measurement: the radial and parallel flow chambers. The radial flow chamber methods involve flowing fluid in a chamber over adhered cells on a stationary substrate where a wide range of radially dependent shear stresses are applied [161] to detach a population of cells in a single experiment. The fluid flow is directed outwards from the center of a circular chamber, impinges on the surface of interest, and flows radially outward over a substrate seeded with cells. The inlet flow is directed outwards from the center of the chamber and the shear stress of the fluid decreases with increasing radial distance in a nonlinear fashion [75]. The radial flow chamber is also known as a stagnation point flow chamber or confined impinging jet [162]. The linear fluid velocity and shear stress will decreased radially across the disk as the flow duct cross-sectional area increases with the radius, and this method provides a continuous range of shear force in a single experiment [162].

The radial flow chamber technique has been used by Cozens-Roberts *et al.* [163] to investigate the effects of ligand and receptor densities and the influence of the pH and ionic strength of the medium on the cell-surface interactions. DiMilla *et al.* [164] studied the human smooth muscle cells (*HSMC*s) initial attachment strength and migration speed on a range of fibronectin and collagen type IV concentration. Their finding's suggested that cell-substrate initial attachment strength is a central variable governing cell migration speed and the cell's maximal migration occurs at an intermediate level of cell-substrate adhesiveness [164]. A number of studies have been carried out to study the adhesion strength of murine *3T3* fibroblasts on the fibronectin-coated of self-assembled monolayers on glass surfaces [161,165–167]. The adhesion of *MC3T3-E1* cells to multilayer polyallylamine hydrochloride (PAH) heparin films was analyzed in order to evaluate biocompatibility of various film chemistries [168]. Rezania *et al.* [169–173] carried out studies on the adhesion of osteoblasts and osteoblast-like cells to RGD peptides [169–171] and the adhesion of endothelial cells to interpenetrating polymer networks [173] for the application of orthopaedic implants. A study by Brown *et al.* [174] showed that low trypsin concentrations could improve cell adhesion and promote stronger endothelial adhesion while high trypsin concentrations significantly reduced the number of functional integrins available on the membrane. Beside human cells, the radial flow method has also been used to understand the adhesion kinetics of bacteria *Escherichia coli* K12-D21 in the in the mid-exponential and stationary growth phases under flow conditions. Chinese hamster ovary cells (*CHO*) were used to explore the adhesion behavior towards different substrates with different treatments [175].

The parallel plate flow chamber consists of a bottom plate and an upper plate separated by a distance of the channel's height to form a rectangular flow channel. Cells are grown on a coverslip and positioned in the flow chamber, constructed by sandwiching a thin rubber gasket between two plates and mounted on a microscope to allow direct observation of the cells during experiments [75]. The flow is often driven using hydrostatic pressure from a raised reservoir or with an automated pump to independently drive the flow [176–178]. The fluid's shear stress can be adjusted by varying the flow's rate of the perfusate, the fluid viscosity, or the channel's height and width [179].

The parallel flow technique was first introduced to study endothelial cell adhesion [178,179] and has been further explored in the adhesion studies of biotinylated endothelial cells adhered on glass with fibronectin or RGD peptide functionalization [180–183]. Tapered height chamber have been developed to produce linear variations of shear stresses along channels at a single flow rate [184]. Cao *et al.* [185] modified the chamber to developed a side-view chamber system where side-view images of cellular deformation and adhesion to various adhesive surfaces under dynamic flow conditions could be observed. Cellular adhesion to surfaces functionalized with artificial ECM proteins and polymer surfaces treated by plasma using different gaseous substances have also been studied [186,187]. The parallel flow technique has been been used to investigate the adhesion potential of various cell types to a range of different materials, including poly-L-lactide (PLL) films [187–189], polyelectrolyte multilayer films [190], polyethylene films [191], and numerous glass-treated surfaces [177,178,192–194]. Interestingly, the flow chamber can be used to observed cells' vascular adhesion potential to the endothelial monolayer, which represents the human endothelial vein system. Gerszten *et al.* [176] were able to study the adhesion of monocytes to vascular cell adhesion molecule-1 (*VCAM-1*) transduced human endothelial cells under physiological flow conditions. Palange *et al.* [195] investigated the extravasation ability of circulating tumor cells (CTCs) to the endotheial cell layer under flow and the potential of natural compounds and curcumin treatment to attenuate the cell's metastasis potential.

Microfluidics. Microfluidic lab-on-a-chip technologies represent a revolution of the flow chamber in laboratory experimentation, bringing the benefit of miniaturization, integration, and automation to many research-based industries. These greatly reduce the size of the devices and make many portable instruments affordable with quick data read-outs. The use of small sample volumes leads to greater efficiency of chemical reagents, straightforward construction and operation processes, and low production costs per device, thereby allowing for disposability and fast sampling times. The ability for real-time observation makes microfluidics bring high promises for cell adhesion studies. In recent years, cell adhesion studies have been carried out in a miniature form of the traditional parallel plate flow chambers as discussed above, using flow in rectangular microchannels to apply shear stresses to cells. These devices are typically constructed from optically transparent PDMS bonded to glass using the soft lithography rapid prototyping process that allows many nearly identical devices to be manufactured in a short amount of time [196]. The optically transparent criteria is important to enable the use of different real-time microscopy techniques to explore cell behaviors under diverse experimental conditions [197]. Small dimensions associated with micrometer-sized channels ensure laminar flow even at very high linear fluid velocities, which is often required when large shear stresses are generated [198].

A microfluidic device consisting of eight parallel channels has been used to assess the effect of varying collagen and fibronectin concentrations on the adhesion strength of endothelial cells [197]. A series of microfluidic channels have been constructed to investigate the adhesion strength of fibroblast cells adhered to fibronectin-coated glass surfaces [198]. In another case, microchannel assays were used to examine the adhesion of various cell types on surfaces with various coatings, including collagen, glutaraldehyde, and silane [199]. Kwon *et al.* [19] used a microfluidic shear device consisting of four parallel channels with different surface topography patterns to separate cancer cells mixed in a population of healthy cells based on adhesion strength. A shear stress-dependent cell detachment from a temperature-responsive cell culture using a microfluidic device has been developed to quantitatively estimate the interaction between cells (*NIH/3T3* mouse fibroblast and bovine aeortic endothelial cells) and materials [200]. Recently, microfluidic technology has moved forward to the studies of single cells. Honarmandi *et al.* [24] integrated microfluidics with optical tweezers for the study of mechanotransduction and focal adhesion of single endothelial cells. Microfluidic devices were `used to provide convenient means of positioning a cell into a specific location in the channel with controlled physiological conditions while the optical tweezers were used to detach the adhered cells. Christ *et al.* [107] have upgraded the application of microfluidics from the study of population cell adhesion to single cell studies. A rectangular microchannel was used to analyze the adhesion strength of single *NIH3T3* fibroblast cells that had been allowed to adhere for 24 h on the collagen and fibronectin coatings on glass [107]. In this work, single cells were imaged throughout the detachment process, and the relationship between adhesion strength and cell geometry was investigated.

3. Advantages and Limitations of the Techniques Used in Cell Adhesion Studies

Single cell approaches allow for precise measurements of the separation of the cell from the substrate. Specialized equipment, which is bulky and expensive, is often required for manipulation and alignment of the probe and testing can be time-intensive. The single cell adhesion measurement approach provides more precise measurement of the individual cell when compared to the population cell approach. The single cell measurement approach allows the system to image biomolecules at nanometer-scale resolution, to have a dynamic range of forces able to be applied to cells, and to process samples in their physiological medium and aqueous buffer [120]. It has been widely used in the cells' teether (adhesion process) formation [126] and in the rupture forces [125,128,201] of the molecular adhesion bonds that couldn't be measured by most of the cell adhesion strength measurement methods. However, beside the precision of the techniques, limitations such as low-throughput measurement, high equipment cost, time consumption, the need of a skilled operator, and other operator variables in the data obtained are unavoidable. These restrictions underscore the need for developing additional simple techniques that do not require expensive equipment, and are able to measure changes in cell adhesion properties associated with diseases or specific physiological perturbations. Some of the methods require computational processing and high-end confocal microscopes, which are not available in most laboratories. Table 2 summarizes the advantages and limitations of techniques used to study cell adhesion. The population cell approach provides data from the average response of a group of cells. Even though this approach could not provide precise data on the characteristic of an individual cell, the approach was still widely used to study cell adhesion until recently. The importance of the approach can not be denied as it provides essential information in the medical field for disease treatments, tissue engineering, and biomaterial compatibility.

Table 2. Comparison of advantages and limitations in the techniques used for cell adhesion studies.

Method	Strength	Weaknesses	References
Polyacylamide-traction Force Microscopy (PA-TFM)	Real time observation; No special and expensive equipment needed for fabrication Inexpensive; Flexible to chemical and mechanical adjustment; Adaptable to a large variety of cells	Needs to record both unstressed and stressed state of the substrate; Suffers from uncertainties in tracking beads' position	[47–54]
Micropatterning (Micropost array/micropillar)	Real-time observation; Force quantification easier and more reliable than PA-TFM; The micropillar stiffness is manipulated by its geometry; Gives good precision over surface chemical properties on micrometer scale	Substrate can alter cell's behavior; Requires sophisticated equipment to fabricate; Needs skilled operator; Sensitivity of the microposts to the particular cell type needs to be optimized	[26,55–60]

Table 2. *Cont.*

Method	Strength	Weaknesses	References
Three Dimensional Traction Force Quantification (3D-TFM)	Real-time observation; Flexible to chemical and mechanical adjustment; Adaptable to a large variety of cells; Flexible to chemical and mechanical adjustment; Adaptable to a large variety of cells	Needs high-end confocal microscope; Needs high computational processing; Needs to record both unstressed and stressed state of the substrate; Suffers from uncertainties in tracking beads position	[61–70]
Wash Assay	Simple	Not a quantitative data, needs further analysis to obtain quantitative data; Poor reproducibility; Insensitive	[71–74]
Resonance Frequency	Real-time observation; Real-time measurement	Poor reproducibility	[36,76–95]
Microfluidics	Straightforward construction and operation; Real-time observation and measurement; Convenience in size (compatible with cell sizes); Fast and simple to operate; Non-invasive to cell	Low detachment force; Restricted to short-term adhesion	*Attachment events* [96–103,107]; *Detachment Events* [19,24,107]; [197–200]
Cytodetachment	Real-time observation; Quick detachment of cell; Range of force produced is high and applicable to long-term adhesion	Alignment of probe and cell; Time-consuming; Needs highly skilled (experienced) operator; Operator variable; Cell damage (hard contact); Expensive equipment; Not real-time measurement	[104–106]
Micropipette Aspiration	Real-time observation and measurement; Common lab equipments	Alignment of probe and cell; High skilled (experienced) operator; Operator variable; Cell damage (hard contact)	[18,108–109]
SCFS-AFM probe	Real-time observation Precise data for short term adhesion studies	Alignment of probe and cell require micromanipulator; Time consuming; Need skilled operator; Operator variable; Cell damage (hard contact); Expensive equipments; Not real-time measurement	[105]; [116–124]

Table 2. *Cont.*

Method	Strength	Weaknesses	References
SCFS-Biomembrane Probe	Real-time observation; Precise data for short term adhesion studies	Low maximum force (pN); Restricted to short term adhesion; High skilled (experienced) operator; Operator variable; Probe variable (fluctuation of probe due to thermal excitation)	[125–128]
SCFS-Optical Tweezer	Real-time observation; Precise data for short term adhesion studies; Compatible with microfluidic device	Low maximum force (pN); Restricted to short term adhesion; High skilled (experienced) operator; Operator variable; Cell damage	[129–138]
Centrifugation	Many analysis can be examined in parallel; Common lab equipments	Low maximum force (uncomplete detachment); Only a single force can be applied per experiment; Nota real-time analysis	[139–147]
Spinning Disk	A range of stresses able to be applied in single experiment; High stresses	Not a real-time analysis; Custom-made apparatuses	[148–160]
Flow chamber: Radial flow; Parallel flow	Radial flow: Ranges of stresses applicable in single experiment; Real-time cell detachment observation; Paralel flow: Simple fabrication; Straightforward operation; Real-time cell detachment observation	Radial flow: Low detachment force; Restricted to short term adhesion; Paralel flow: Low detachment force; Restricted to short term adhesion	[163–175]; [176–195]

4. Summary

Studying human diseases from a biomechanical perspective can lead to a better understanding of the pathophysiology and pathogenesis of a variety of illnesses because changes occurring at the molecular level will affect, and can be correlated to, changes at the macroscopic level. Research on biomechanics at the cellular and molecular levels not only leads to a better elucidation of the mechanisms behind disease progression, it can also lead to new methods for early disease detection, thus providing important knowledge in the fight and treatments against the diseases. Sickle cell disease (SCD) could be characterized by observing the red blood cells' (RBC) adhesiveness and deformability [98,202–205]. Cell adhesiveness was found to be reduced in human cancers. Diseased cells' properties have been found to be physically different from that of healthy cells [206]. The adhesion strength of cancer cells was found to be lower than the normal cells [19] and decreased in line with their increased "metastatic potential" [18]. Reduced intercellular adhesiveness allows cancer cells to disobey the social order, resulting in the destruction of the histological structure, which is the morphological hallmark of malignant tumors [8]. Polymorphonuclear leukocytes (PMNs) migrate from the bloodstream to the inflammation sites by adhering to the surface of the endothelium during infection and tissue injury [207].

The knowledge obtained can also be useful in the development of new and improved assays and diagnostic devices, and the techniques are not only sensitive enough for the early detection of diseases, but they are also highly accurate, so it is possible to detect diseases when the symptoms or signs are hardly discernable. Determining chronic diseases in their initial stages is promising in curing the illness and saving lives, thus improving the quality of human health.

Figure 6 summarizes the importance of cell adhesion studies and its applications categorized by attachment and detachment events and grouped by single cell and population studies. Cell adhesion studies cover a wide range of important applications from the fundamental single cell adhesion behavior (morphology, migration, kinetics) and understanding the cell signaling pathway to how the physiological factors (temperature, pH, fluid flow), treatments (chemical, drugs, toxic, different substrate) and conditions affect cell adhesion and cancer metastasis studies as well as tissue engineering and biocompatibility studies for implants. This essential information obtained from the adhesion studies leads to the development of the computational model for further studying and understanding cell adhesion [27,96,208–217]. The future potential of single cell adhesion characterization is especially significant for early disease diagnosis. This emerging field can lead to the development of biomarkers for chronic diseases and cancers in their early stage at the cellular level. Furthermore, the new techniques or devices will bring high promises in the search of suitable treatments for those who have diseases in their early stages. Beside the importance of single cell adhesion, the cell adhesion population approach plays an important role in and brings high promises for the development of biomaterials in tissue engineering for implantable bioMEMs/biosensors, tissue scaffold production, and the applications of artificial bones as well as tooth replacement. Cell adhesion population studies are also essential in analyzing the potential of drug treatments, improving drug delivery systems, attenuating cancer metastasis development, understanding the dynamic mechanism of cell adhesion in many important biological processes, and finding a cure for many diseases or human health-realated problems.

5. Conclusions and Future Directions

Cell adhesion is a very important process in the human biological system. Studying both cell attachment and detachment events provides essential knowledge in understanding many important functional processes in the human body, which lead us to find the causes and problems that trigger certain diseases and thus develop the strategy for curing and improving them. Many different techniques and adhesion assays have been developed to study cell adhesion applicable to a wide range of fields. Every method is unique and was developed for specific important and independent purposes, which makes them difficult to compare in finding the best method applicable for cell adhesion studies. Choosing an appropriate technique is highly dependent on the purpose of the information that a person desires to obtain. Both single cell and population studies are equally important and required to fully understand how cells behave and function in the human system.

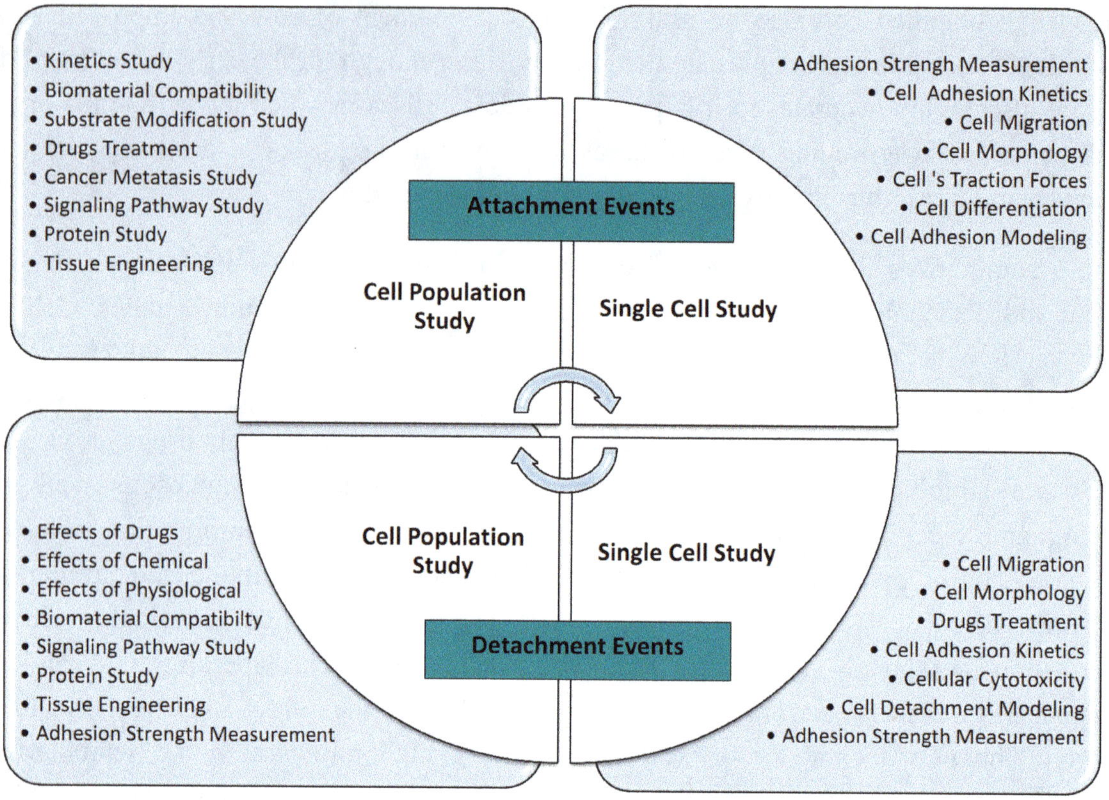

Figure 6. Summary of the importance of adhesion studies and their applications.

Acknowledgments

The research was supported by the Ministry of Higher Education Malaysia (grant Nos. 4L640 and 4F351) and Universiti Teknologi Malaysia, (grant Nos. 02G46, 03H82 and 03H80). Thanks to them for funding this project and for their endless support.

Author Contributions

Amelia Ahmad Khalili and Mohd Ridzuan Ahmad wrote and edited the review, respectively.

References

1. Sagvolden, G.; Giaever, I.; Pettersen, E.O.; Feder, J. Cell adhesion force microscopy. *Proc. Natl. Acad. Sci. USA* **1999**, *96*, 471–476.
2. Dembo, M.; Torney, D.; Saxman, K.; Hammer, D. The kinetics of membrane-to-surface adhesion and detachment. *Proc. R. Soc.* **1988**, *234*, 55–83.
3. Shen, Y.; Nakajima, M.; Kojima, S.; Homma, M.; Kojima, M.; Fukuda, T. Single cell adhesion force measurement for cell viability identification using an AFM cantilever-based micro putter. *Meas. Sci. Technol.* **2011**, *22*, 115802.

4. Huang, S.; Ingber, D.E. The structural and mechanical complexity of cell-growth control. *Nat. Cell Biol.* **1999**, *1*, E131.
5. Lasky, L.A.; Singer, M.S.; Dowbenko, D.; Imai, Y.; Henzel, W.J.; Grimley, C.; Fennie, C.; Gillett, N.; Watson, S.R.; Rosent, S.D. An endothelial ligand for L-Selectin is a novel mucin-like molecule. *Cell* **1992**, *69*, 927–938.
6. Szekanecz, Z.; Koch, A.E. Cell-cell interactions in synovitis: Endothelial cells and immune cell migration. *Arthritis Res.* **2000**, *2*, 368–373.
7. Okegawa, T.; Pong, R.-C.; Li, Y.; Hsieh, J.-T. The role of cell adhesion molecule in cancer progression and its application in cancer therapy. *Acta Biochim. Pol.* **2004**, *51*, 445–457.
8. Hirohashi, S.; Kanai, Y. Cell adhesion system and human cancer morphogenesis. *Cancer Sci.* **2003**, *94*, 575–581.
9. Perinpanayagam, H.; Zaharias, R.; Stanford, C.; Keller, J.; Schneider, G.; Brand, R. Early cell adhesion events differ between osteoporotic and non-osteoporotic osteoblasts. *J. Orthop. Res.* **2001**, *19*, 993–1000.
10. Cho, P.; Schneider, G.B.; Kellogg, B.; Zaharias, R.; Keller, J.C. Effect of glucocorticoid-induced osteoporotic-like conditions on osteoblast cell attachment to implant surface microtopographies. *Implant Dent.* **2006**, *15*, 377–385.
11. Serhan, C.N.; Savill, J. Resolution of inflammation: The beginning programs the end. *Nat. Immunol.* **2005**, *6*, 1191–1197.
12. Simon, S.; Green, C.E. Molecular mechanics and dynamics of leukocyte recruitment during inflammation. *Annu. Rev. Biol.* **2005**, *7*, 151–185.
13. Spangenberg, C.; Lausch, E.U.; Trost, T.M.; Prawitt, D.; May, A.; Keppler, R.; Fees, S.A; Reutzel, D.; Bell, C.; Schmitt, S.; *et al.* ERBB2-mediated transcriptional up-regulation of the α5β1 integrin fibronectin receptor promotes tumor cell survival under adverse conditions. *Cancer Res.* **2006**, *66*, 3715–3725.
14. Zou, J.X.; Liu, Y.; Pasquale, E.B.; Ruoslahti, E. Activated Src oncogene phosphorylates R-ras and suppresses integrin activity. *J. Biol. Chem.* **2002**, *277*, 1824–1827.
15. Mierke, C.T. Cancer cells regulate biomechanical properties of human microvascular endothelial cells. *J. Biol. Chem.* **2011**, *286*, 40025–40037.
16. Huang, W.; Anvari, B.; Torres, J.H.; LeBaron, R.G.; Athanasiou, K.A. Temporal effects of cell adhesion on mechanical characteristics of the single chondrocyte. *J. Orthop. Res.* **2003**, *21*, 88–95.
17. Wang, C.-C.; Hsu, Y.C.; Su, F.C.; Lu, S.C.; Lee, T.M. Effects of passivation treatments on titanium alloy with nanometric scale roughness and induced changes in fibroblast initial adhesion evaluated by a cytodetacher. *J. Biomed. Mater. Res.* **2009**, *88*, 370–383.
18. Palmer, C.P.; Mycielska, M.E.; Burcu, H.; Osman, K.; Collins, T.; Beckerman, R.; Perrett, R.; Johnson, H.; Aydar, E.; Djamgoz, M.B.A. Single cell adhesion measuring apparatus (SCAMA): Application to cancer cell lines of different metastatic potential and voltage-gated Na^+ channel expression. *Eur. Biophys. J.* **2008**, *37*, 359–368.
19. Kwon, K.W.; Choi, S.S.; Lee, S.H.; Kim, B.; Lee, S.N.; Park, M.C.; Kim, P.; Hwang, S.Y.; Suh, K.Y. Label-free, microfluidic separation and enrichment of human breast cancer cells by adhesion difference. *Lab Chip* **2007**, *7*, 1461–1468.

20. Saif, M.A.T.; Sager, C.R.; Coyer, S. Functionalized biomicroelectromechanical systems sensors for force response study at local adhesion sites of single living cells on substrates. *Ann. Biomed. Eng.* **2003**, *31*, 950–961.
21. Horwitz, A.F. Integrins and health. *Sci. Am.* **1997**, *276*, 68–75.
22. Chen, C.S. Geometric control of cell life and death. *Science* **1997**, *276*, 1425–1428.
23. Hwang, W.C.; Waugh, R.E. Energy of dissociation of lipid bilayer from the membrane skeleton of red blood cells. *Biophys. J.* **1997**, *72*, 2669–2678.
24. Honarmandi, P.; Lee, H.; Lang, M.J.; Kamm, R.D. A microfluidic system with optical laser tweezers to study mechanotransduction and focal adhesion recruitment. *Lab Chip* **2011**, *11*, 684–694.
25. Beningo, K.A.; Dembo, M.; Kaverina, I.; Small, J.V.; Wang, Y. Nascent focal adhesions are responsible for the generation of strong propulsive forces in migrating fibroblasts. *J. Cell Biol.* **2001**, *153*, 881–887.
26. Tan, J.L.; Tien, J.; Pirone, D.M.; Gray, D.S.; Bhadriraju, K.; Chen, C.S. Cells lying on a bed of microneedles: An approach to isolate mechanical force. *Proc. Natl. Acad. Sci. USA* **2003**, *100*, 1484–1489.
27. Zhong, Y.; He, S.; Ji, B. Mechanics in mechanosensitivity of cell adhesion and its roles in cell migration. *Int. J. Comp. Mater. Sci. Eng.* **2012**, *01*, 1250032.
28. Dumbauld, D.W.; Lee, T.T.; Singh, A.; Scrimgeour, J.; Gersbach, C.A.; Zamir, E.A.; Fu, J.; Chen, C.S.; Curtis, J.E.; Craig, S.W.; *et al.*, How vinculin regulates force transmission. *Proc. Natl. Acad. Sci. USA* **2013**, *110*, 9788–9793.
29. Geiger, B.; Bershadsky, A.; Pankov, R.; Yamada, K.M. Transmembrane crosstalk between the extracellular matrix and the cytoskeleton. *Nat. Rev. Mol. Cell Biol.* **2001**, *2*, 793–805.
30. Hynes, R.O. Integrins: A family of cell surface receptors. *Cell* **1987**, *48*, 549–554.
31. Van der Flier, A.; Sonnenberg, A. Function and interactions of integrins. *Cell Tissue Res.* **2001**, *305*, 285–298.
32. DiMilla, P.A.; Barbee, K.; Douglas, A.; Lauffenburger, D.A. Mathematical model for the effects of adhesion and mechanics on cell migration speed. *Biophys. J.* **1991**, *60*, 15–37.
33. Burridge, K.; Wennerberg, K.; Hill, C.; Carolina, N. Rho and Rac take center stage. *Cell* **2004**, *116*, 167–179.
34. Hall, A. Rho GTPases and the actin cytoskeleton. *Science* **1998**, *279*, 509–514.
35. LeBaron, R.G.; Athanasiou, K.A. *Ex vivo* synthesis of articular cartilage. *Biomaterials* **2000**, *21*, 2575–2587.
36. Hong, S.; Ergezen, E.; Lec, R.; Barbee, K.A. Real-time analysis of cell-surface adhesive interactions using thickness shear mode resonator. *Biomaterials* **2006**, *27*, 5813–5820.
37. Taubenberger, A.; Cisneros, D.; Friedrichs, J.; Puech, P.-H.; Muller, D.J.; Franz, C.M. Revealing early steps of alpha2beta1 integrin-mediated adhesion to collagen type I by using single-cell force spectroscopy. *Mol. Biol. Cell* **2007**, *18*, 1634–1644.
38. Helenius, J.; Heisenberg, C.-P.; Gaub, H.E.; Muller, D.J. Single-cell force spectroscopy. *J. Cell Sci.* **2008**, *121*, 1785–1791.
39. McEver, R.P.; Zhu, C. Rolling cell adhesion. *Ann. Rev. Cell Dev. Biol.* **2010**, *26*, 363–396.

40. Green, C.E.; Pearson, D.N.; Camphausen, R.T.; Staunton, D.E.; Simon, S.I. Shear-dependent capping of L-selectin and P-selectin glycoprotein ligand 1 by E-selectin signals activation of high-avidity β2-integrin on neutrophils. *J. Immunol.* **2004**, *172*, 7780–7790.
41. Belloni, P.; Tressler, R. Microvascular endothelial cell heterogeneity: Interactions with leukocytes and tumor cells. *Cancer Metastasis Rev.* **1990**, *8*, 353–389.
42. Honn, K.; Tang, D. Adhesion molecules and tumor cell interaction with endothelium and subendothelial matrix. *Cancer Metastasis Rev.* **1992**, *11*, 353–375.
43. Ley, K.; Laudanna, C.; Cybulsky, M.I.; Nourshargh, S. Getting to the site of inflammation: The leukocyte adhesion cascade updated. *Nat. Rev. Immunol.* **2007**, *7*, 678–689.
44. Petri, B.; Phillipson, M.; Kubes, P. The physiology of leukocyte recruitment: An *in vivo* perspective. *J. Immunol.* **2008**, *180*, 6439–6446.
45. Imhof, B.A.; Aurrand-Lions, M. Adhesion mechanisms regulating the migration of monocytes. *Nat. Rev. Immunol.* **2004**, *4*, 432–444.
46. Muller, W.A. Leukocyte–endothelial-cell interactions in leukocyte transmigration and the inflammatory response. *Trends Immunol.* **2015**, *24*, 326–333.
47. Dembo, M.; Wang, Y. Stresses at the cell-to-substrate interface during locomotion of fibroblasts. *Biophys. J.* **1999**, *76*, 2307–2316.
48. Kraning-Rush, C.M.; Califano, J.P.; Reinhart-King, C.A. Cellular traction stresses increase with increasing metastatic potential. *PLoS ONE* **2012**, *7*, e32572.
49. Huynh, J.; Bordeleau, F.; Kraning-Rush, C.M.; Reinhart-King, C.A. Substrate stiffness regulates PDGF-induced circular dorsal ruffle formation through MLCK. *Cell. Mol. Bioeng.* **2014**, *6*, 1–16.
50. Reinhart-King, C.A.; Dembo, M.; Hammer, D.A. Endothelial cell traction forces on RGD-derivatized polyacrylamide substrata. *Langmuir* **2003**, *19*, 1573–1579.
51. Sabass, B.; Gardel, M.L.; Waterman, C.M.; Schwarz, U.S. High resolution traction force microscopy based on experimental and computational advances. *Biophys. J.* **2008**, *94*, 207–220.
52. Ng, M.R.; Besser, A.; Danuser, G.; Brugge, J.S. Substrate stiffness regulates cadherin-dependent collective migration through myosin-II contractility. *J. Cell Biol.* **2012**, *199*, 545–563.
53. Wen, J.H.; Vincent, L.G.; Fuhrmann, A.; Choi, Y.S.; Hribar, K.C.; Taylor-weiner, H.; Chen, S.; Engler, A.J. Interplay of matrix stiffness and protein tethering in stem cell differentiation. *Nat. Mater.* **2014**, *13*, 979–984.
54. Califano, J.P.; Reinhart-King, C.A. Substrate stiffness and cell area predict cellular traction stresses in single cells and cells in contact. *Cell. Mol. Bioeng.* **2011**, *3*, 68–75.
55. Zhou, D.W.; García, A.J. Measurement systems for cell adhesive forces. *J. Biomech. Eng.* **2015**, *137*, 020908.
56. Théry, M. Micropatterning as a tool to decipher cell morphogenesis and functions. *J. Cell Sci.* **2010**, *123*, 4201–4213.
57. D'Arcangelo, E.; McGuigan, A.P. Micropatterning strategies to engineer controlled cell and tissue architecture *in vitro*. *Biotechniques* **2014**, *58*, 13–23.
58. Mandal, K.; Balland, M.; Bureau, L. Thermoresponsive micropatterned substrates for single cell studies. *PLoS ONE* **2012**, *7*, e37548.

59. Polio, S.R.; Rothenberg, K.E.; Stamenović, D.; Smith, M.L. A micropatterning and image processing approach to simplify measurement of cellular traction forces. *Acta Biomater.* **2012**, *8*, 82–88.
60. Ting, L.H.; Jahn, J.R.; Jung, J.I.; Shuman, B.R.; Feghhi, S.; Han, S.J.; Rodriguez, M.L.; Sniadecki, N.J. Flow mechanotransduction regulates traction forces, intercellular forces, and adherens junctions. *Am. J. Physiol.* **2012**, *302*, H2220–H2229.
61. Santos, E.; Hernández, R.M.; Pedraz, J.L.; Orive, G. Novel advances in the design of three-dimensional bio-scaffolds to control cell Fate: Translation from 2D to 3D. *Trends Biotechnol.* **2012**, *30*, 331–341.
62. Baker, B.M.; Chen, C.S. Deconstructing the third dimension: How 3D culture microenvironments alter cellular cues. *J. Cell Sci.* **2012**, *125*, 3015–3024.
63. Legant, W.R.; Miller, J.S.; Blakely, B.L.; Cohen, D.M.; Genin, G.M.; Chen, C.S. Measurement of mechanical tractions exerted by cells in three-dimensional matrices. *Nat. Methods* **2010**, *7*, 969–971.
64. Khetan, S.; Guvendiren, M.; Legant, W.R.; Cohen, D.M.; Chen, C.S.; Burdick, J.A. Degradation-mediated cellular traction directs stem cell fate in covalently crosslinked three-dimensional hydrogels. *Nat. Mater.* **2013**, *12*, 458–465.
65. Meseke, M.; Förster, E. A 3D-matrigel/microbead assay for the visualization of mechanical tractive forces at the neurite-substrate interface of cultured neurons. *J. Biomed. Mater. Res. A* **2013**, *101A*, 1726–1733.
66. Fraley, S.I.; Feng, Y.; Krishnamurthy, R.; Kim, D.-H.; Celedon, A.; Longmore, G.D.; Wirtz, D. A distinctive role for focal adhesion proteins in three-dimensional cell motility. *Nat. Cell Biol.* **2010**, *12*, 598–604.
67. Kraning-Rush, C.M.; Carey, S.P.; Califano, J.P.; Smith, B.N.; Reinhart-King, C.A. The role of the cytoskeleton in cellular force generation in 2D and 3D environments. *Phys. Biol.* **2011**, *8*, 015009.
68. Zaman, M.H.; Trapani, L.M.; Sieminski, A.L.; MacKellar, D.; Gong, H.; Kamm, R.D.; Wells, A.; Lauffenburger, D.A.; Matsudaira, P. Migration of tumor cells in 3D matrices is governed by matrix stiffness along with cell-matrix adhesion and proteolysis. *Proc. Natl. Acad. Sci. USA* **2006**, *103*, 10889–10894.
69. Bonakdar, N.; Butler, J.P.; Fabry, B.; Koch, T.M.; Mu, S. 3D traction forces in cancer cell invasion. *PLoS ONE* **2012**, *7*, e33476.
70. Kutys, M.L.; Yamada, K.M. An extracellular matrix-specific GEF-GAP interaction regulates Rho GTPase crosstalk for 3D collagen migration. *Nat. Cell Biol.* **2015**, *16*, 909–917.
71. Park, J.Y.; Arnaout, M.A.; Gupta, V. A simple, no-wash cell adhesion-based high-throughput assay for the discovery of small-molecule regulators of the integrin CD11b/CD18. *J. Biomol. Screen.* **2011**, *12*, 406–417.
72. Garcia, A.J.; Gallant, N.D. Stick and grip: Measurement systems and quantitative analyses of integrin-mediated cell adhesion strength. *Cell Biochem. Biophys.* **2003**, *39*, 61–73.
73. Chen, Y.; Lu, B.; Yang, Q.; Fearns, C.; Yates, J.R.; Lee, J.D. Combined integrin phosphoproteomic analyses and siRNA-based functional screening identified key regulators for cancer cell adhesion and migration. *Cancer Res.* **2010**, *69*, 3713–3720.
74. Mianabadi, M.; Yazdanparast, R. Inhibition of substrate-tumor cell adhesion under the effect of gnidilatimonoein purified from daphne mucronata. *Am. J. Chin. Med.* **2004**, *32*, 369–376.

75. Christ, K.V; Turner, K.T. Methods to measure the strength of cell adhesion to substrates. *J. Adhes. Sci. Technol.* **2010**, *24*, 37–41.
76. Ferreira, G.N.M.; da-Silva, A.-C.; Tomé, B. Acoustic wave biosensors: Physical models and biological applications of quartz crystal microbalance. *Trends Biotechnol.* **2009**, *27*, 689–697.
77. Wegener, J.; Janshoff, A. Cell adhesion monitoring using a quartz crystal microbalance: Comparative analysis of different mammalian cell lines. *Eur. Biophys. J.* **1998**, *28*, 26–37.
78. Heitmann, V.; Wegener, J. Monitoring cell adhesion by piezoresonators: Impact of increasing oscillation amplitudes. *Anal. Chem.* **2007**, *79*, 3392–3400.
79. Xi, J.; Chen, J.Y.; Garcia, M.P.; Penn, L.S. Quartz crystal microbalance in cell biology studies. *J. Biochips Tissue Chipissue Chip.* **2013**, *S5*, 1–9.
80. Li, F.; Wang, J.H.-C.; Wang, Q.-M. Monitoring cell adhesion by using thickness shear mode acoustic wave sensors. *Biosens. Bioelectron.* **2007**, *23*, 42–50.
81. Fohlerová, Z.; Skládal, P.; Turánek, J. Adhesion of eukaryotic cell lines on the gold surface modified with extracellular matrix proteins monitored by the piezoelectric sensor. *Biosens. Bioelectron.* **2007**, *22*, 1896–1901.
82. Heitmann, V.; Reiβ, B.; Wegener, J. The quartz crystal microbalance in cell biology: Basics and applications. *Springer Ser. Chem. Sens. Biosens.* **2007**, *5*, 303–338.
83. Wegener, J.; Seebach, J.; Janshoff, A.; Galla, H.J. Analysis of the composite response of shear wave resonators to the attachment of mammalian cells. *Biophys. J.* **2000**, *78*, 2821–2833.
84. Modin, C.; Stranne, A.-L.; Foss, M.; Duch, M.; Justesen, J.; Chevallier, J.; Andersen, L.K.; Hemmersam, A.G.; Pedersen, F.S.; Besenbacher, F. QCM-D studies of attachment and differential spreading of pre-osteoblastic cells on Ta and Cr surfaces. *Biomaterials* **2006**, *27*, 1346–1354.
85. Lord, M.S.; Modin, C.; Foss, M.; Duch, M.; Simmons, A.; Pedersen, F.S.; Milthorpe, B.K.; Besenbacher, F. Monitoring cell adhesion on tantalum and oxidised polystyrene using a quartz crystal Microbalance with dissipation. *Biomaterials* **2006**, *27*, 4529–4537.
86. Le Guillou-Buffello, D.; Hélary, G.; Gindre, M.; Pavon-Djavid, G.; Laugier, P.; Migonney, V. Monitoring cell adhesion processes on bioactive polymers with the quartz crystal resonator technique. *Biomaterials* **2005**, *26*, 4197–4205.
87. Zhu, Y.; Qiu, H.; Trzeciakowski, J.P.; Sun, Z.; Li, Z.; Hong, Z.; Hill, M.A; Hunter, W.C.; Vatner, D.E.; Vatner, S.F.; *et al.* Temporal analysis of vascular smooth muscle cell elasticity and adhesion reveals oscillation waveforms that differ with aging. *Aging Cell* **2012**, *11*, 741–750.
88. Hayward, L.; Thompson, M. Acoustic waves and the study of biochemical macromolecules and cells at the sensor—Liquid interface critical review. *Analyst* **1999**, *124*, 1405–1420.
89. Da-Silva, A.-C.; Soares, S.S.; Ferreira, G.N.M. Acoustic detection of cell Adhesion to a coated quartz crystal microbalance—Implications for studying the biocompatibility of polymers. *Biotechnol. J.* **2013**, *8*, 690–698.
90. Marx, K.A.; Zhou, T.; Montrone, A.; McIntosh, D.; Braunhut, S.J. Quartz crystal microbalance biosensor study of endothelial cells and their extracellular matrix following cell removal: Evidence for transient cellular stress and viscoelastic changes during detachment and the elastic behavior of the pure matrix. *Anal. Biochem.* **2005**, *343*, 23–34.

91. Marx, K.A.; Zhou, T.; Montrone, A.; McIntosh, D.; Braunhut, S.J. A comparative study of the cytoskeleton binding drugs nocodazole and taxol with a mammalian cell quartz crystal microbalance biosensor: Different dynamic responses and energy dissipation effects. *Anal. Biochem.* **2007**, *361*, 77–92.
92. Marx, K.A.; Zhou, T.; Montrone, A.; Schulze, H.; Braunhut, S.J. A quartz crystal microbalance cell biosensor: Detection of microtubule alterations in living cells at nM nocodazole concentrations. *Biosens. Bioelectron.* **2001**, *16*, 773–782.
93. Zhou, T.; Marx, K.A.; Warren, M.; Schulze, H.; Braunhut, S.J. The quartz crystal microbalance as a continuous monitoring tool for the study of endothelial cell surface attachment and growth. *Biotechnol. Prog.* **2000**, *16*, 268–277.
94. Warrick, J.W.; Young, E.W.K.; Schmuck, E.G.; Saupe, K.W.; Beebe, D.J. High-content adhesion assay to address limited cell samples. *Integr. Biol.* **2013**, *5*, 720–727.
95. Hartmann, A.; Stamp, M.; Kmeth, R.; Buchegger, S.; Stritzker, B.; Saldamli, B.; Burgkart, R.; Schneider, M.F.; Wixforth, A. A novel tool for dynamic cell adhesion studies-the De-Adhesion number investigator DANI. *Lab Chip* **2014**, *14*, 542–546.
96. Gupta, V.K.; Sraj, I.A.; Konstantopoulos, K.; Eggleton, C.D. Multi-scale simulation of L-selectin-PSGL-1-dependent homotypic leukocyte binding and rupture. *Biomech. Model. Mechanobiol.* **2010**, *9*, 613–627.
97. Rupprecht, P.; Golé, L.; Rieu, J.-P.; Vézy, C.; Ferrigno, R.; Mertani, H.C.; Rivière, C. A tapered channel microfluidic device for comprehensive cell adhesion analysis, using measurements of detachment kinetics and shear stress-dependent motion. *Biomicrofluidics* **2012**, *6*, 14107–1410712.
98. Alapan, Y.; Little, J.A.; Gurkan, U.A. Heterogeneous red blood cell adhesion and deformability in sickle cell disease. *Sci. Rep.* **2014**, *4*, 7173.
99. Chaw, K.C.; Manimaran, M.; Tay, E.H.; Swaminathan, S. Multi-step microfluidic device for studying cancer metastasis. *Lab Chip* **2007**, *7*, 1041–1047.
100. Nalayanda, D.D.; Wang, Q.; Fulton, W.B.; Wang, T.-H.; Abdullah, F. Engineering an artificial alveolar-capillary membrane: A novel continuously perfused model within microchannels. *J. Pediatr. Surg.* **2010**, *45*, 45–51.
101. Song, J.W.; Cavnar, S.P.; Walker, A.C.; Luker, K.E.; Gupta, M.; Tung, Y.-C.; Luker, G.D.; Takayama, S. Microfluidic endothelium for studying the intravascular adhesion of metastatic breast cancer cells. *PLoS ONE* **2009**, *4*, e5756.
102. Fu, Y.; Kunz, R.; Wu, J.; Dong, C. Study of local hydrodynamic environment in cell-substrate adhesion using side-view μPIV technology. *PLoS ONE* **2012**, *7*, e30721.
103. Riahi, R.; Yang, Y.L.; Kim, H.; Jiang, L.; Wong, P.K.; Zohar, Y. A microfluidic model for organ-specific extravasation of circulating tumor cells. *Biomicrofluidics* **2014**, *8*, 024103.
104. Yamamoto, A.; Mishima, S.; Maruyama, N.; Sumita, M. Quantitative evaluation of cell attachment to glass, polystyrene, and fibronectin- or collagen-coated polystyrene by measurement of cell adhesive shear force and cell detachment energy. *J. Biomed. Mater. Res.* **2000**, *50*, 114–124.
105. Lee, C. The technique for measurement of cell adhesion force. *J. Med. Biol. Eng.* **2004**, *24*, 51–56.
106. Yang, S.-P.; Yang, C.-Y.; Lee, T.-M.; Lui, T.-S. Effects of calcium-phosphate topography on osteoblast mechanobiology determined using a cytodetacher. *Mater. Sci. Eng.* **2012**, *32*, 254–262.

107. Christ, K.V.; Williamson, K.B.; Masters, K.S.; Turner, K.T. Measurement of single-cell adhesion strength using a microfluidic assay. *Biomed. Microdevices* **2010**, *12*, 443–455.
108. Hochmuth, R.M. Micropipette aspiration of living cells. *J. Biomech.* **2000**, *33*, 15–22.
109. Gao, Z.; Wang, S.; Zhu, H.; Su, C.; Xu, G.; Lian, X. Using selected uniform cells in round shape with a micropipette to measure cell adhesion strength on silk fibroin-based materials. *Mater. Sci. Eng. C* **2008**, *28*, 1227–1235.
110. Rotsch, C.; Jacobson, K.; Radmacher, M. Dimensional and mechanical dynamics of active and stable edges in motile fibroblasts investigated by using atomic force microscopy. *Proc. Natl. Acad. Sci. USA* **1999**, *96*, 921–926.
111. Muller, D.J. AFM: A nanotool in membrane biology. *Biochemistry* **2008**, *47*, 7896–7898.
112. Engel, A.; Gaub, H.E. Structure and mechanics of membrane proteins. *Annu. Rev. Biochem.* **2008**, *77*, 127–148.
113. Hansma, H.G.; Kasuya, K.O.E. Atomic force microscopy imaging and pulling of nucleic acids. *Curr. Opin. Struct. Biol.* **2004**, *14*, 380–385.
114. Hirano, Y.; Takahashi, H.; Kumeta, M.; Hizume, K.; Hirai, Y.; Otsuka, S.; Yoshimura, S.H.; Takeyasu, K. Nuclear architecture and chromatin dynamics revealed by atomic force microscopy in combination with biochemistry and cell biology. *Pflügers Archiv* **2008**, *456*, 139–153.
115. Pesen, D.; Hoh, J.H. Micromechanical architecture of the endothelial cell cortex. *Biophys. J.* **2005**, *88*, 670–679.
116. Collinsworth, A.M.; Zhang, S.; Kraus, W.E.; Truskey, G.A. Apparent elastic modulus and hysteresis of skeletal muscle cells throughout differentiation. *Am. J. Physiol. Cell Physiol.* **2002**, *283*, C1219–C1227.
117. Solon, J.; Levental, I.; Sengupta, K.; Georges, P.C.; Janmey, P.A. Fibroblast adaptation and stiffness matching to soft elastic substrates. *Biophys. J.* **2007**, *93*, 4453–4461.
118. Rotsch, C.; Braet, F.; Wisse, E.; Radmacher, M. AFM imaging and elasticity measurements on living rat liver macrophages. *Cell Biol. Int.* **1997**, *21*, 685–696.
119. Wu, C.-C.; Su, H.-W.; Lee, C.-C.; Tang, M.-J.; Su, F.-C. Quantitative measurement of changes in adhesion force involving focal adhesion kinase during cell attachment, spread, and migration. *Biochem. Biophys. Res. Commun.* **2005**, *329*, 256–265.
120. Kim, W. *A Micro-Aspirator Chip Using Vacuum Expanded Microchannels for High-Throughput Mechanical Characterization of Biological Cells*; The Royal Society of Chemistry Publishing: London, UK, 2010; pp. 1–86.
121. Puech, P.-H.; Taubenberger, A.; Ulrich, F.; Krieg, M.; Muller, D.J.; Heisenberg, C.-P. Measuring cell adhesion forces of primary gastrulating cells from zebrafish using atomic force microscopy. *J. Cell Sci.* **2005**, *118*, 4199–4206.
122. Weder, G.; Vörös, J.; Giazzon, M.; Matthey, N.; Heinzelmann, H.; Liley, M. Measuring cell adhesion forces during the cell cycle by force spectroscopy. *Biointerphases* **2009**, *4*, 27–34.
123. Hoffmann, S.C.; Cohnen, A.; Ludwig, T.; Watzl, C. 2B4 engagement mediates rapid LFA-1 and actin-dependent NK cell adhesion to tumor cells as measured by single cell force spectroscopy. *J. Immunol.* **2011**, *186*, 2757–2764.

124. Beaussart, A.; El-Kirat-Chatel, S.; Sullan, R.M.A; Alsteens, D.; Herman, P.; Derclaye, S.; Dufrêne, Y.F. Quantifying the forces guiding microbial cell adhesion using single-cell force spectroscopy. *Nat. Protoc.* **2014**, *9*, 1049–1055.
125. Gourier, C.; Jegou, A.; Husson, J.; Pincet, F.A Nanospring named erythrocyte. The biomembrane force probe. *Cell. Mol. Bioeng.* **2008**, *1*, 263–275.
126. Evans, E.; Ritchie, K.; Merkel, R. Sensitive force technique to probe molecular adhesion and structural linkages at biological interfaces. *Biophys. J.* **1995**, *68*, 2580–2587.
127. Evans, E.; Berk, D.; Leung, A. Detachment of agglutinin-bonded red blood cells. I. Forces to rupture molecular-point attachments. *Biophys. J.* **1991**, *59*, 838–848.
128. Evans, E.; Heinrich, V.; Leung, A.; Kinoshita, K. Nano- to microscale dynamics of P-selectin detachment from leukocyte interfaces. I. Membrane separation from the cytoskeleton. *Biophys. J.* **2005**, *88*, 2288–2298.
129. Zhang, H.; Liu, K.-K. Optical tweezers for single cells. *J. R. Soc. Interface* **2008**, *5*, 671–690.
130. Ashkin, A. Optical trapping and manipulation of neutral particles using lasers. *Proc. Natl. Acad. Sci. USA* **1997**, *94*, 4853–4860.
131. Thoumine, O.; Kocian, P.; Kottelat, A; Meister, J.J. Short-term binding of fibroblasts to fibronectin: Optical tweezers experiments and probabilistic analysis. *Eur. Biophys. J.* **2000**, *29*, 398–408.
132. Tsai, J.-W.; Liao, B.-Y.; Huang, C.-C.; Hwang, W.-L.; Wang, D.-W.; Chiou, A.E.T.; Lin, C.-H. Applications of optical tweezers and an integrated force measurement module for biomedical research. *Int. Soc. Opt. Photonics* **2000**, *4082*, 213–221.
133. Curtis, J.E.; Spatz, J.P. Getting a grip: Hyaluronan-mediated cellular adhesion. *Int. Soc. Opt. Photonics* **2004**, *5514*, 455–466.
134. Castelain, M.; Pignon, F.; Piau, J.-M.; Magnin, A.; Mercier-Bonin, M.; Schmitz, P. Removal forces and adhesion properties of saccharomyces cerevisiae on glass substrates probed by optical tweezer. *J. Chem. Phys.* **2007**, *127*, 135104.
135. Castelain, M.; Rouxhet, P.G.; Pignon, F.; Magnin, A.; Piau, J.-M. Single-cell adhesion probed in-situ using optical tweezers: A case study with saccharomyces cerevisiae. *J. Appl. Phys.* **2012**, *111*, 114701.
136. Schwingel, M.; Bastmeyer, M. Force mapping during the formation and maturation of cell adhesion sites with multiple optical tweezers. *PLoS ONE* **2013**, *8*, e54850.
137. Askenasy, N.; Farkas, D.L. Optical imaging of PKH-labeled hematopoietic cells in recipient bone marrow *in vivo*. *Stem Cells* **2002**, *20*, 501–513.
138. Castelain, M.; Pignon, F.; Piau, J.-M.; Magnin, A. The initial single yeast cell adhesion on glass via optical trapping and Derjaguin–Landau–Verwey–Overbeek predictions. *J. Chem. Phys.* **2008**, *128*, 135101.
139. Koo, L.Y.; Irvine, D.J.; Mayes, A.M.; Lauffenburger, D.A.; Griffith, L.G. Co-regulation of cell adhesion by nanoscale RGD organization and mechanical stimulus. *J. Cell Sci.* **2002**, *115*, 1423–1433.
140. Giacomello, E.; Neumayer, J.; Colombatti, A.P.R. Centrifugal assay for fluorescence-based cell adhesion adapted to the analysis of *ex vivo* cells and capable of determining relative binding strengths. *Biotechniques* **1999**, *26*, 764–766.

141. Channavajjala, L.S.; Eidsath, A.; Saxinger, W.C.A. Simple method for measurement of cell-substrate attachment forces: Application to HIV-1 Tat. *J. Cell Sci.* **1997**, *110*, 249–256.
142. Capadona, J.R.; Collard, D.M.; García, A.J. Fibronectin adsorption and cell adhesion to mixed monolayers of tri(ethylene glycol)- and methyl-terminated alkanethiols. *Langmuir* **2002**, *19*, 1847–1852.
143. Keselowsky, B.G.; García, A.J. Quantitative methods for analysis of integrin binding and focal adhesion formation on biomaterial surfaces. *Biomaterials* **2005**, *26*, 413–418.
144. Reyes, C.D.; García, A.J. A centrifugation cell adhesion assay for high-throughput screening of biomaterial surfaces. *J. Biomed. Mater. Res. A* **2003**, *67*, 328–333.
145. Reyes, C.D.; Petrie, T.A.; García, A.J. Mixed extracellular matrix ligands synergistically modulate integrin adhesion and signaling. *J. Cell. Physiol.* **2008**, *217*, 450–458.
146. Harbers, G.M.; Healy, K.E. The effect of ligand type and density on osteoblast adhesion, proliferation, and matrix mineralization. *J. Biomed. Mater. Res. A* **2005**, *75*, 855–869.
147. Harbers, G.M.; Gamble, L.J.; Irwin, E.F.; Castner, D.G.; Healy, K.E. Development and characterization of a high-throughput system for assessing cell-surface receptor-ligand engagement. *Langmuir* **2005**, *21*, 8374–8384.
148. Reutelingsperger, C.P.M.; van Gool, R.G.J.; Heijnen, V.; Frederik, P.; Lindhout, T. The rotating disc as a device to study the adhesive properties of endothelial cells under differential shear stresses. *J. Mater. Sci. Mater. Med.* **1994**, *5*, 361–367.
149. Gallant, N.D.; Michael, K.E.; Garcı, J. Cell adhesion strengthening: Contributions of adhesive area , integrin binding , and focal adhesion assembly. *Mol. Biol. Cell* **2005**, *16*, 4329–4340.
150. García, A.J.; Ducheyne, P.; Boettiger, D. Quantification of cell adhesion using a spinning disc device and application to surface-reactive materials. *Biomaterials* **1997**, *18*, 1091–1098.
151. García, A.J.; Ducheyne, P.; Boettiger, D. Effect of surface reaction stage on fibronectin-mediated adhesion of osteoblast-like cells to bioactive glass. *J. Biomed. Mater. Res.* **1998**, *40*, 48–56.
152. Deligianni, D.D.; Katsala, N.D.; Koutsoukos, P.G.; Missirlis, Y.F. Effect of surface roughness of hydroxyapatite on human bone marrow cell adhesion, proliferation, differentiation and detachment strength. *Biomaterials* **2000**, *22*, 87–96.
153. Deligianni, D.; Korovessis, P.; Porte-Derrieu, M.C.; Amedee, J. Fibronectin preadsorbed on hydroxyapatite together with rough surface structure increases osteoblasts' adhesion "*in vitro*": The theoretical usefulness of fibronectin preadsorption on hydroxyapatite to increase permanent stability and longevity in spine implant. *J. Spinal Disord. Tech.* **2005**, *18*, 257–262.
154. Lee, M.H.; Adams, C.S.; Boettiger, D.; DeGrado, W.F.; Shapiro, I.M.; Composto, R.J.; Ducheyne, P. Adhesion of MC3T3-E1 cells to RGD peptides of different flanking residues: Detachment strength and correlation with long-term cellular function. *J. Biomed. Mater. Res. A* **2007**, *81A*, 150–160.
155. Michael, K.E.; Dumbauld, D.W.; Burns, K.L.; Hanks, S.K.; García, A.J. Focal adhesion kinase modulates cell adhesion strengthening via integrin activation. *Mol. Biol. Cell* **2009**, *20*, 2508–2519.
156. Shi, Q.; Boettiger, D. A novel mode for integrin-mediated signaling: Tethering is required for phosphorylation of FAK Y397. *Mol. Biol. Cell* **2003**, *14*, 4306–4315.
157. Datta, A.; Huber, F.; Boettiger, D. Phosphorylation of B3 integrin controls ligand binding strength. *J. Biol. Chem.* **2002**, *277*, 3943–3949.

158. Lynch, L.; Vodyanik, P.I.; Boettiger, D.; Guvakova, M.A. Insulin-like growth factor I controls adhesion strength mediated by α5β1 integrins in motile carcinoma cells. *Mol. Biol. Cell* **2005**, *16*, 51–63.
159. Miller, T.; Boettiger, D. Control of intracellular signaling by modulation of fibronectin conformation at the cell-materials interface. *Langmuir* **2002**, *19*, 1723–1729.
160. García, A.J.; Takagi, J.; Boettiger, D. Two-stage activation for α5β1 integrin binding to surface-adsorbed fibronectin. *J. Biol. Chem.* **1998**, *273*, 34710–34715.
161. Goldstein, A.S.; DiMilla, P.A. Application of fluid mechanic and kinetic models to characterize mammalian cell detachment in a radial-flow chamber. *Biotechnol. Bioeng.* **1997**, *55*, 616–629.
162. Detry, J.G.; Deroanne, C.; Sindic, M. Hydrodynamic systems for assessing surface fouling, soil adherence and cleaning in laboratory installations. *Biotechnol. Agron. Soc. Environ.* **2009**, *13*, 427–439.
163. Cozens-roberts, C.; Quinn, J.A.; Lauffenburger, D.A. Receptor-mediated adhesion phenomena model studies with the radial-flow detachment assay. *Biophys. J.* **1990**, *58*, 107–125.
164. Dimilla, P.A.; Stone, J.A.; Albelda, S.M.; Biology, S. Maximal migration of human smooth muscle cells on fibronectin and type IV collagen occurs at an intermediate attachment strength. *J. Cell Biol.* **1993**, *122*, 729–737.
165. Goldstein, A.S.; DiMilla, P.A. Examination of membrane rupture as a mechanism for mammalian cell detachment from fibronectin-coated biomaterials. *J. Biomed. Mater. Res. A* **2003**, *67A*, 658–666.
166. Goldstein, A.S.; DiMilla, P.A. Effect of adsorbed fibronectin concentration on cell adhesion and deformation under shear on hydrophobic surfaces. *J. Biomed. Mater. Res.* **2002**, *59*, 665–675.
167. Goldstein, A.S.; DiMilla, P.A. Comparison of converging and diverging radial flow for measuring cell adhesion. *AIChE J.* **1998**, *44*, 465–473.
168. Kreke, M.R.; Badami, A.S.; Brady, J.B.; Michael Akers, R.; Goldstein, A.S. Modulation of protein adsorption and cell adhesion by poly(allylamine hydrochloride) heparin films. *Biomaterials* **2005**, *26*, 2975–2981.
169. Rezania, A.; Healy, K.E. Biomimetic peptide surfaces that regulate adhesion, spreading, cytoskeletal organization, and mineralization of the matrix deposited by osteoblast-like cells. *Biotechnol. Prog.* **1999**, *15*, 19–32.
170. Rezania, A.; Healy, K.E. Integrin subunits responsible for adhesion of human osteoblast-like cells to biomimetic peptide surfaces. *J. Orthop. Res.* **1999**, *17*, 615–623.
171. Rezania, A.; Thomas, C.H.; Branger, A.B.; Waters, C.M.; Healy, K.E. The detachment strength and morphology of bone cells contacting materials modified with a peptide sequence found within bone sialoprotein. *J. Biomed. Mater. Res.* **1997**, *37*, 9–19.
172. Rezania, A.; Thomas, C.; Healy, K. A probabilistic approach to measure the strength of bone cell adhesion to chemically modified surfaces. *Ann. Biomed. Eng.* **1997**, *25*, 190–203.
173. Bearinger, J.P.; Castner, D.G.; Golledge, S.L.; Rezania, A.; Hubchak, S.; Healy, K.E. P(AAm-co-EG) interpenetrating polymer networks grafted to oxide surfaces: Surface characterization, protein adsorption, and cell detachment studies. *Langmuir* **1997**, *13*, 5175–5183.

174. Brown, M.A.; Wallace, C.S.; Anamelechi, C.C.; Clermont, E.; Reichert, W.M.; Truskey, G.A. The use of mild trypsinization conditions in the detachment of endothelial cells to promote subsequent endothelialization on synthetic surfaces. *Biomaterials* **2007**, *28*, 3928–3935.
175. Sordel, T.; Kermarec-Marcel, F.; Garnier-Raveaud, S.; Glade, N.; Sauter-Starace, F.; Pudda, C.; Borella, M.; Plissonnier, M.; Chatelain, F.; Bruckert, F.; *et al.*, Influence of glass and polymer coatings on CHO cell morphology and adhesion. *Biomaterials* **2007**, *28*, 1572–1584.
176. Gerszten, R.E.; Lim, Y.-C.; Ding, H.T.; Snapp, K.; Kansas, G.; Dichek, D.A.; Cabañas, C.; Sánchez-Madrid, F.; Gimbrone, M.A.; Rosenzweig, A.; *et al.* Adhesion of monocytes to vascular cell adhesion molecule-1–transduced human endothelial cells: Implications for atherogenesis. *Circ. Res.* **1998**, *82*, 871–878.
177. Truskey, G.A.; Pirone, J.S. The effect of fluid shear stress upon cell adhesion to fibronectin-treated surfaces. *J. Biomed. Mater. Res.* **1990**, *24*, 1333–1353.
178. Van Kooten, T.G.; Schakenraad, J.M.; van der Mei, H.C.; Busscher, H.J. Development and use of a parallel-plate flow chamber for studying cellular adhesion to solid surfaces. *J. Biomed. Mater. Res.* **1992**, *26*, 725–738.
179. Lane, W.O.; Jantzen, A.E.; Carlon, T.A; Jamiolkowski, R.M.; Grenet, J.E.; Ley, M.M.; Haseltine, J.M.; Galinat, L.J.; Lin, F.-H.; Allen, J.D.; *et al.* Parallel-plate flow chamber and continuous flow circuit to evaluate endothelial progenitor cells under laminar flow shear stress. *J. Vis. Exp.* **2012**, 1–12.
180. Xiao, Y.; Truskey, G.A. Effect of receptor-ligand affinity on the strength of endothelial cell adhesion. *Biophys. J.* **1996**, *71*, 2869–2884.
181. Bhat, V.D.; Truskey, G.A.; Reichert, W.M. Using avidin-mediated binding to enhance initial endothelial cell attachment and spreading. *J. Biomed. Mater. Res.* **1998**, *40*, 57–65.
182. Chan, B.P.; Bhat, V.D.; Yegnasubramanian, S.; Reichert, W.M.; Truskey, G.A. An equilibrium model of endothelial cell adhesion via integrin-dependent and integrin-independent ligands. *Biomaterials* **1999**, *20*, 2395–2403.
183. Bhat, V.D.; Truskey, G.A.; Reichert, W.M. Fibronectin and avidin–biotin as a heterogeneous ligand system for enhanced endothelial cell adhesion. *J. Biomed. Mater. Res.* **1998**, *41*, 377–385.
184. Usami, S.; Chen, H.-H.; Zhao, Y.; Chien, S.; Skalak, R. Design and construction of a linear shear stress flow chamber. *Ann. Biomed. Eng.* **1993**, *21*, 77–83.
185. Cao, J.; Usami, S.; Dong, C. Development of a side-view chamber for studying cell-surface adhesion under flow conditions. *Ann. Biomed. Eng.* **1997**, *25*, 573–580.
186. Heilshorn, S.C.; DiZio, K.A.; Welsh, E.R.; Tirrell, D.A. Endothelial cell adhesion to the fibronectin CS5 domain in artificial extracellular matrix proteins. *Biomaterials* **2003**, *24*, 4245–4252.
187. Wan, Y.; Yang, J.; Yang, J.; Bei, J.; Wang, S. Cell adhesion on gaseous plasma modified poly-(l-lactide) surface under shear stress field. *Biomaterials* **2003**, *24*, 3757–3764.
188. Renshaw, K.M.; Orr, D.E.; Burg, K.J.L. Design and evaluation of a novel flow chamber for measuring cell adhesion to absorbable polymer films. *Biotechnol. Prog.* **2005**, *21*, 538–545.
189. Yang, J.; Wan, Y.; Yang, J.; Bei, J.; Wang, S. Plasma-treated, collagen-anchored polylactone: Its cell affinity evaluation under shear or shear-free conditions. *J. Biomed. Mater. Res. A* **2003**, *67A*, 1139–1147.

190. Boura, C.; Muller, S.; Vautier, D.; Dumas, D.; Schaaf, P.; Claude Voegel, J.; François Stoltz, J.; Menu, P. Endothelial cell—Interactions with polyelectrolyte multilayer films. *Biomaterials* **2005**, *26*, 4568–4575.
191. Lee, J.H.; Lee, S.J.; Khang, G.; Lee, H.B. The effect of fluid shear stress on endothelial cell adhesiveness to polymer surfaces with wettability gradient. *J. Colloid Interf. Sci.* **2000**, *230*, 84–90.
192. Van Kooten T.G.; Schakenraad J.M.; van der Mei, H.C.; Busscher, H.J. Influence of Substratum Wettability on the Strength of Adhesion of Human Fibroblasts. *Biomaterials* **1992**, *13*, 897–904.
193. Truskey, G.A.; Proulx, T.L. Relationship between 3T3 cell spreading and the strength of adhesion on glass and silane surfaces. *Biomaterials* **1993**, *14*, 243–254.
194. Vogel, J.; Bendas, G.; Bakowsky, U.; Hummel, G.; Schmidt, R.R.; Kettmann, U.; Rothe, U. The role of glycolipids in mediating cell adhesion: A flow chamber Study 1. *Biochim. Biophys. Acta* **1998**, *1372*, 205–215.
195. Palange, A.L.; di Mascolo, D.; Singh, J.; de Franceschi, M.S.; Carallo, C.; Gnasso, A.; Decuzzi, P. Modulating the vascular behavior of Mmetastatic breast cancer cells by curcumin treatment. *Front. Oncol.* **2012**, *2*, 161.
196. Xia, Y.; Whitesides, G.M. Soft lithography. *Angew. Chem. Int. Ed.* **1998**, *37*, 550–575.
197. Young, E.W.K.; Wheeler, A.R.; Simmons, C.A. Matrix-dependent adhesion of vascular and valvular endothelial cells in microfluidic channels. *Lab Chip* **2007**, *7*, 1759–1766.
198. Lu, H.; Koo, L.Y.; Wang, W.M.; Lauffenburger, D.A; Griffith, L.G.; Jensen, K.F. Microfluidic shear devices for quantitative analysis of cell adhesion. *Anal. Chem.* **2004**, *76*, 5257–5264.
199. Zhang, X.; Jones, P.; Haswell, S. Attachment and detachment of living cells on modified microchannel surfaces in a microfluidic-based lab-on-a-chip system. *Chem. Eng. J.* **2008**, *135*, S82–S88.
200. Tang, Z.; Akiyama, Y.; Itoga, K.; Kobayashi, J.; Yamato, M.; Okano, T. Shear stress-dependent cell detachment from temperature-responsive cell culture surfaces in a microfluidic device. *Biomaterials* **2012**, *33*, 7405–7411.
201. Merkel, R.; Nassoy, P.; Leung, A.; Ritchie, K.; Evans, E. Energy landscapes of receptor-ligand bonds explored with dynamic force spectroscopy. *Nature* **1999**, *397*, 50–53.
202. An, X.; Mohandas, N. Disorders of red cell membrane. *Br. J. Haematol.* **2008**, *141*, 367–375.
203. Mohandas, N.; Gallagher, P.G. Red cell membrane: Past, present, and future. *Blood* **2008**, *112*, 3939–3949.
204. Correlates, M. Sickle erythrocyte adherence to vascular endothelium morphologic correlates and the requirement for divalent cations and collagen-binding plasma proteins. *J. Clin. Investig.* **1985**, *76*, 1605–1612.
205. Hebbel, R.P.; Boogaerts, M.A.B.; Eaton, J.W.; Steinberg, M.H. Erythrocyte adherence to endothelium in sickle-cell anemia. *N. Engl. J. Med.* **1980**, *302*, 992–995.
206. Lee, G.Y.H.; Lim, C.T. Biomechanics approaches to studying human diseases. *Trends Biotechnol.* **2007**, *25*, 111–118.
207. Girdhar, G.; Shao, J.-Y. Simultaneous tether extraction from endothelial cells and leukocytes: Observation, mechanics, and significance. *Biophys. J.* **2007**, *93*, 4041–4052.

208. Schlüter, D.K.; Ramis-Conde, I.; Chaplain, M.A.J. Multi-scale modelling of the dynamics of cell colonies: Insights into cell-adhesion forces and cancer invasion from in silico simulations. *J. R. Soc. Interface* **2014**, *12*, 20141080.
209. Deshpande, V.; Mrksich, M.; Mcmeeking, R.; Evans, A. A bio-mechanical model for coupling cell contractility with focal adhesion formation. *J. Mech. Phys. Solids* **2008**, *56*, 1484–1510.
210. Cheung, L.S.-L.; Zheng, X.; Wang, L.; Baygents, J.C.; Guzman, R.; Schroeder, J.A.; Heimark, R.L.; Zohar, Y. Adhesion dynamics of circulating tumor cells under shear flow in a bio-functionalized microchannel. *J. Micromech. Microeng.* **2011**, *21*, 054033.
211. Kong, D.; Ji, B.; Dai, L. Nonlinear mechanical modeling of cell adhesion. *J. Theor. Biol.* **2008**, *250*, 75–84.
212. Hodges, S.R.; Jensen, O.E. Spreading and peeling dynamics in a model of cell adhesion. *J. Fluid Mech.* **2002**, *460*, 381–409.
213. Kong, D.; Ji, B.; Dai, L. Stability of adhesion clusters and cell reorientation under lateral cyclic tension. *Biophys. J.* **2008**, *95*, 4034–4044.
214. Nicolas, A.; Besser, A.; Safran, S.A. Dynamics of cellular focal adhesions on deformable substrates: Consequences for cell force microscopy. *Biophys. J.* **2008**, *95*, 527–539.
215. Gallant, N.D.; García, A.J. Model of integrin-mediated cell adhesion strengthening. *J. Biomech.* **2007**, *40*, 1301–1309.
216. Ronan, W.; Deshpande, V.S.; McMeeking, R.M.; McGarry, J.P. Numerical investigation of the active role of the actin cytoskeleton in the compression resistance of cells. *J. Mech. Behav. Biomed. Mater.* **2012**, *14*, 143–157.
217. Kong, D.; Ji, B.; Dai, L. Stabilizing to disruptive transition of focal adhesion response to mechanical forces. *J. Biomech.* **2010**, *43*, 2524–2529.

10

Digital Microfluidics for Manipulation and Analysis of a Single Cell

Jie-Long He, An-Te Chen, Jyong-Huei Lee and Shih-Kang Fan *

Department of Mechanical Engineering, National Taiwan University, Taipei 10617, Taiwan; E-Mails: D93b47202@ntu.edu.tw (J.-L.H.); andy80103@hotmail.com (A.-T.C.); b00505004@ntu.edu.tw (J.-H.L.)

* Author to whom correspondence should be addressed; E-Mail: skfan@fan-tasy.org

Academic Editors: Fan-Gang Tseng and Tuhin Subhra Santra

Abstract: The basic structural and functional unit of a living organism is a single cell. To understand the variability and to improve the biomedical requirement of a single cell, its analysis has become a key technique in biological and biomedical research. With a physical boundary of microchannels and microstructures, single cells are efficiently captured and analyzed, whereas electric forces sort and position single cells. Various microfluidic techniques have been exploited to manipulate single cells through hydrodynamic and electric forces. Digital microfluidics (DMF), the manipulation of individual droplets holding minute reagents and cells of interest by electric forces, has received more attention recently. Because of ease of fabrication, compactness and prospective automation, DMF has become a powerful approach for biological application. We review recent developments of various microfluidic chips for analysis of a single cell and for efficient genetic screening. In addition, perspectives to develop analysis of single cells based on DMF and emerging functionality with high throughput are discussed.

Keywords: microfluidic chips; digital microfluidics; single cell analysis; genetic screening; assisted reproductive technologies

---

1. Introduction

Multicellular organisms are composed of varied cells grouped into specialized tissues and organs, which typically comprise cells of diverse types present in widely varying abundance. The single cell is the basic structural and functional unit of a living organism. The most critical knowledge for biological and biomedical science is constructed from research on cell biology [1,2]; this information about cell functionality and behavior has been applied in many clinical and biomedical applications [3], such as drug development, disease diagnosis, cancer research and assisted reproductive technology (ART).

Traditional cell assays to study cell differentiation, gene expression and drug response were focused mainly on a population of cells. For an average result of multiple cells, the outcome of all cells is assumed to be homogeneous, but several authors have found cellular heterogeneity or multi-modal distributions in a cell population [4–6]. The heterogeneity might arise through genetic or non-genetic processes, which induce distinct cellular decision-making [7]. If an average response of multiple cells is taken to be representative of a typical population, cellular heterogeneity might result in a misleading interpretation [8]. Analysis of a single cell indicated that individual cells that form a cell population might have a complicated distribution. This information is ignored in a traditional cell assay, which emphasizes evaluating the mean of a cell population [9]. To understand the variability from cell to cell and to improve the clinical and biomedical applicability, development of new approaches to isolate, to manipulate, to treat and to analyze single cells are required. Analysis of single cells has become a key technique in biological and biomedical research, as one can thereby analyze individual cells within a cell population [1,7].

Various techniques for analysis of a single cell have been developed, such as flow cytometry and microfluidic chips [10]. Flow cytometry, which has been under development for many years, has become a mature technique for single-cell analysis. In particular, fluorescence-activated cell sorting (FACS), builds upon flow cytometry in order to sort single cells that are tagged with specific fluorescent markers [11,12]. Flow cytometry is, however, a complicated and costly system: it cannot support analysis of cells in real time in their natural environment. Furthermore, integration of an entire assay based on a single cell typically entails processes such as cell manipulation, treatment and final detection. To overcome these problems, integrating various methods of cell manipulation with microfluidic lab-on-a-chip (LOC) platforms, also known as microfluidic chips, has become a major activity in assays of single cells [8,11,13–15]. Microfluidic chips could provide efficient genetic screening through a well-controlled microenvironment for analysis and treatment of a single cell. On these platforms, isolation of a single cell, purification of mRNA and subsequent multiplex quantitative polymerase chain reaction (qPCR) in real time must be effected on a chip [16]. Single-cell genomics, transcriptomics, epigenomics and proteomics will allow many enduring questions in biological and biomedical sciences to be answered [17–20]. Digital microfluidics (DMF), the manipulation of individual droplets by electric forces, is one of the particularly important techniques for single cell manipulation and analysis on a microfluidic chip. In this paper, we review recent developments in DMF chips for analysis of a single cell and efficient genetic screening, which involve manipulation of a single cell, next-generation sequencing (NGS) and assisted-reproductive applications. We conclude with views on the future development of DMF chips for analysis of a single cell and discuss how this emerging efficient tool can advance biological and biomedical research.

2. Microfluidic Chips for Analysis of a Single Cell

In an organism, a single cell is small: its volume is about 1 pL. Cells constituting a tissue or organ are complicated and diverse. An extracellular matrix (ECM) provides structural and biochemical support of the structural connection. To analyze a single cell on a chip, a microfluidic chip typically entails integration of four functions: (1) cell sorting, through microwells [21–25], traps [26–29], optical tweezers [30–33] or dielectrophoresis (DEP) [34–37] to isolate a single cell; (2) cell manipulation, through a traditional syringe pump [38] or an electrowetting-on-dielectric (EWOD) technique [39]; (3) cell lysis, through a mechanical [40], chemical [24] or electrical [23,41,42] process; and (4) analysis of an individual cell, such as with PCR [21,43] or NGS [44].

Microfluidic chips have been applied in biological and biomedical research, including culturing, sorting, patterning and genetic screening of single cells for clinical diagnostics [45–49]. These platforms have become a promising tool for efficient genetic screening through analysis of single cells, because these microfluidic chips can provide rapid, real-time and automated analysis. Gossett *et al.* [50] demonstrated an automated microfluidic technique capable of probing single cells. A rapid assay of the deformability of native populations of leukocytes and malignant cells in pleural effusions has been enabled on this chip. Guan *et al.* [51] introduced a new microfluidic chip with real-time feedback control to evaluate single-cell deformability, which was used to discriminate different kinds of cells for cancer diagnosis [30]. Guo *et al.* [52] produced a microfluidic chip to distinguish red blood cells containing parasitic *Plasmodium falciparum* from uninfected cells. Several microfluidic chips have been generated to capture single cells and to measure the impedance of the cells, such as human cervical epithelioid carcinoma (HeLa) cells [53,54] or circulating tumor cells (CTCs) from blood [55,56]. Kurz *et al.* [57] reported a microfluidic chip to trap single cells and to measure the impedance for the monitoring of sub-toxic effects on cell membranes.

The method most frequently used to isolate a single cell is physical separation. At designed physical boundaries, an individual cell is isolated, captured and sorted with mechanical structures on a chip. Capturing an individual cell with microwells is an attractive strategy, because it is simple and easily operated. Jen *et al.* [23,24] reported microfluidic chips with arrays of microwells that isolated individual cells and provided chemical and electric lysis of a single cell with high throughput (Figure 1a). Lindstrom *et al.* [21,22,58,59] developed a novel microplate with microwells for efficient analyses of single cells. This platform allowed each single cell to be cultivated and analyzed individually for reprogramming factor evaluation on stem cells [22], PCR amplification and genetic analysis [21] (Figure 1b).

Microfluidic hydrodynamic traps combine dynamic cell isolation with prospective high throughput on a chip [60,61]. Di Carlo *et al.* [26,62] produced a dynamic platform that allows culture of a single cell with a consistent environment and dynamic control of individual cells (Figure 2a). Kobel *et al.* [60] reported a microfluidic chip with efficiency of trapping a single cell enhanced up to 97% (Figure 2b).

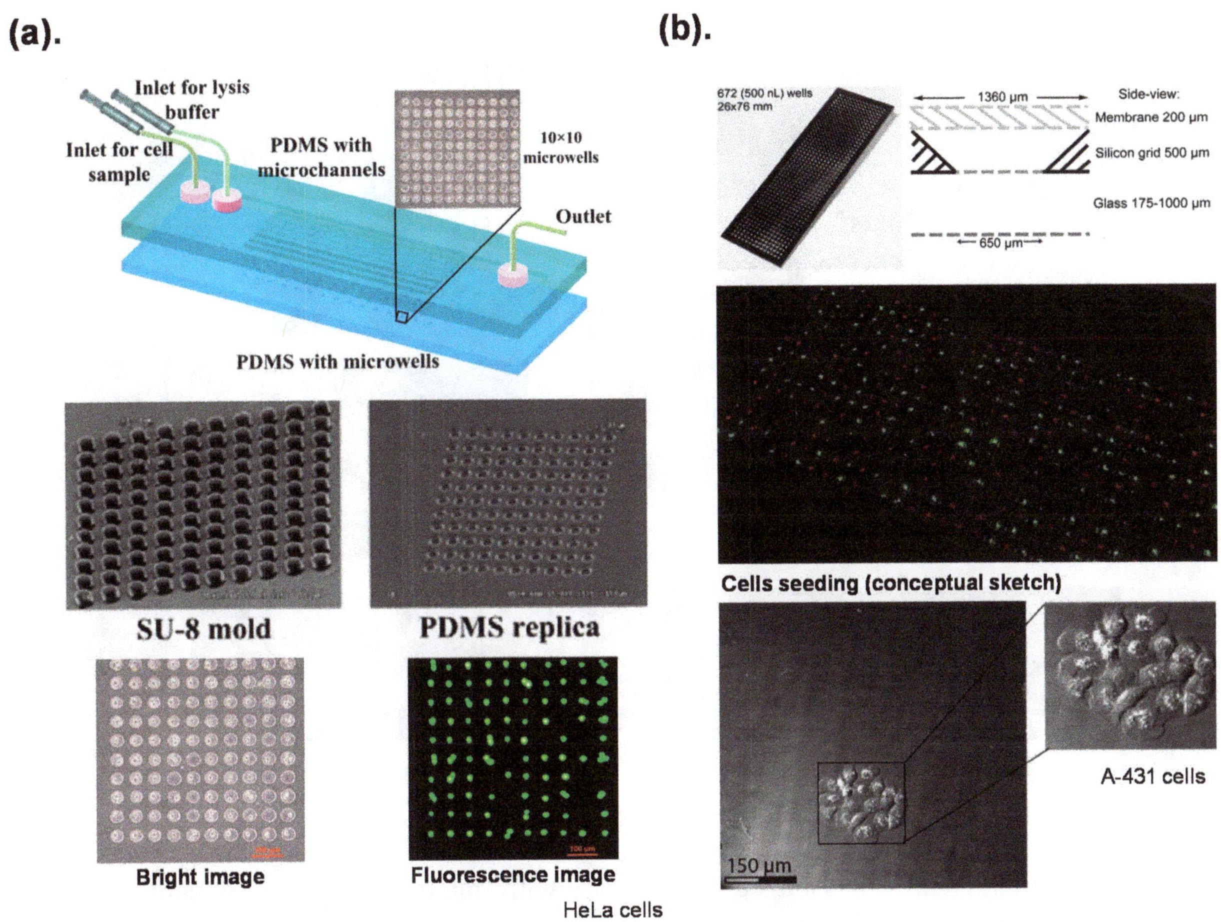

Figure 1. Individual cells isolated on a chip with microwells described in: (**a**) Jen *et al.*, 2012 [24]; (**b**) Lindstrom *et al.*, 2009 [21]. Reproduction of the figures has been made with permission from Multidisciplinary Digital Publishing Institute and The Royal Society of Chemistry.

Compared with use of mechanical structures, isolating a cell with an optical or electric force is a contact-free method that eliminates deformation and damage of cells [63,64]. Here we focus more on the electric forces, including DEP and EWOD. The DEP force generated with a non-uniform electric field interacting with polarized, suspended particles [65–67] has been widely used to manipulate cells [39,63,68,69]. The DEP force is classified as positive DEP (pDEP) and negative DEP (nDEP) based on the polarizability of the particles or cells and the liquid. Fan *et al.* [39] used DEP forces to concentrate suspended particles in a liquid droplet with dielectric-coated electrodes patterned on a plate (Figure 3a). Creating two droplets with mammalian cells and polystyrene beads at distinct concentrations was achieved with DEP and EWOD (Figure 3b).

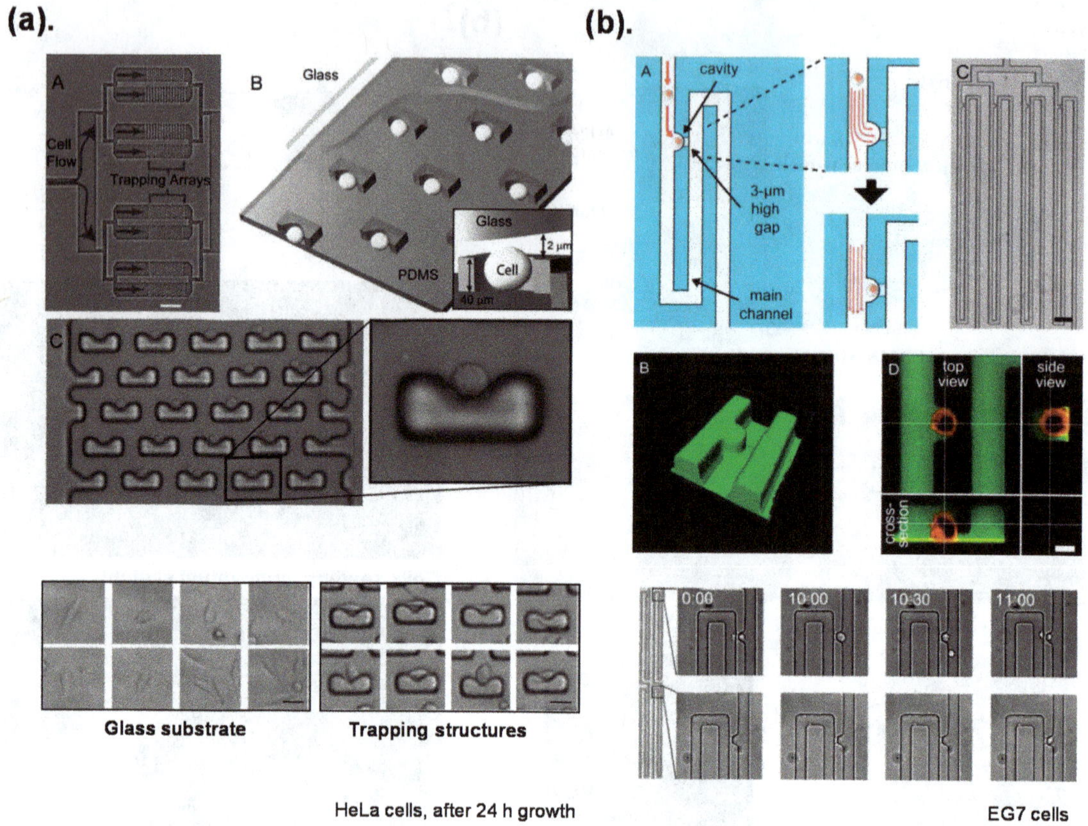

Figure 2. Individual cell isolated on a chip with microfluidic hydrodynamic traps described in: (**a**) Di Carlo *et al.*, 2006 [26]: (**A**) A photograph of the cell trapping device; (**B**) A diagram of the device and mechanism of trapping; (**C**) A high resolution micrograph of the trapping device; (**b**) Kobel *et al.*, 2010 [60]: (**A**) Schematic illustration of the single cell trap; (**B**) A three-dimensional reconstruction image of the cell trap; (**C**) An array of the single cell traps; (**D**) An orthogonal view of a fluorescently labeled single cell. Reproduction of the figures has been made with permission from The Royal Society of Chemistry.

3. Digital Microfluidic Chips and Biological Application

Digital microfluidics (DMF), according to which tiny droplets are manipulated with electric forces, including EWOD and DEP, have been confirmed to be a powerful platform for reagent addition, droplet transmission, solution mixing, splitting and dispensing for biological application [70–80]. For a simple configuration of device and easy modular interfaces, portable and wearable DMF systems with assembled modules for continuous actuation of droplets were demonstrated [70] (Figure 3c).

A DMF chip with arrays of electrodes is typically fabricated on indium tin oxide (ITO) glass plates using photolithography and wet etching. The patterned ITO electrodes are then covered with a dielectric and a hydrophobic layer [16,39,73,81]. The advantages of DMF include ease of fabrication, simple device structure, small consumption of reagents, easy integration with analytical instruments and prospective automation. Thus, DMF has become highly suitable for biological application. Barbulovic-Nad *et al.* [80] introduced a DMF chip to implement cell-based assays; the platform was demonstrated to be advantageous for cell-based assays because of potential for automated manipulation of multiple reagents. Vergauwe *et al.* [78] reported a DMF chip for homogeneous and heterogeneous bio-assays with great analytical performance capable of medical applications. Kumar *et al.* [75] demonstrated the first use of a DMF technique for individual protoplasts from *Arabidopsis thaliana* plants. Shih *et al.* developed the first DMF chip capable of cell impedance sensing [76]; they also integrated droplet-in-channel microfluidics with DMF to develop a novel chip to perform complicated assays [81].

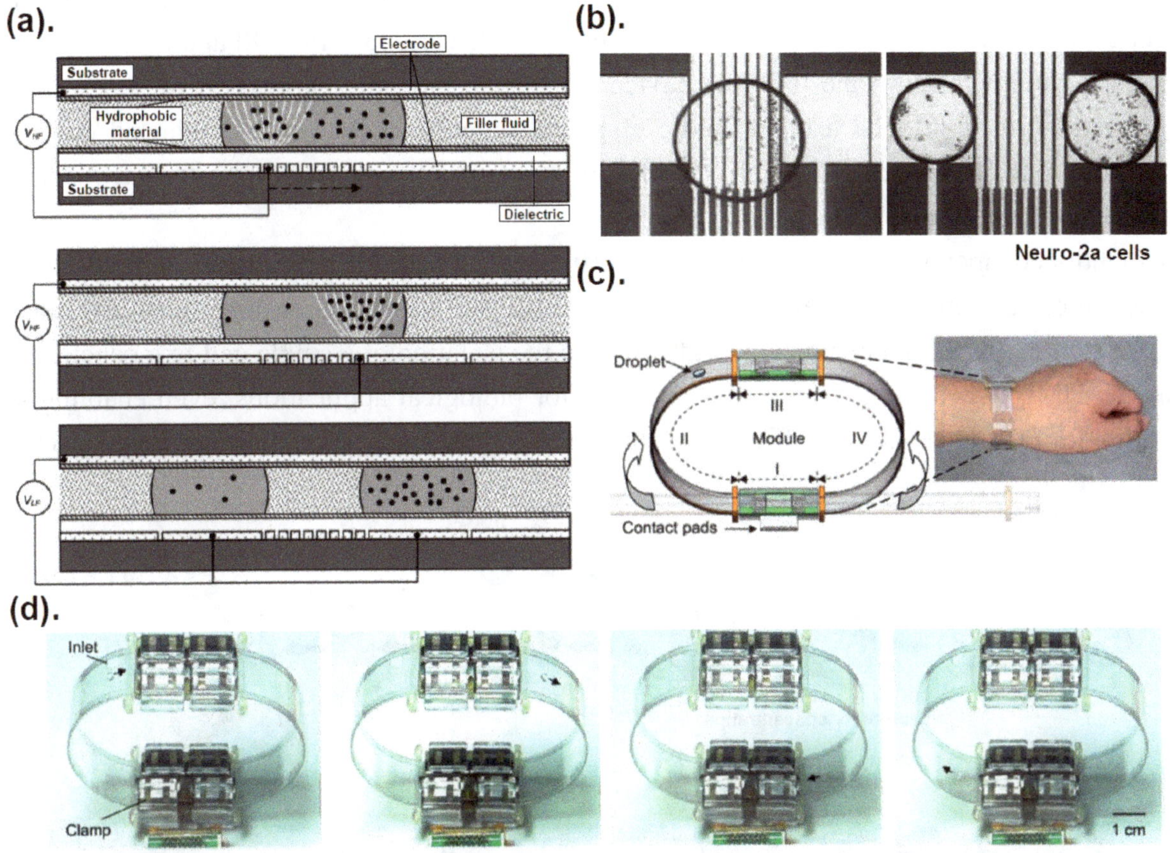

Figure 3. Dielectrophoresis (DEP) forces exerting on the suspended particles described by Fan *et al.*, 2008 [39]. (**a**) A parallel-plate device with square and strip electrodes to manipulate droplets and a particle; (**b**) DEP forces exerted on suspended particles including mammalian cells (Neuro-2a); (**c,d**) Prototype of EWOD-based, continuously microfluidic module for a portable system reported by Fan *et al.*, 2011 [70]. Reproduction of the figures has been made with permission from The Royal Society of Chemistry.

This work demonstrates that DMF chips would be a generic and powerful platform for the biological assays, including drug screening, immunoassays, analysis of single cells and digital PCR. This promising new technique might allow the efficient genetic screening based on a single cell to become a reality.

4. Digital Microfluidic Chips for Genetic Screening

Investigating gene expression and developing genetic screening at a level of a single cell provides an important capability to resolve the problem of disease etiology, cancer pathology and other biomedical applications [82]. Traditional methods of genetic screening require a large amount of sample for an analysis, which typically decreases the sensitivity and accuracy on analysis of only a single cell [83,84]. Various microfluidic techniques have been developed to address this problem. Digital polymerase chain reaction (digital PCR) platforms have measured DNA or cDNA of a single cell [85,86], but challenges persist in treating the integration of varied programs for genetic analyses of a single cell on a device, including cell sorting, manipulation, lysis, PCR and genetic screening. Within the past decade, microfluidic chips have become one of the most powerful platforms to achieve efficient genetic screening at the level of a single cell [16,87]. Toriello *et al.* [88] and Bontoux *et al.* [89] reported a polydimethylsiloxane (PDMS) device to analyze the gene expression of single cells, and Marcus *et al.* [90] developed a microfluidic chip with integrated flow that conducted cell capture, lysis, mRNA purification, cDNA synthesis and purification but with a complicated auxiliary system for control.

As mentioned above, the DMF chip has advantages of ease of fabrication, simple supporting instrumentation and prospective automation. Rival *et al.* [16] described an integrated and automated EWOD system to perform a complete workflow from the isolation of a single cell to a genetic analysis (Figure 4). DMF is becoming a powerful approach for biological applications, even enabling sample preparation for PCR to develop an efficient genetic screening platform based on a single cell [16,91,92].

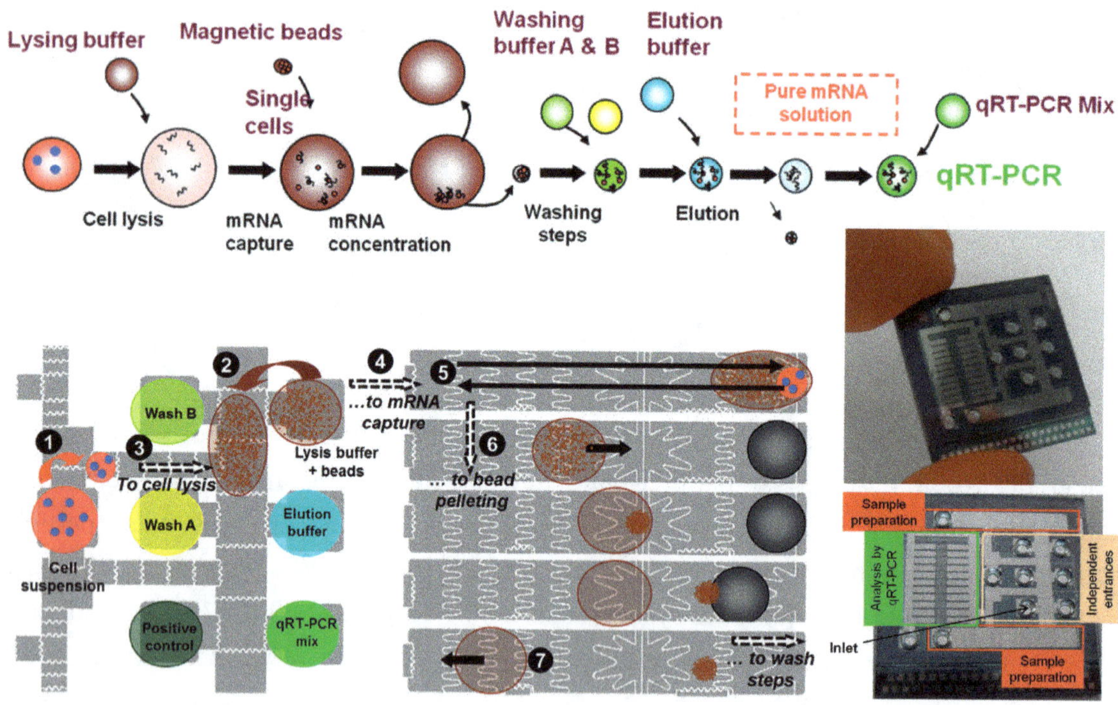

Figure 4. An integrated EWOD system for genetic analysis based on a single cell described by Rival *et al.*, 2014 [16]. The lysis and mRNA capture steps: (1) A droplet containing a few cells is dispensed; (2) A droplet of lysis buffer and magnetic beads is dispensed; (3) A droplet with the cells and a droplet with the beads are merged for cell lysis; (4) This merged droplet is moved to the "sample preparation" electrodes for operating magnetic beads; (5) The droplet is moved back and forth to enable bead mixing and mRNA capture; (6) The droplet is moved towards the magnet, the beads are concentrated to result in bead extraction; (7) An empty droplet is moved towards the waist. Reproduction of the figures has been made with permission from The Royal Society of Chemistry.

5. Efficient Genetic Screening for Assisted Reproductive Techniques

The early development of a mammalian embryo is a complicated process involving an upheaval of a transcriptional architecture [93–95]. In applications, human *in vitro* fertilization (IVF) is an important scientific achievement in the twentieth century, but until recently there has been little knowledge of regulatory mechanisms in genes of early mammalian embryos. The early embryo undergoes cleavage divisions in a series from two cell, four cells, eight cells, morula, even to blastocyst. A platform for genetic screening based on a single cell could provide critical knowledge to clarify regulatory mechanisms of genes in early mammalian embryos [93]. This emerging efficient technique would benefit a biomedical approach, such as assisted reproductive technology (ART).

The microfluidic ART platforms under development are focused on simulation of the Fallopian tube to optimize the IVF, especially the early embryo culture *in vitro* [96–98]. Pre-implantation genetic diagnosis (PGD) has been recently developed to detect genetic diseases [99,100]. The current methods of sorting single cells include taking trophectoderm cells from blastocyst, blastomeres from embryos at a cleavage stage and polar bodies from the oocyte or zygote [101]. The advantages of a DMF chip include simple accompanying instrumentation and prospective automation for sorting single cells from an early embryo. Huang *et al.* [98] demonstrated a EWOD-based microfluidic device for culture of early mammalian embryos *in vitro*, presaging future clinical application. Although a DMF chip applied in efficient genetic diagnosis is still at an initial stage, we believe that this powerful platform will have a major impact on the ART field.

6. Conclusions

Recent developments of microfluidic chips for single cell analysis were reviewed; microfluidic techniques provide numerous advantages for biological and biomedical research, including ease for modularity, small sample requirement, potential of automation, and high-throughput. Perspectives on DMF in analysis of a single cell and efficient genetic screening were particularly focused on and described. We expect that these novel single-cell techniques on microfluidic chips will be important in biomedical areas.

Acknowledgments

Jie-Long He thanks the Ministry of Science and Technology, Taiwan for providing a post-doctoral fellowship.

Author Contributions

Jie-Long He and Shih-Kang Fan wrote the manuscript. An-Te Chen and Jyong-Huei Lee assisted in collecting and integrating the references.

References

1. Svahn, H.A.; van den Berg, A. Single cells or large populations? *Lab Chip* **2007**, *7*, 544–546.
2. Bahcall, O.G. Single cell resolution in regulation of gene expression. *Mol. Syst. Biol.* **2005**, *1*, 2005.0015, doi:10.1038/msb4100020.
3. Dittrich, P.S.; Tachikawa, K.; Manz, A. Micro total analysis systems. Latest advancements and trends. *Anal. Chem.* **2006**, *78*, 3887–3908.
4. Graf, T.; Stadtfeld, M. Heterogeneity of embryonic and adult stem cells. *Cell Stem Cell* **2008**, *3*, 480–483.
5. Zhou, L.; Shen, Y.; Jiang, L.; Yin, D.; Guo, J.; Zheng, H.; Sun, H.; Wu, R.; Guo, Y. Systems mapping for hematopoietic progenitor cell heterogeneity. *PLoS ONE* **2015**, *10*, e0126937.
6. Donati, G.; Watt, F.M. Stem cell heterogeneity and plasticity in epithelia. *Cell Stem Cell* **2015**, *16*, 465–476.
7. Perkins, T.J.; Swain, P.S. Strategies for cellular decision-making. *Mol. Syst. Biol.* **2009**, *5*, 326.
8. Yin, H.; Marshall, D. Microfluidics for single cell analysis. *Curr. Opin. Biotechnol.* **2012**, *23*, 110–119.
9. Wang, D.; Bodovitz, S. Single cell analysis: The new frontier in "omics". *Trends Biotechnol.* **2010**, *28*, 281–290.
10. Shah, P.; Zhu, X.; Chen, C.; Hu, Y.; Li, C.Z. Lab-on-chip device for single cell trapping and analysis. *Biomed. Microdevices* **2014**, *16*, 35–41.
11. Cheung, K.C.; di Berardino, M.; Schade-Kampmann, G.; Hebeisen, M.; Pierzchalski, A.; Bocsi, J.; Mittag, A.; Tarnok, A. Microfluidic impedance-based flow cytometry. *Cytom. Part A J. Int. Soc. Anal. Cytol.* **2010**, *77*, 648–666.
12. Ning, B.; Yihong, Z.; Chang, L. Microfluidic electroporative flow cytometry for studying single-cell biomechanics. *Anal. Chem.* **2008**, *80*, 7714–7719.
13. Shah, P.; Kaushik, A.; Zhu, X.; Zhang, C.; Li, C.Z. Chip based single cell analysis for nanotoxicity assessment. *Analyst* **2014**, *139*, 2088–2098.
14. Lindstrom, S.; Andersson-Svahn, H. Miniaturization of biological assays—Overview on microwell devices for single-cell analyses. *Biochim. Biophys. Acta* **2011**, *1810*, 308–316.
15. Le Gac, S.; van den Berg, A. Single cells as experimentation units in lab-on-a-chip devices. *Trends Biotechnol.* **2010**, *28*, 55–62.
16. Rival, A.; Jary, D.; Delattre, C.; Fouillet, Y.; Castellan, G.; Bellemin-Comte, A.; Gidrol, X. An EWOD-based microfluidic chip for single-cell isolation, mRNA purification and subsequent multiplex qPCR. *Lab Chip* **2014**, *14*, 3739–3749.
17. Shapiro, E.; Biezuner, T.; Linnarsson, S. Single-cell sequencing-based technologies will revolutionize whole-organism science. *Nat. Rev. Genet.* **2013**, *14*, 618–630.
18. Ullal, A.V.; Peterson, V.; Agasti, S.S.; Tuang, S.; Juric, D.; Castro, C.M.; Weissleder, R. Cancer cell profiling by barcoding allows multiplexed protein analysis in fine-needle aspirates. *Sci. Transl. Med.* **2014**, *6*, 219ra219.
19. Salehi-Reyhani, A.; Burgin, E.; Ces, O.; Willison, K.R.; Klug, D.R. Addressable droplet microarrays for single cell protein analysis. *Analyst* **2014**, *139*, 5367–5374.
20. Kalisky, T.; Quake, S.R. Single-cell genomics. *Nat. Methods* **2011**, *8*, 311–314.

21. Lindstrom, S.; Hammond, M.; Brismar, H.; Andersson-Svahn, H.; Ahmadian, A. PCR amplification and genetic analysis in a microwell cell culturing chip. *Lab Chip* **2009**, *9*, 3465–3471.
22. Lindstrom, S.; Eriksson, M.; Vazin, T.; Sandberg, J.; Lundeberg, J.; Frisen, J.; Andersson-Svahn, H. High-density microwell chip for culture and analysis of stem cells. *PLoS ONE* **2009**, *4*, e6997.
23. Jen, C.P.; Amstislavskaya, T.G.; Liu, Y.H.; Hsiao, J.H.; Chen, Y.H. Single-cell electric lysis on an electroosmotic-driven microfluidic chip with arrays of microwells. *Sensors (Basel)* **2012**, *12*, 6967–6977.
24. Jen, C.P.; Hsiao, J.H.; Maslov, N.A. Single-cell chemical lysis on microfluidic chips with arrays of microwells. *Sensors (Basel)* **2012**, *12*, 347–358.
25. Osada, K.; Hosokawa, M.; Yoshino, T.; Tanaka, T. Monitoring of cellular behaviors by microcavity array-based single-cell patterning. *Analyst* **2014**, *139*, 425–430.
26. Di Carlo, D.; Wu, L.Y.; Lee, L.P. Dynamic single cell culture array. *Lab Chip* **2006**, *6*, 1445–1449.
27. Skelley, A.M.; Kirak, O.; Suh, H.; Jaenisch, R.; Voldman, J. Microfluidic control of cell pairing and fusion. *Nat. Methods* **2009**, *6*, 147–152.
28. Lee, S.-W.; Kang, J.Y.; Lee, I.-H.; Ryu, S.-S.; Kwak, S.-M.; Shin, K.-S.; Kim, C.; Jung, H.-I.; Kim, T.-S. Single-cell assay on CD-like lab chip using centrifugal massive single-cell trap. *Sens. Actuators A Phys.* **2008**, *143*, 64–69.
29. Liu, W.; Dechev, N.; Foulds, I.G.; Burke, R.; Parameswaran, A.; Park, E.J. A novel permalloy based magnetic single cell micro array. *Lab Chip* **2009**, *9*, 2381–2390.
30. Remmerbach, T.W.; Wottawah, F.; Dietrich, J.; Lincoln, B.; Wittekind, C.; Guck, J. Oral cancer diagnosis by mechanical phenotyping. *Cancer Res.* **2009**, *69*, 1728–1732.
31. Guck, J.; Schinkinger, S.; Lincoln, B.; Wottawah, F.; Ebert, S.; Romeyke, M.; Lenz, D.; Erickson, H.M.; Ananthakrishnan, R.; Mitchell, D.; *et al.* Optical deformability as an inherent cell marker for testing malignant transformation and metastatic competence. *Biophys. J.* **2005**, *88*, 3689–3698.
32. Wang, M.M.; Tu, E.; Raymond, D.E.; Yang, J.M.; Zhang, H.; Hagen, N.; Dees, B.; Mercer, E.M.; Forster, A.H.; Kariv, I.; *et al.* Microfluidic sorting of mammalian cells by optical force switching. *Nat. Biotechnol.* **2005**, *23*, 83–87.
33. Yang, T.; Paie, P.; Nava, G.; Bragheri, F.; Martinez Vazquez, R.; Minzioni, P.; Veglione, M.; di Tano, M.; Mondello, C.; Osellame, R.; *et al.* An integrated optofluidic device for single-cell sorting driven by mechanical properties. *Lab Chip* **2015**, *15*, 1262–1266.
34. Taff, B.M.; Voldman, J. A scalable addressable positive-dielectrophoretic cell-sorting array. *Anal. Chem.* **2005**, *77*, 7976–7983.
35. Hu, X.; Bessette, P.H.; Qian, J.; Meinhart, C.D.; Daugherty, P.S.; Soh, H.T. Marker-specific sorting of rare cells using dielectrophoresis. *Proc. Natl. Acad. Sci. USA* **2005**, *102*, 15757–15761.
36. Shah, G.J.; Ohta, A.T.; Chiou, E.P.; Wu, M.C.; Kim, C.J. EWOD-driven droplet microfluidic device integrated with optoelectronic tweezers as an automated platform for cellular isolation and analysis. *Lab Chip* **2009**, *9*, 1732–1739.
37. Park, S.; Wijethunga, P.A.; Moon, H.; Han, B. On-chip characterization of cryoprotective agent mixtures using an EWOD-based digital microfluidic device. *Lab Chip* **2011**, *11*, 2212–2221.

38. Rosenbluth, M.J.; Lam, W.A.; Fletcher, D.A. Analyzing cell mechanics in hematologic diseases with microfluidic biophysical flow cytometry. *Lab Chip* **2008**, *8*, 1062–1070.
39. Fan, S.K.; Huang, P.W.; Wang, T.T.; Peng, Y.H. Cross-scale electric manipulations of cells and droplets by frequency-modulated dielectrophoresis and electrowetting. *Lab Chip* **2008**, *8*, 1325–1331.
40. Di Carlo, D.; Jeong, K.H.; Lee, L.P. Reagentless mechanical cell lysis by nanoscale barbs in microchannels for sample preparation. *Lab Chip* **2003**, *3*, 287–291.
41. Zheng, S.; Lin, H.; Liu, J.Q.; Balic, M.; Datar, R.; Cote, R.J.; Tai, Y.C. Membrane microfilter device for selective capture, electrolysis and genomic analysis of human circulating tumor cells. *J. Chromatogr. A* **2007**, *1162*, 154–161.
42. Bao, N.; Kodippili, G.C.; Giger, K.M.; Fowler, V.M.; Low, P.S.; Lu, C. Single-cell electrical lysis of erythrocytes detects deficiencies in the cytoskeletal protein network. *Lab Chip* **2011**, *11*, 3053–3056.
43. Ginsberg, S.D. RNA amplification strategies for small sample populations. *Methods* **2005**, *37*, 229–237.
44. Feng, X.; Du, W.; Luo, Q.; Liu, B.F. Microfluidic chip: Next-generation platform for systems biology. *Anal. Chim. Acta* **2009**, *650*, 83–97.
45. Koster, S.; Angile, F.E.; Duan, H.; Agresti, J.J.; Wintner, A.; Schmitz, C.; Rowat, A.C.; Merten, C.A.; Pisignano, D.; Griffiths, A.D.; *et al.* Drop-based microfluidic devices for encapsulation of single cells. *Lab Chip* **2008**, *8*, 1110–1115.
46. Clausell-Tormos, J.; Lieber, D.; Baret, J.C.; El-Harrak, A.; Miller, O.J.; Frenz, L.; Blouwolff, J.; Humphry, K.J.; Koster, S.; Duan, H.; *et al.* Droplet-based microfluidic platforms for the encapsulation and screening of mammalian cells and multicellular organisms. *Chem. Biol.* **2008**, *15*, 427–437.
47. Brouzes, E.; Medkova, M.; Savenelli, N.; Marran, D.; Twardowski, M.; Hutchison, J.B.; Rothberg, J.M.; Link, D.R.; Perrimon, N.; Samuels, M.L. Droplet microfluidic technology for single-cell high-throughput screening. *Proc. Natl. Acad. Sci. USA* **2009**, *106*, 14195–14200.
48. Yu, L.; Chen, M.C.; Cheung, K.C. Droplet-based microfluidic system for multicellular tumor spheroid formation and anticancer drug testing. *Lab Chip* **2010**, *10*, 2424–2432.
49. Pekin, D.; Skhiri, Y.; Baret, J.C.; le Corre, D.; Mazutis, L.; Salem, C.B.; Millot, F.; El Harrak, A.; Hutchison, J.B.; Larson, J.W.; *et al.* Quantitative and sensitive detection of rare mutations using droplet-based microfluidics. *Lab Chip* **2011**, *11*, 2156–2166.
50. Gossett, D.R.; Tse, H.T.; Lee, S.A.; Ying, Y.; Lindgren, A.G.; Yang, O.O.; Rao, J.; Clark, A.T.; di Carlo, D. Hydrodynamic stretching of single cells for large population mechanical phenotyping. *Proc. Natl. Acad. Sci. USA* **2012**, *109*, 7630–7635.
51. Guan, G.; Chen, P.C.Y.; Peng, W.K.; Bhagat, A.A.; Ong, C.J.; Han, J. Real-time control of a microfluidic channel for size-independent deformability cytometry. *J. Micromech. Microeng.* **2012**, *22*, 105037.
52. Guo, Q.; Reiling, S.J.; Rohrbach, P.; Ma, H. Microfluidic biomechanical assay for red blood cells parasitized by *Plasmodium falciparum*. *Lab Chip* **2012**, *12*, 1143–1150.
53. Jang, L.S.; Wang, M.H. Microfluidic device for cell capture and impedance measurement. *Biomed. Microdevices* **2007**, *9*, 737–743.

54. Malleo, D.; Nevill, J.T.; Lee, L.P.; Morgan, H. Continuous differential impedance spectroscopy of single cells. *Microfluid. Nanofluid.* **2010**, *9*, 191–198.
55. Han, K.H.; Han, A.; Frazier, A.B. Microsystems for isolation and electrophysiological analysis of breast cancer cells from blood. *Biosens. Bioelectron.* **2006**, *21*, 1907–1914.
56. Chung, J.; Issadore, D.; Ullal, A.; Lee, K.; Weissleder, R.; Lee, H. Rare cell isolation and profiling on a hybrid magnetic/size-sorting chip. *Biomicrofluidics* **2013**, *7*, 54107.
57. Kurz, C.M.; Buth, H.; Sossalla, A.; Vermeersch, V.; Toncheva, V.; Dubruel, P.; Schacht, E.; Thielecke, H. Chip-based impedance measurement on single cells for monitoring sub-toxic effects on cell membranes. *Biosens. Bioelectron.* **2011**, *26*, 3405–3412.
58. Lindstrom, S.; Mori, K.; Ohashi, T.; Andersson-Svahn, H. A microwell array device with integrated microfluidic components for enhanced single-cell analysis. *Electrophoresis* **2009**, *30*, 4166–4171.
59. Lindstrom, S.; Larsson, R.; Svahn, H.A. Towards high-throughput single cell/clone cultivation and analysis. *Electrophoresis* **2008**, *29*, 1219–1227.
60. Kobel, S.; Valero, A.; Latt, J.; Renaud, P.; Lutolf, M. Optimization of microfluidic single cell trapping for long-term on-chip culture. *Lab Chip* **2010**, *10*, 857–863.
61. Bow, H.; Pivkin, I.V.; Diez-Silva, M.; Goldfless, S.J.; Dao, M.; Niles, J.C.; Suresh, S.; Han, J. A microfabricated deformability-based flow cytometer with application to malaria. *Lab Chip* **2011**, *11*, 1065–1073.
62. Di Carlo, D.; Lee, L.P. Dynamic single-cell analysis for quantitative biology. *Anal. Chem.* **2006**, *78*, 7918–7925.
63. Lan, K.C.; Jang, L.S. Integration of single-cell trapping and impedance measurement utilizing microwell electrodes. *Biosens. Bioelectron.* **2011**, *26*, 2025–2031.
64. Huang, N.T.; Zhang, H.L.; Chung, M.T.; Seo, J.H.; Kurabayashi, K. Recent advancements in optofluidics-based single-cell analysis: Optical on-chip cellular manipulation, treatment, and property detection. *Lab Chip* **2014**, *14*, 1230–1245.
65. Rahman, A.R.A.; Price, D.T.; Bhansali, S. Effect of electrode geometry on the impedance evaluation of tissue and cell culture. *Sens. Actuators B Chem.* **2007**, *127*, 89–96.
66. Fan, S.K.; Chiu, C.P.; Hsu, C.H.; Chen, S.C.; Huang, L.L.; Lin, Y.H.; Fang, W.F.; Chen, J.K.; Yang, J.T. Particle chain display—An optofluidic electronic paper. *Lab Chip* **2012**, *12*, 4870–4876.
67. Fan, S.K.; Chiu, C.P.; Huang, P.W. Transmittance tuning by particle chain polarization in electrowetting-driven droplets. *Biomicrofluidics* **2010**, *4*, 43011.
68. Choi, H.; Kim, K.B.; Jeon, C.S.; Hwang, I.; Lee, S.; Kim, H.K.; Kim, H.C.; Chung, T.D. A label-free DC impedance-based microcytometer for circulating rare cancer cell counting. *Lab Chip* **2013**, *13*, 970–977.
69. Mernier, G.; Duqi, E.; Renaud, P. Characterization of a novel impedance cytometer design and its integration with lateral focusing by dielectrophoresis. *Lab Chip* **2012**, *12*, 4344–4349.
70. Fan, S.K.; Yang, H.; Hsu, W. Droplet-on-a-wristband: Chip-to-chip digital microfluidic interfaces between replaceable and flexible electrowetting modules. *Lab Chip* **2011**, *11*, 343–347.
71. Fan, S.K.; Hsieh, T.H.; Lin, D.Y. General digital microfluidic platform manipulating dielectric and conductive droplets by dielectrophoresis and electrowetting. *Lab Chip* **2009**, *9*, 1236–1242.

72. Fan, S.K.; Chen, W.J.; Lin, T.H.; Wang, T.T.; Lin, Y.C. Reconfigurable liquid pumping in electric-field-defined virtual microchannels by dielectrophoresis. *Lab Chip* **2009**, *9*, 1590–1595.
73. Fan, S.K.; Yang, H.; Wang, T.T.; Hsu, W. Asymmetric electrowetting—Moving droplets by a square wave. *Lab Chip* **2007**, *7*, 1330–1335.
74. Choi, K.; Ng, A.H.; Fobel, R.; Wheeler, A.R. Digital microfluidics. *Annu. Rev. Anal. Chem.* **2012**, *5*, 413–440.
75. Kumar, P.T.; Toffalini, F.; Witters, D.; Vermeir, S.; Rolland, F.; Hertog, M.L.A.T.M.; Nicolai, B.M.; Puers, R.; Geeraerd, A.; Lammertyn, J. Digital microfluidic chip technology for water permeability measurements on single isolated plant protoplasts. *Sens. Actuators B Chem.* **2014**, *199*, 479–487.
76. Shih, S.C.; Barbulovic-Nad, I.; Yang, X.; Fobel, R.; Wheeler, A.R. Digital microfluidics with impedance sensing for integrated cell culture and analysis. *Biosens. Bioelectron.* **2013**, *42*, 314–320.
77. Witters, D.; Vergauwe, N.; Vermeir, S.; Ceyssens, F.; Liekens, S.; Puers, R.; Lammertyn, J. Biofunctionalization of electrowetting-on-dielectric digital microfluidic chips for miniaturized cell-based applications. *Lab Chip* **2011**, *11*, 2790–2794.
78. Vergauwe, N.; Witters, D.; Ceyssens, F.; Vermeir, S.; Verbruggen, B.; Puers, R.; Lammertyn, J. A versatile electrowetting-based digital microfluidic platform for quantitative homogeneous and heterogeneous bio-assays. *J. Micromech. Microeng.* **2011**, *21*, 054026.
79. Vergauwe, N.; Witters, D.; Atalay, Y.T.; Verbruggen, B.; Vermeir, S.; Ceyssens, F.; Puers, R.; Lammertyn, J. Controlling droplet size variability of a digital lab-on-a-chip for improved bio-assay performance. *Microfluid. Nanofluid.* **2011**, *11*, 25–34.
80. Barbulovic-Nad, I.; Yang, H.; Park, P.S.; Wheeler, A.R. Digital microfluidics for cell-based assays. *Lab Chip* **2008**, *8*, 519–526.
81. Shih, S.C.; Gach, P.C.; Sustarich, J.; Simmons, B.A.; Adams, P.D.; Singh, S.; Singh, A.K. A droplet-to-digital (D2D) microfluidic device for single cell assays. *Lab Chip* **2015**, *15*, 225–236.
82. Thompson, A.M.; Gansen, A.; Paguirigan, A.L.; Kreutz, J.E.; Radich, J.P.; Chiu, D.T. Self-digitization microfluidic chip for absolute quantification of mRNA in single cells. *Anal. Chem.* **2014**, *86*, 12308–12314.
83. Thompson, A.M.; Paguirigan, A.L.; Kreutz, J.E.; Radich, J.P.; Chiu, D.T. Microfluidics for single-cell genetic analysis. *Lab Chip* **2014**, *14*, 3135–3142.
84. Lee, T.M.; Hsing, I.M. DNA-based bioanalytical microsystems for handheld device applications. *Anal. Chim. Acta* **2006**, *556*, 26–37.
85. Sanders, R.; Huggett, J.F.; Bushell, C.A.; Cowen, S.; Scott, D.J.; Foy, C.A. Evaluation of digital PCR for absolute DNA quantification. *Anal. Chem.* **2011**, *83*, 6474–6484.
86. White, R.A., 3rd; Quake, S.R.; Curr, K. Digital PCR provides absolute quantitation of viral load for an occult RNA virus. *J. Virol. Methods* **2012**, *179*, 45–50.
87. Bennett, M.R.; Hasty, J. Microfluidic devices for measuring gene network dynamics in single cells. *Nat. Rev. Genet.* **2009**, *10*, 628–638.

88. Toriello, N.M.; Douglas, E.S.; Thaitrong, N.; Hsiao, S.C.; Francis, M.B.; Bertozzi, C.R.; Mathies, R.A. Integrated microfluidic bioprocessor for single-cell gene expression analysis. *Proc. Natl. Acad. Sci. USA* **2008**, *105*, 20173–20178.
89. Bontoux, N.; Dauphinot, L.; Vitalis, T.; Studer, V.; Chen, Y.; Rossier, J.; Potier, M.C. Integrating whole transcriptome assays on a lab-on-a-chip for single cell gene profiling. *Lab Chip* **2008**, *8*, 443–450.
90. Marcus, J.S.; Anderson, W.F.; Quake, S.R. Microfluidic single-cell mRNA isolation and analysis. *Anal. Chem.* **2006**, *78*, 3084–3089.
91. Schell, W.A.; Benton, J.L.; Smith, P.B.; Poore, M.; Rouse, J.L.; Boles, D.J.; Johnson, M.D.; Alexander, B.D.; Pamula, V.K.; Eckhardt, A.E.; *et al.* Evaluation of a digital microfluidic real-time PCR platform to detect DNA of *Candida albicans* in blood. *Eur. J. Clin. Microbiol. Infect. Dis.* **2012**, *31*, 2237–2245.
92. Hua, Z.; Rouse, J.L.; Eckhardt, A.E.; Srinivasan, V.; Pamula, V.K.; Schell, W.A.; Benton, J.L.; Mitchell, T.G.; Pollack, M.G. Multiplexed real-time polymerase chain reaction on a digital microfluidic platform. *Anal. Chem.* **2010**, *82*, 2310–2316.
93. Xue, Z.; Huang, K.; Cai, C.; Cai, L.; Jiang, C.Y.; Feng, Y.; Liu, Z.; Zeng, Q.; Cheng, L.; Sun, Y.E.; *et al.* Genetic programs in human and mouse early embryos revealed by single-cell RNA sequencing. *Nature* **2013**, *500*, 593–597.
94. Niakan, K.K.; Han, J.; Pedersen, R.A.; Simon, C.; Pera, R.A. Human pre-implantation embryo development. *Development* **2012**, *139*, 829–841.
95. Zeng, F.; Baldwin, D.A.; Schultz, R.M. Transcript profiling during preimplantation mouse development. *Dev. Biol.* **2004**, *272*, 483–496.
96. Swain, J.E.; Smith, G.D. Advances in embryo culture platforms: Novel approaches to improve preimplantation embryo development through modifications of the microenvironment. *Hum. Reprod. Update* **2011**, *17*, 541–557.
97. Smith, G.D.; Swain, J.E.; Bormann, C.L. Microfluidics for gametes, embryos, and embryonic stem cells. *Semin. Reprod. Med.* **2011**, *29*, 5–14.
98. Huang, H.Y.; Shen, H.H.; Tien, C.H.; Li, C.J.; Fan, S.K.; Liu, C.H.; Hsu, W.S.; Yao, D.J. Digital microfluidic dynamic culture of mammalian embryos on an electrowetting on dielectric (EWOD) chip. *PLoS ONE* **2015**, *10*, e0124196.
99. Egozcue, J.; Santalo, J.; Gimenez, C.; Durban, M.; Benet, J.; Navarro, J.; Vidal, F. Preimplantation genetic screening and human implantation. *J. Reprod. Immunol.* **2002**, *55*, 65–72.
100. Yan, L.; Wei, Y.; Huang, J.; Zhu, X.; Shi, X.; Xia, X.; Yan, J.; Lu, C.; Lian, Y.; Li, R.; *et al.* Advances in preimplantation genetic diagnosis/screening. *Sci. China Life Sci.* **2014**, *57*, 665–671.
101. Harper, J.C.; Harton, G. The use of arrays in preimplantation genetic diagnosis and screening. *Fertil. Steril.* **2010**, *94*, 1173–1177.

11

Microfluidic Impedance Flow Cytometry Enabling High-Throughput Single-Cell Electrical Property Characterization

Jian Chen [1], Chengcheng Xue [1], Yang Zhao [1], Deyong Chen [1], Min-Hsien Wu [2,*] and Junbo Wang [1,*]

[1] State Key Laboratory of Transducer Technology, Institute of Electronics, Chinese Academy of Sciences, Beijing 100190, China; E-Mails: chenjian@mail.ie.ac.cn (J.C.); xuechengcheng13@mails.ucas.ac.cn (C.X.); zhaoyang110@mails.ucas.ac.cn (Y.Z.); dychen@mail.ie.ac.cn (D.C.)

[2] Graduate Institute of Biochemical and Biomedical Engineering, Chang Gung University, Taoyuan 333, Taiwan

* Authors to whom correspondence should be addressed; E-Mails: mhwu@mail.cgu.edu.tw (M.-H.W.); jbwang@mail.ie.ac.cn (J.W.)

Academic Editor: Fan-Gang Tseng

Abstract: This article reviews recent developments in microfluidic impedance flow cytometry for high-throughput electrical property characterization of single cells. Four major perspectives of microfluidic impedance flow cytometry for single-cell characterization are included in this review: (1) early developments of microfluidic impedance flow cytometry for single-cell electrical property characterization; (2) microfluidic impedance flow cytometry with enhanced sensitivity; (3) microfluidic impedance and optical flow cytometry for single-cell analysis and (4) integrated point of care system based on microfluidic impedance flow cytometry. We examine the advantages and limitations of each technique and discuss future research opportunities from the perspectives of both technical innovation and clinical applications.

Keywords: microfluidics; single-cell electrical property analysis; impedance flow cytometry; high throughput

---

1. Introduction

Single-cell electrical properties (e.g., membrane capacitance or cytoplasm resistance) can be utilized as cellular biophysical markers to evaluate cellular status in a label-free manner [1,2]. They have been demonstrated to classify various types of tumor cells [3–5], stem cells [6] and blood cells [7–11].

Conventionally there are mainly three techniques capable of characterizing single-cell electrical properties: dielectrophoresis, patch clamping and electrorotation [12]. Dielectrophoresis is demonstrated to quantify cellular electrical properties by curve fitting of the Clausius–Mossotti factor spectra or cell count spectra. However, since the spectra are not from the measurements of the same cells, only average electrical properties of a cell population can be obtained [3,4,12–17]. Patch-clamp devices characterize the activities of cellular ion channels by sucking a portion of cell membrane into a micropipette tip to form a high electrical resistance seal, enabling the quantification of specific membrane capacitance of single cells (an intrinsic size-independent electrical parameter of cells) [18–24]. In electro-rotation, a rotating electric field is exerted to rotate a suspended single cell as a result of Maxwell-Wagner polarization. By measuring the rotating rate as a function of the applied frequency, this method is capable of collecting membrane permittivity and cytoplasm conductivity of single cells [25–32]. However, patch clamping and electrorotation rely on the precise manipulation and positioning of pipettes (patch clamping) or cells (electrorotation) which is time-consuming and labor-intensive [12,33–36]. This could greatly affect the measurement efficiency and therefore hamper the wide application of using these techniques to acquire statistically-meaningful data.

Microfluidics is the science and technology on the processing and manipulation of small amounts of fluids (10^{-9} to 10^{-18} liters) in channels with dimensions of tens of micrometers [37–39]. The micrometer dimension well matches with the size of typical biological calls, making microfluidics an ideal platform for cell studies [40–44]. Based on the advantageous features of microfluidic technologies, microfluidics has been used for characterizing the biochemical (e.g., gene and protein) and/or biophysical properties (mechanical and electrical) of cells at the single-cell level [45–51].

Microfluidics-based devices for the characterization of single-cell electrical properties have been proposed, in which two major approaches, the micro electrical impedance spectroscopy (μ-EIS) and microfluidic impedance flow cytometry [12,35,36], are commonly used. μ-EIS is a non-invasive approach to characterize immobilized single cells between two electrodes relying on hydrodynamic fluid trapping [52–58], vacuum aspiration [59–65], dielectrophoretic forces [66–69] or surface modifications [70–72]. Although this technique can conduct spectroscopy sweeping on the trapped single cells, it normally suffers from limited throughput and thus might not be suitable for collecting data from large amounts of cells [12,33–36].

Meanwhile, microfluidic impedance flow cytometry has also been demonstrated where single cells are pushed to continuously flow through two microelectrodes in which the impedance data of cells at multiple frequencies are measured [35,36]. Compared to the conventional coulter counters which rely on DC or low-frequency signal for cell size characterization [73–76], the multiple-frequency-based impedance data obtained from the microfluidic impedance flow cytometry enable the characterization of cellular sizes, membrane capacitance and cytoplasm resistance in a high-throughput manner [35,36].

In this review, we focus on the recent advances of the four perspectives of microfluidics-based flow cytometers for single-cell electrical property characterization: (1) early developments of microfluidic impedance flow cytometry for single-cell electrical property characterization; (2) microfluidic impedance flow cytometry with enhanced sensitivity; (3) microfluidic impedance and optical flow cytometry for single-cell analysis and (4) integrated point of care system based on microfluidic impedance flow cytometry (see Table 1).

Table 1. Key developments in the field of microfluidic impedance flow cytometry enabling high-throughput cellular electrical property characterization.

Techniques	Quantified Parameters	Classified Objects and Key Observations	References
Coplanar microelectrodes	Two-frequency impedance data (1.7 and 15.0 MHz)	Polymer beads of 5 and 8 μm, normal erythrocytes and their ghost counterparts	[77]
Coplanar microelectrodes	One-frequency impedance data (100 kHz)	Liver tumor cells at normal, apoptotic and necrotic status, leukemia cells	[78]
Coplanar microelectrodes	One-frequency impedance data (2.0 MHz)	Different stages of *P. falciparum* infected red blood cells and uninfected red blood cells	[9]
Parallel microelectrodes	Two-frequency impedance opacity $\|Z_{high}\|/\|Z_{ref}\|$ (f_{ref} = 602 kHz, f_{high} = 350 kHz–20.0 MHz)	Polymer beads of 5, 6 μm, red blood cells and their fixed counterparts	[79]
Parallel microelectrodes	Two-frequency impedance opacity $\|Z_{high}\|/\|Z_{ref}\|$ (f_{ref} = 500 kHz, f_{high} = 0.5–250.0 MHz)	Wild-type yeasts and a mutant with different sizes and distribution of vacuoles in the intracellular fluid	[80]
Parallel microelectrodes + insulating fluid focusing	One-frequency impedance data (503 kHz)	Polymer beads of 1, 2 μm, and *E coli*	[81]
Parallel microelectrodes + resonance	Two-frequency impedance data (87.2 and 89.2 MHz)	*E. coli*, *B. subtilis* and polymer beads of 2 μm	[82]
Constriction channel	One-frequency impedance data (100 kHz)	Size-comparable tumor cells and their more malignant counterparts	[83]
Constriction channel	One-frequency impedance data (100 kHz)	Adult red blood cells and neonatal red blood cells	[84]
Constriction channel	Four-frequency impedance data (50 kHz, 250 kHz, 500 kHz and 1.0 MHz)	Polymer beads of 20 μm, undifferentiated stem cells and differentiated stem cells	[6]
Constriction channel + equivalent circuit model	Specific membrane capacitance and cytoplasm conductivity	Characterization of size-independent intrinsic cellular electrical properties from hundreds of single cells	[85]
Constriction channel + equivalent circuit model	Specific membrane capacitance and cytoplasm conductivity	Paired high- and low-metastatic cancer cells, and tumor cells with single oncogenes under regulation	[5]

Table 1. *Cont.*

Techniques	Quantified Parameters	Classified Objects and Key Observations	References
Parallel microelectrodes + optical lens	Two-frequency impedance data (503 kHz and 1.7 MHz) and fluorescent signals	lymphocytes, monocytes and neutrophils	[10]
Parallel microelectrodes + optical lens	Two-frequency impedance data (503 kHz and 10.0 MHz) and fluorescent signals	Lymphocytes, lymphocytes + CD4 beads, granulocytes, monocytes and monocytes + CD4	[11]
Parallel microelectrodes + on-chip optical fibers	One-frequency impedance data (1.0 MHz), fluorescent signals, and side scattered light	Microbeads (10 and 15 μm diameter fluorescent, 20 and 25 μm diameter plain)	[86]
Parallel microelectrodes + on-chip waveguides	Two-frequency impedance data (500 kHz and 2.0 MHz), fluorescent signals, and side scattered light	Lymphocytes, granulocytes, monocytes, neutrophils and CD4 labelled white blood cells	[87]
Parallel microelectrodes + sample pretreatment module	Two-frequency impedance data (500 kHz and 1.7 MHz)	Lymphocytes, monocytes, neutrophils, red blood cells and platelets	[88]
Parallel microelectrodes + sample pretreatment module	Two-frequency impedance data (303 kHz and 1.7 MHz)	$CD4^+$ and $CD8^+$ lymphocytes	[7]

2. Early Development of Microfluidic Flow Cytometry for Single-Cell Electrical Property Characterization

Renaud *et al.* are the pioneers in the field of microfluidic impedance flow cytometry [77,79,89–93]. In 2001, Renaud *et al.* proposed the first microfluidics-based impedance flow cytometry for high-throughput single-cell electrical property characterization [77]. As shown in Figure 1a, a microfluidic chip with channels integrated with a differential pair of coplanar microelectrodes was used to characterize electrical properties of single cells. The cells were flushed through the measurement area in a high-throughput manner with the impedance data measured at two given frequencies. In this study, an equivalent circuit model for microfluidic impedance flow cytometry was developed where C_m, R_c, R_{sol} and C_{dl} represent cell membrane capacitance, cytoplasm resistance, buffer solution resistance and electrical double layer capacitance, respectively (see Figure 1a).

In addition, complex impedance spectrum of a cell as simulated using an equivalent circuit model was shown in Figure 1b. Based on simulation results, the authors suggested that the impedance data for frequencies lower than 100 kHz, between 100 kHz–1 MHz, 2–5 MHz and 10–100 MHz reflect the electrical double layer, cellular size, membrane capacitance and cytoplasm resistance, respectively. Note that this impedance spectrum has served as the guiding rule of frequency choice in the subsequent development of microfluidic impedance flow cytometry.

To demonstrate its applications, the microfluidic device was used to differentiate latex beads of 5 and 8 μm at 1.72 MHz. The result confirmed that impedance data at ~1 MHz does reflect particle sizes (see Figure 1c). Furthermore, normal erythrocytes and erythrocyte ghost cells (namely the erythrocytes with cytoplasm replaced with phosphate buffer solution) were characterized and differentiated. The impedance data for these two types of cells were found similar at 1.72 MHz indicating comparable cell sizes whereas, significantly different at 15 MHz suggesting differences in cytoplasm conductivity (see Figure 1d).

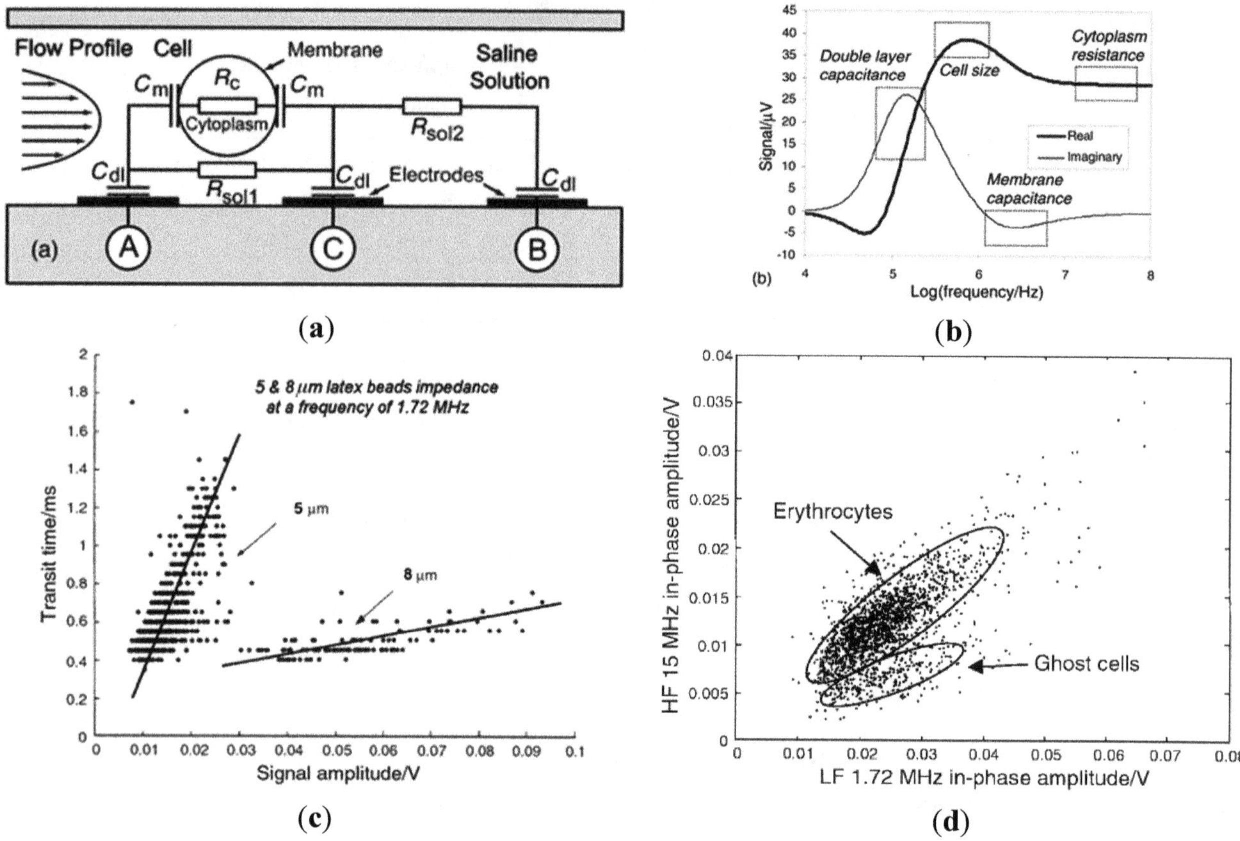

Figure 1. (**a**) The first-generation microfluidic impedance flow cytometry where a microfluidic chip with integrated channels and a differential pair of coplanar microelectrodes were proposed to quantify two-frequency impedance data of single cells flushed through the measurement area in a high-throughput manner; (**b**) The complex impedance spectrum of a cell is simulated using an equivalent circuit model where impedance data at various frequency domains indicate the electrical double layer, cellular size, membrane capacitance and cytoplasm resistance, respectively; (**c**) Impedance amplitude difference of 5 and 8 μm latex beads, confirming that impedance data at ~1 MHz can reflect particle sizes. Note that "transit time" indicates the traveling velocity of latex beads which were also obtained from impedance data; (**d**) Normal erythrocytes and erythrocyte ghost cells were characterized, with comparable low-frequency impedance data indicating size comparability and significant differences at high-frequency impedance data suggesting cytoplasm conductivity differences [77].

In 2005, Renaud *et al.* proposed the second-generation microfluidic impedance flow cytometry [79] where the parallel overlap microelectrodes were used to replace the previously reported coplanar microelectrode, enabling the production of more homogeneous current density around the single cells under measurement (see Figure 2a). Furthermore, systematic experiments were conducted to classify polystyrene beads (5 and 6 μm), normal red blood cells and fixed red blood cells based on the impedance data at the frequency of 602 kHz and 10 MHz (see Figure 2b).

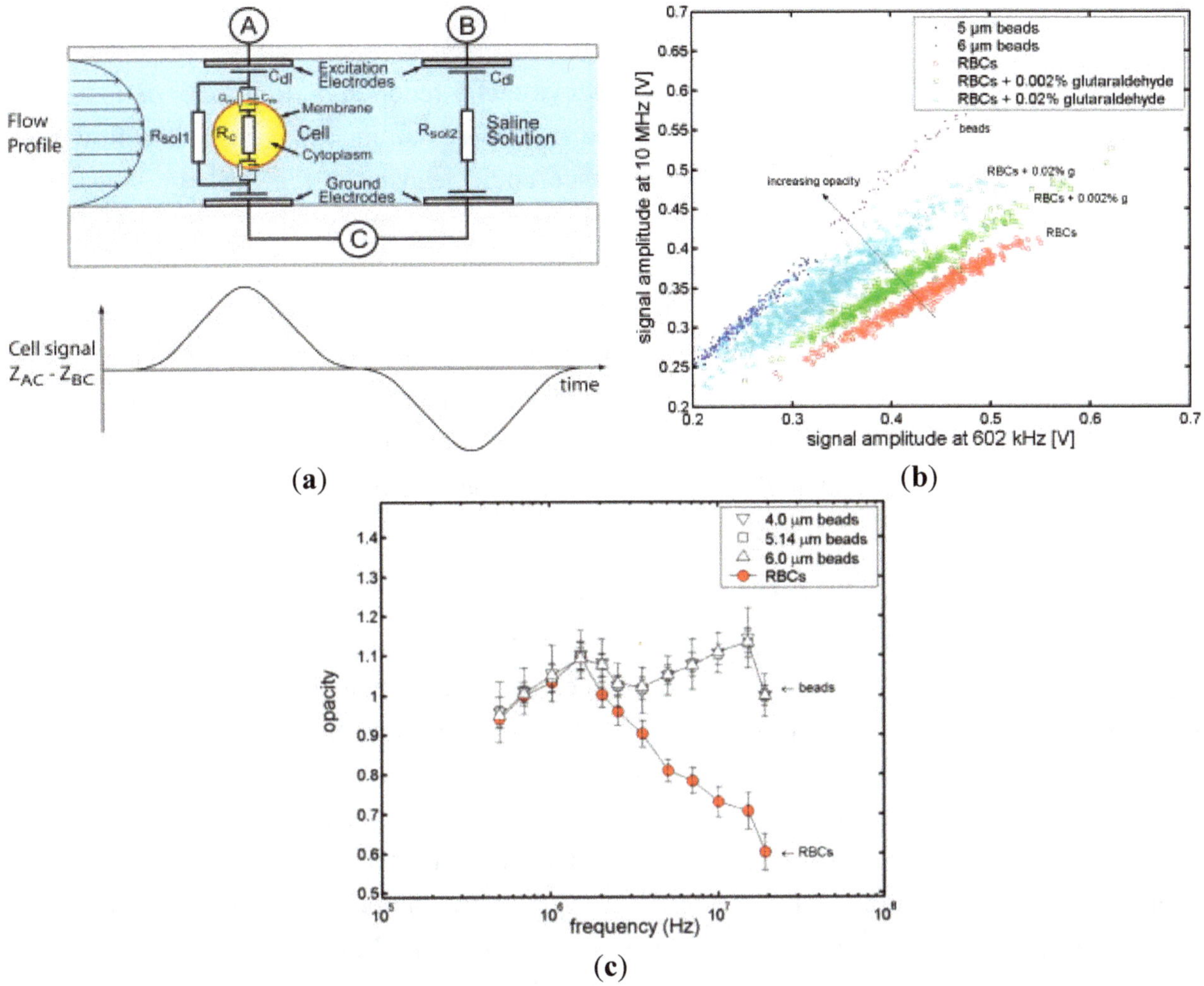

Figure 2. (**a**) The second-generation microfluidic impedance flow cytometry where the parallel overlap micro electrodes were used to replace the previously reported coplanar micro electrodes; (**b**) Two-frequency impedance data of polystyrene beads, normal red blood cells and fixed red blood cells, which can be classified to an extent based on opacity defined as $|Z_{high}|/|Z_{ref}|$; (**c**) Opacity spectrum of red blood cells and polystyrene beads where no significant difference was noticed among the opacity spectra for polystyrene beads of different diameters, confirming that opacity can be used to normalize the particle size. In addition, a decrease in opacity at the high frequency domain of red blood cells compared to polystyrene beads was observed, confirming that the cytoplasm of red blood cells is more conductive than polystyrene beads [79].

In this study, opacity was defined as $|Z_{high}|/|Z_{ref}|$ to partially remove the dependence of the impedance data on particle sizes. As shown Figure 2c, no significant difference was noticed among the opacity spectrum (f_{ref} = 602 kHz)) for polystyrene beads of 4.0, 5.1, and 6.0 μm diameters, confirming that, to an extent, opacity is insensitive to particle sizes. In addition, a decrease in opacity at the high frequency domain of red blood cells compared to polystyrene beads was observed, confirming that the cytoplasm of red blood cells is more conductive than polystyrene beads. As a valuable impedance parameter, opacity has been commonly used in the subsequent development of microfluidic impedance flow cytometry to evaluate electrical properties of single cells.

3. Microfluidic Impedance Flow Cytometry with Enhanced Sensitivity

The drawback of the microfluidic impedance flow cytometry reported by Renaud *et al.* is the lack of close contact between cells and electrodes when the cells were continuously flushed to flow through the detection area between two electrodes. This issue could lead to current leakage where electric signals circumvent the cells under measurement by travelling through solutions surrounding the cells. In addition, the relative positions of travelling single cells between two facing electrodes (*i.e.*, in the middle of two facing electrodes *vs.* alongside the boundary of one detecting electrode) can also lead to issues of low detection stability and repeatability. In order to address these issues, the detection area of the microfluidic impedance flow cytometry needs to be further reduced. Two approaches have been developed to this end: sandwiching cells between two insulating fluid layers (e.g., insulating fluid flow [81,94]) or confining cells within solid constriction channels (cross sectional area smaller than biological cells) [63,83,84].

As the first demonstration, Morgan *et al.* developed a microfluidic impedance flow cytometer which utilized an insulating fluid to hydrodynamically focus a sample stream of cells suspended in electrolyte through the sensing area between two microelectrodes [81] (see Figure 3a). The focusing technique enhanced the measurement sensitivity without reducing the dimensions of the microfluidic channels so that channel blockage can be avoided. This microfluidic platform was used to successfully classify polystyrene beads with diameters of 1 and 2 μm based on impedance amplitudes at the frequency of 503 kHz (see Figure 3b). As to the classification of 2 μm diameter polystyrene beads and *E. coli* (~2 μm in length and 0.5 μm in width), a significant overlap in the impedance amplitude histogram was observed, which may result from the comparable sizes between 2 μm beads and *E. coli* (see Figure 3b). In this study, only one frequency at 503 kHz was used, which was previously demonstrated as the frequency enabling particle size quantification [77,79]. More frequencies higher than 503 kHz may be further used to characterize the electrical properties of *E. coli*.

Although this technique can, to an extent, address the current leakage problem by sandwiching the detection solution between two insulating fluid flows, this type of sandwiching is only one dimensional and produces a vertical conductive sheet of cells (see Figure 3a, top view and side view), which still suffers from the current leakage. In addition, the impedance data of cells can also be affected by the z-direction position of single cells during the measurement [95], which leads to additional concerns on the measurement accuracy in this microfluidic system.

To tackle the technical hurdle, a constriction channel design was put forward to further decrease the current leakage and enable single-cell electrical property characterization [63,65,83,84]. The constriction channel design with a cross-section area smaller than that of biological cells was initially used for the characterization of cellular mechanical properties (e.g., red blood cells [84,96–98], white blood cells [99] and tumor cells [83,100–102]), by which single cells were aspirated into the constriction channel with their entry times adopted as a biophysical marker.

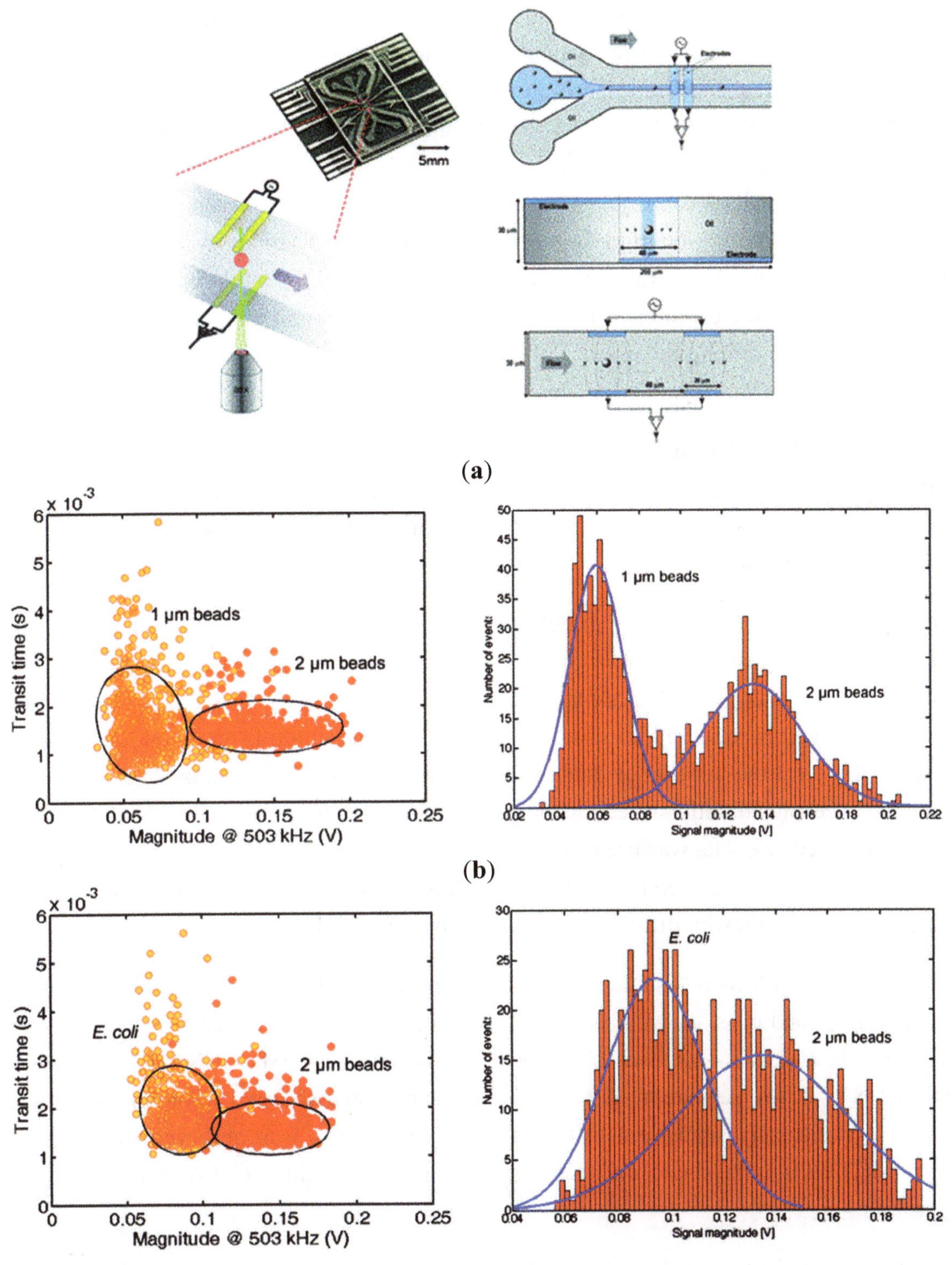

Figure 3. (**a**) A microfluidic impedance flow cytometer uses an insulating fluid to hydrodynamically focus a sample stream of cells suspended in electrolyte through the sensing area of two microelectrodes; (**b**) Successful classification of 1 and 2 µm diameter polystyrene beads based on impedance amplitudes at 503 kHz; (**c**) As to the classification of 2 µm diameter polystyrene beads and *E coli* (~2 µm in length and 0.5 µm in width), a significant overlap in the impedance amplitude histogram at 503 kHz was observed [81].

In 2011, Sun *et al.* proposed the first microfluidic impedance flow cytometry based on the constriction channel design, where single cells were continuously aspirated through the constriction channel while cell elongations and single-frequency impedance profiles were measured simultaneously (see Figure 4a) [83]. When a cell is aspirated through the constriction channel, it blocks electric fields and leads to a higher impedance amplitude value, which is used as an indicator of cellular electrical properties (see Figure 4b). This technique was used to classify two types of bone cells (osteoblasts *vs.* osteocytes) using a constriction channel of 6 μm × 6 μm in dimensions (at 100 kHz). To quantify the overall impedance of the cell, an impedance amplitude ratio was adopted which is defined as the ratio between the highest impedance amplitude value captured while cells are squeezed through the constriction channel and the background impedance amplitude value without cells. Compared with osteocytes, osteoblasts were found to have a larger cell elongation length and a higher impedance amplitude ratio (see Figure 4c).

The constriction channel design (8 μm × 8 μm at 100 kHz) was then used to characterize tumor cells EMT6 and their more malignant counterparts EMT6/AR 1.0, revealing a linear trend between the cell elongation length and the impedance amplitude ratio with different slopes and different *y*-axis intersections (see Figure 4d).

Furthermore, based on equivalent circuit models and two-frequency measurements, these raw impedance data were translated to intrinsic cellular electrical parameters including specific membrane capacitance ($C_{specific\ membrane}$) and cytoplasm conductivity ($\sigma_{cytoplasm}$) [5,85,103,104]. As shown in Figure 5a, when a cell is squeezed into the constriction channel, there is an increase in amplitude and a decrease in phase for the impedance data at the frequency of 1 and 100 kHz. The 1 kHz impedance data was used to evaluate cellular sealing properties with constriction channel walls to obtain R_{leak} while 100 kHz impedance data was used to quantify $C_{membrane}$ and $R_{cytoplasm}$, which were then translated to $C_{specific\ membrane}$ and $\sigma_{cytoplasm}$ (see Figure 5a) [85]. Based on the above translations, the $C_{specific\ membrane}$ and $\sigma_{cytoplasm}$ of tumor cells with different types were quantified [5].

For paired high- and low-metastatic carcinoma strains 95D and 95C cells, significant differences in both $C_{specific\ membrane}$ and $\sigma_{cytoplasm}$ were observed (see Figure 5b). In addition, a statistically significant difference only in $C_{specific\ membrane}$ was observed for 95D cells and 95D CCNY-KD cells with single oncogene CCNY down regulation (CCNY is a membrane-associated protein) (see Figure 5c). Furthermore, a statistically significant difference only in $\sigma_{cytoplasm}$ was observed for A549 cells and A549 CypA-KD cells with single oncogene CypA down regulation (CypA is a cytosolic protein) (see Figure 5d).

Although the combination of impedance flow cytometry with the constriction channel design can adequately tackle the current leakage issue, the use of constriction channel could reduce the detection throughput and may lead to channel blockage. Thus, the detection throughput in the constriction channel-based microfluidic impedance flow cytometry is normally lower than its conventional counterparts [77,79,81].

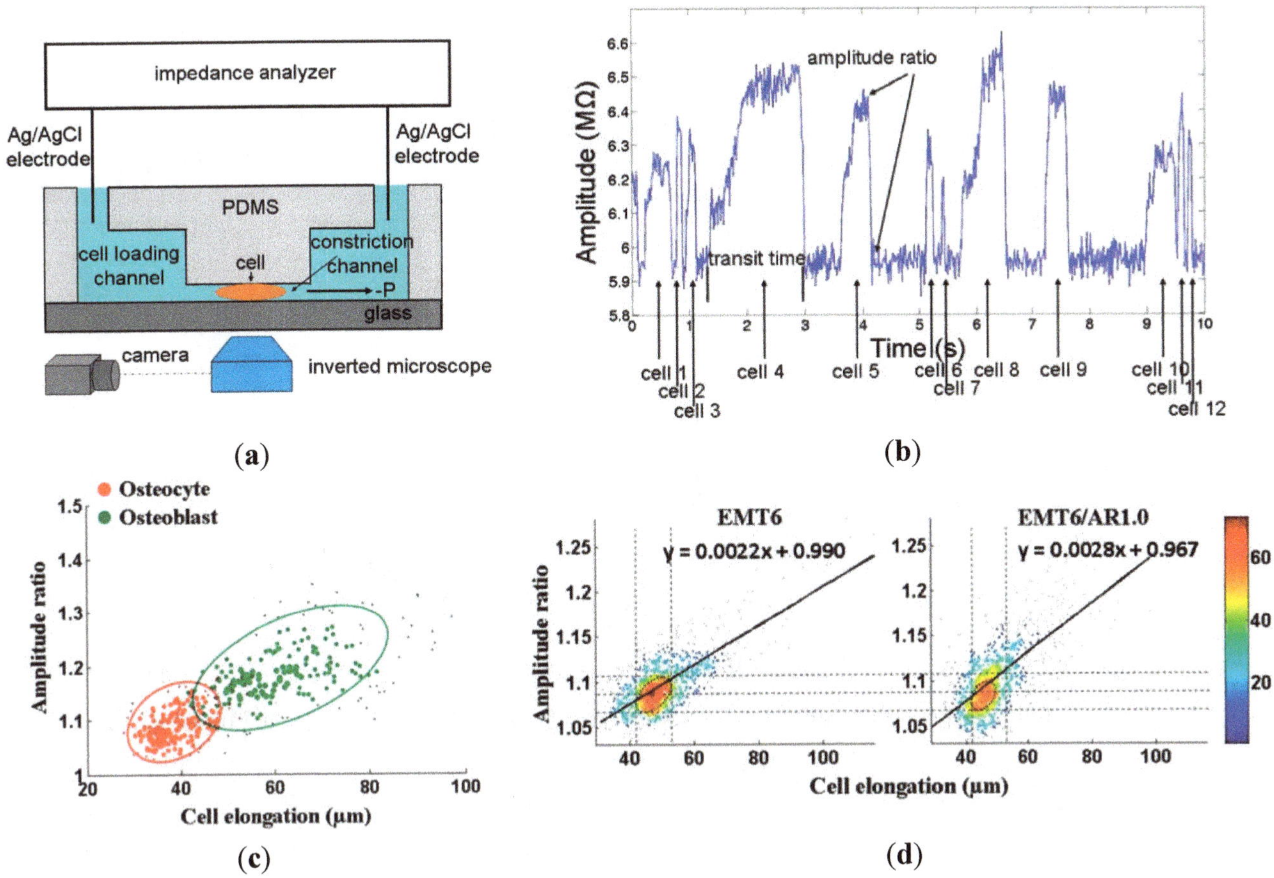

Figure 4. (**a**) The constriction channel based microfluidic impedance flow cytometry where single cells were aspirated through a constriction continuously while cell elongations and single frequency impedance profiles are measured simultaneously; (**b**) Raw impedance data of single cells, recording higher impedance amplitudes during cellular squeezing through the constriction channel; (**c**) The scatter plot of impedance amplitude ratio *vs.* cell elongation length for osteocytes and osteoblasts. Compared with osteocytes, osteoblasts have a larger cell elongation length and a higher impedance amplitude ratio; (**d**) The scatter plot of impedance amplitude ratio *vs.* cell elongation length for tumor cell EMT6 and their more malignant counterparts EMT6/AR 1.0, revealing a linear trend between cell elongation length and impedance amplitude ratio with different slopes and different *y*-axis intersections Reproduced by permission of the Royal Scoeity of Chemistry [83].

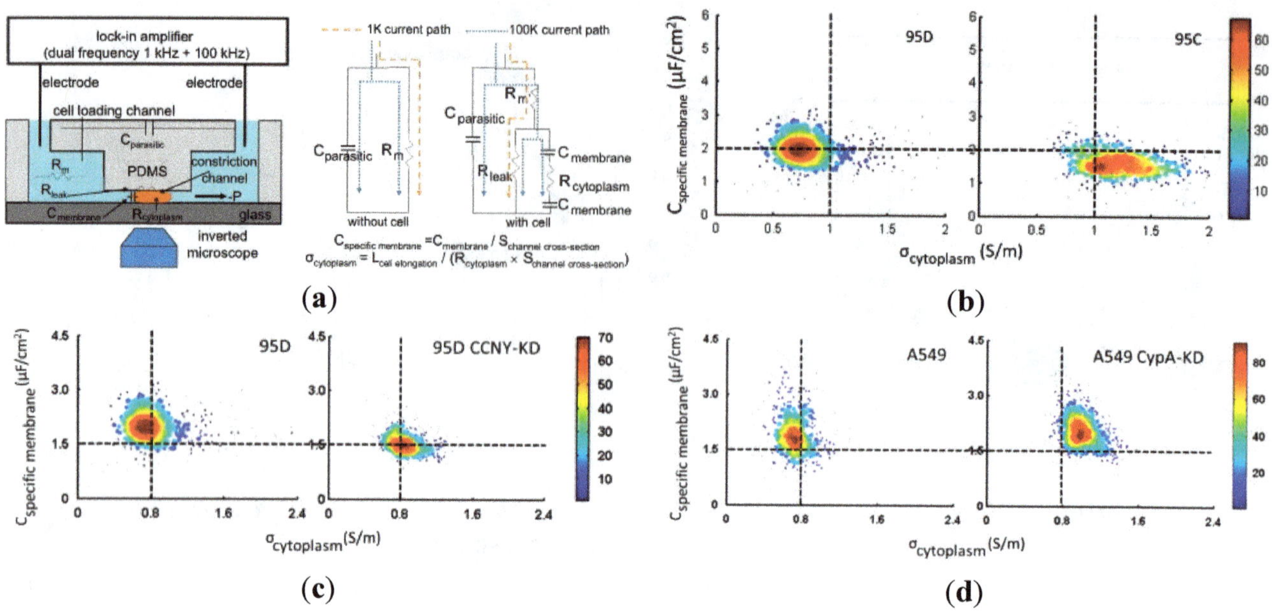

Figure 5. (**a**) The microfluidic impedance flow cytometry for continuous characterization of specific membrane capacitance ($C_{specific\ membrane}$) and cytoplasm conductivity ($\sigma_{cytoplasm}$) of single cells. Cells are aspirated continuously through the constriction channel with impedance data at 1 and 100 kHz measured simultaneously where 1 kHz impedance data were used to evaluate cellular sealing properties with constriction channel walls while 100 kHz impedance data were used to quantify $C_{specific\ membrane}$ and $\sigma_{cytoplasm}$ [85]; (**b**) For paired high- and low-metastatic carcinoma strains 95D and 95C cells, significant differences in both $C_{specific\ membrane}$ and $\sigma_{cytoplasm}$ were observed; (**c**) A statistically significant difference only in $C_{specific\ membrane}$ was observed for 95D cells and 95D CCNY-KD cells with single oncogene *CCNY* down regulation (CCNY is a membrane-associated protein); (**d**) A statistically significant difference only in $\sigma_{cytoplasm}$ was observed for A549 cells and A549 CypA-KD cells with single oncogene *CypA* down regulatio n (CypA is a cytosolic protein) [5].

4. Microfluidic Impedance and Fluorescent Flow Cytometry for Single-Cell Analysis

In order to enhance the functionality of microfluidic impedance flow cytometry and provide a more comprehensive understanding on cellular biophysical and biochemical properties, Morgan *et al.* integrated the functions of impedance measurement and fluorescence detection in a microfluidic impedance and fluorescent flow cytometry [8,10,11,35,81,86–88,95,105–117]. Figure 6a shows the first-generation microfluidic platforms capable of characterizing both cellular electrical and optical property consisting of dual laser excitation, three color detection and dual frequency impedance measurement [10]. As the first demonstration, whole blood cells were successfully classified by the microfluidic impedance and fluorescent flow cytometry (see Figure 6b) [10]. In this study, the lymphocytes were differentiated from monocytes and neutrophils due to their significantly smaller cell sizes based on impedance data at 503 kHz. In addition, the neutrophils were differentiated from monocytes due to their significant differences in membrane capacitance based on impedance data at 1.707 MHz.

Furthermore, the whole blood cells mixed with CD4 antibody coated beads were successfully characterized by the microfluidic impedance (frequency: 503 kHz and 10 MHz) and fluorescent flow cytometry. In this work, the lymphocytes, lymphocytes + CD4 beads, granulocytes & monocytes and monocytes + CD4 beads were classified. This method was found useful for $CD4^+$ T-lymphocyte counting (see Figure 6c) [11]. Note that these impedance based cell type classification were confirmed by the simultaneous fluorescent measurement.

In the first-generation microfluidics-based impedance and fluorescent flow cytometry, laser excitation and fluorescent collection was implemented by using optical lens and thus only fluorescent signals can be obtained while other optical parameters such as side scattered light cannot be acquired. In order to address the limitation, Morgan *et al.* proposed the second-generation impedance and fluorescent flow cytometry where optical fibers were integrated into the microfluidic platform enabling the simultaneous measurement of impedance signals (at two frequencies) and optical signals (e.g., side scattered light and fluorescence) [86,87]. Figure 7a shows the microfluidic impedance flow cytometry with on-chip optical components. More specifically, a groove in SU-8 material holds a fiber to launch incident light perpendicular to the channel, which is focused into a sheet across the width of the channel using an air compound lens. Fluorescent emission is then collected with the fibers placed in two grooves on the same side as the incident light (at 135°). A 7° fiber is used to measure the optical extinction signal and the light loss due when a particle passes through an incident beam. Furthermore, two more collection fibers placed at 22.5° and 45° were designed to measure side scattered light [86]. The reported microfluidic platform was used to classify a mixture of beads (the fluorescent beads with 10 and 15 μm in diameter, and the plain beads with 20 and 25 μm in diameter). Figure 7b shows that the beads with four different sizes can be distinguished from optical side scattered light, but the impedance signals provide much better discrimination between the populations. The fluorescence signals from the 10 and 15 μm beads provide easy discrimination in this platform.

In 2014, Spencer and Morgan proposed a novel microfluidics-based impedance and fluorescent flow cytometry capable of measuring four different parameters, namely fluorescence, large angle side scattered light and dual frequency electrical impedance (electrical volume and opacity) (see Figure 7c) [87]. In this study, on-chip waveguides were used to replace the inserted fibers described in the previous study, which can effectively address the issues of optical fiber misalignments and incident light scattering from multiple interfaces. In addition, a sheath-less particle focusing technique was used and thus hydrodynamic focusing is no longer required. Figure 7d shows a 3-D scatter plot for a CD4 labelled white blood cell sample based on parameters of side scattered light, fluorescence, and two-frequency impedance data. Both side scattered light and low frequency impedance data at 0.5 MHz provide information on cell sizes, which separate smaller lymphocytes from granulocytes. High-frequency impedance data at 2 MHz separate monocytes from neutrophils due to differences in cell membrane capacitance while CD4 labelled white blood cells were distinguished from white blood cells without CD4 labelling based on fluorescent data.

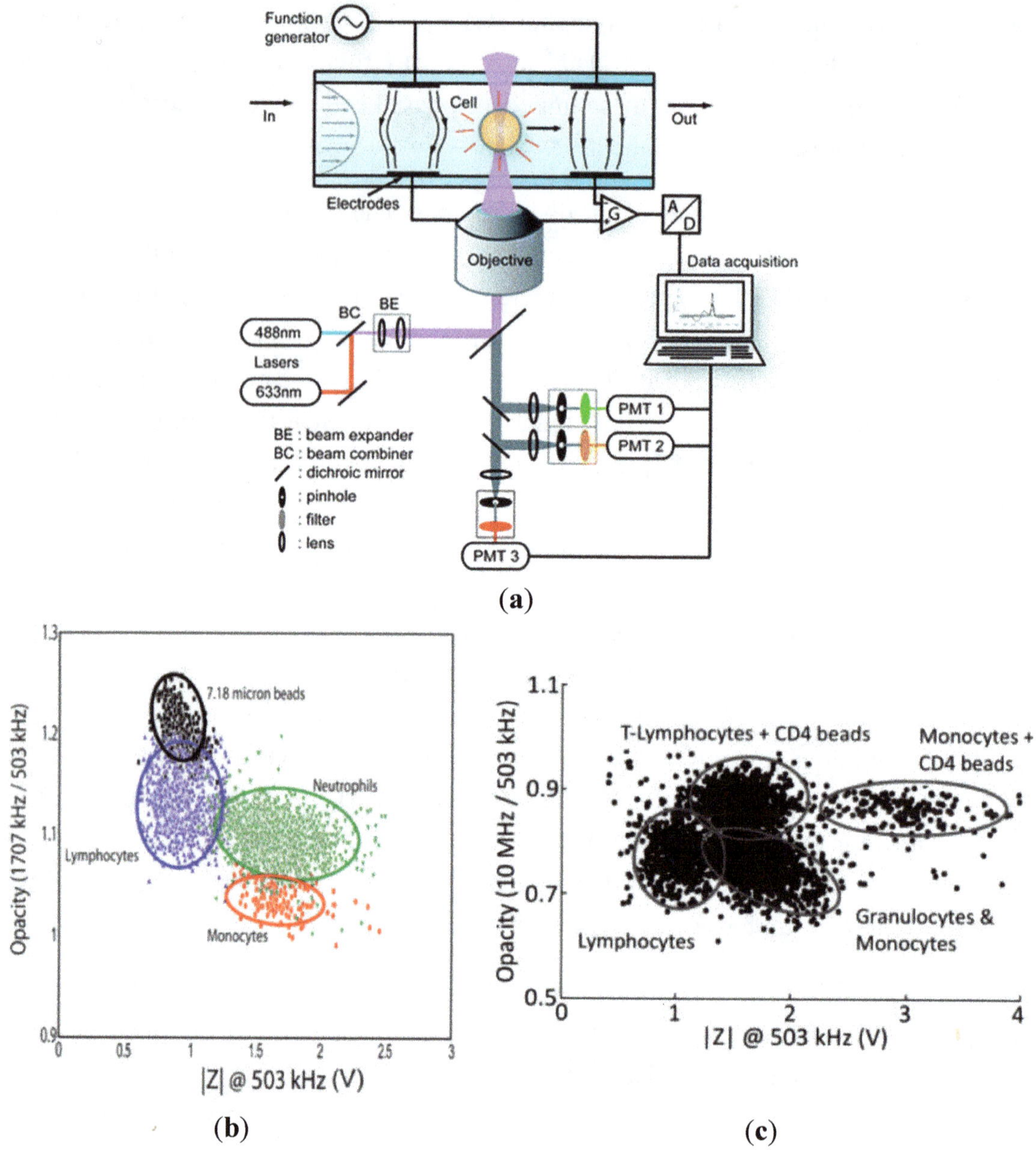

Figure 6. (**a**) The first-generation microfluidic impedance and fluorescent flow cytometry where a cell flows between two pairs of electrodes and the optical detection region composed of dual laser excitation, three color detection and dual frequency impedance measurement; (**b**) Impedance and fluorescent measurement results. Based on low frequency impedance amplitudes, lymphocytes can be differentiated from monocytes and neutrophils due to significantly smaller cell sizes. High frequency impedance amplitudes were used to differentiate neutrophils from monocytes due to significant differences in membrane capacitance. Note that these impedance based classification were validated by the simultaneous fluorescent detection by fluorescently labelling whole blood cells; (**c**) Whole blood cells mixing with CD4 antibody coated beads were characterized by the microfluidic impedance and fluorescent flow cytometry where lymphocytes, lymphocytes + CD4 beads, granulocytes & monocytes and monocytes + CD4 beads were successfully classified and confirmed by simultaneous fluorescent characterization [10,11].

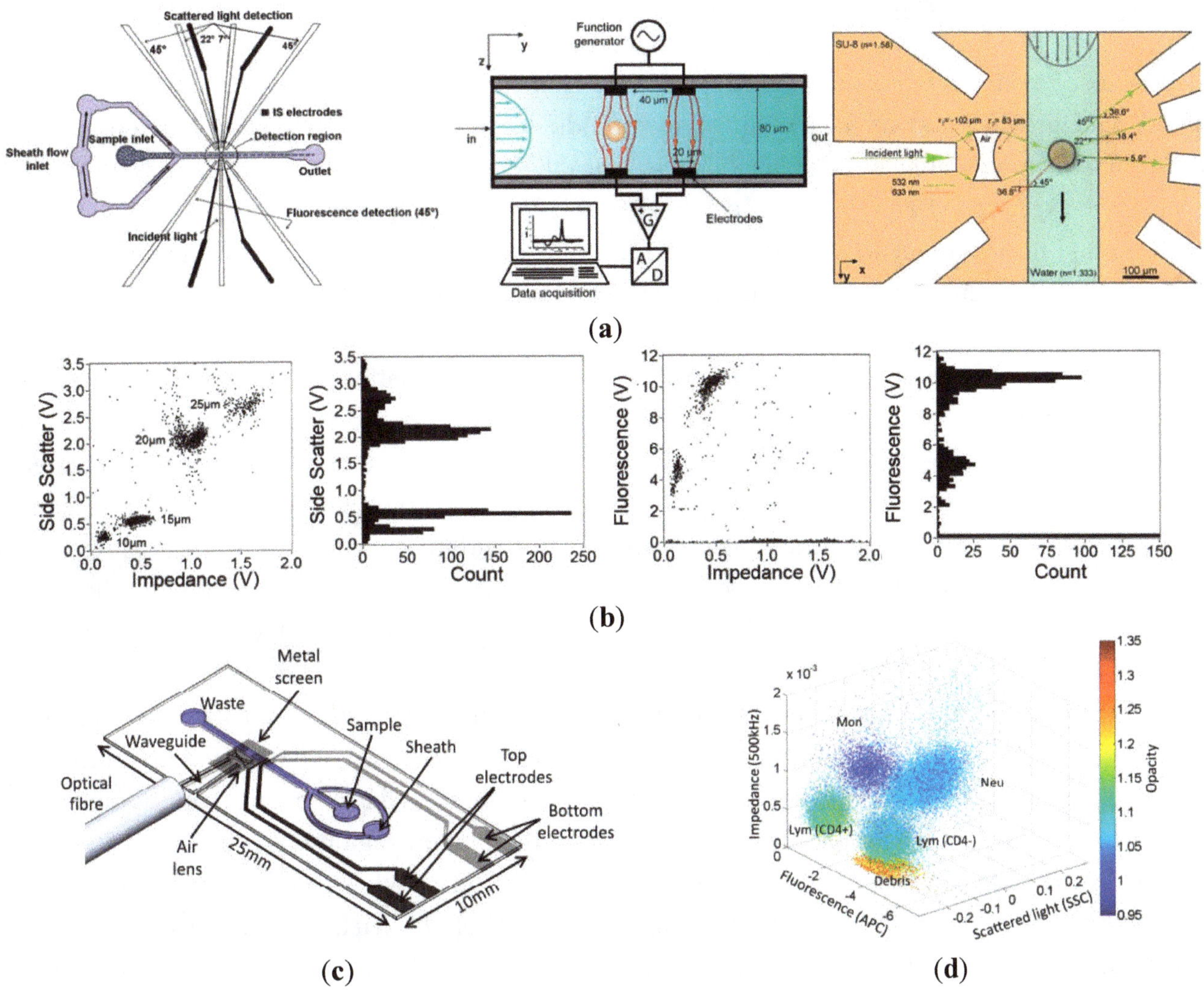

Figure 7. (**a**) The second-generation microfluidic impedance and fluorescent flow cytometry with on-chip optical components where a groove in SU-8 holds a fiber to launch incident light, which is then focused into the channel using an air compound lens. Fibers at various angles are used to collect fluorescence emission, optical extinction signal loss, and side scattered light, respectively; (**b**) Side scattered light, fluorescence and impedance data based classification of a mixture of different beads (10 and 15 µm diameter fluorescent, 20 and 25 µm diameter plain); (**c**) A new microfluidic impedance and fluorescent flow cytometry with on-chip waveguides in a sheath-less manner, which can effectively address misalignment of the optical fibers, incident light scatter from multiple interfaces and signal dependent on particle positions; (**d**) The 3-D scatter plot for CD4 labelled white blood cells based on parameters of side scatter light, fluorescence, and two-frequency impedance data. Both side scattered light and low frequency impedance data provide information on cell sizes, which discriminate smaller lymphocytes from granulocytes. High-frequency impedance data discriminates monocytes from neutrophils due to differences in cell membrane capacitance while CD4 labelled white blood cells were distinguished from white blood cells without CD4 labelling based on fluorescent data [86,87].

5. Integrated Point of Care System Based on Microfluidic Impedance Flow Cytometry

Diagnostic testing at or near the site of patient care is often termed as "near-patient" or "point-of-care" (POC) testing, which can be facilitated by microfluidic technologies [118–123]. Blood cell counting, as the most common clinical indicator of patient health, is one area where microfluidics based POC systems are expected to bring significant advancements [124–126]. Due to the advantages of compactness, low cost and no requirement for optical interfaces, microfluidic impedance flow cytometry has been integrated with sample pretreatment components to enable whole blood cell counting in the POC manner [7,88].

In 2011, Morgan *et al.* proposed an integrated microfluidic platform based on impedance flow cytometry, enabling the counting of 3-part differential leukocytes (granulocyte, lymphocyte and monocyte), as well as erythrocytes and platelets from raw blood samples [8,88]. As shown in Figure 8a, the integrated system consists of two parts: an impedance detection chip and a microfluidic sample preparation block. The microfluidic sample preparation block performs whole blood loading, pre-treatment and dilution into two separate fluid channels for impedance characterization, respectively. The bottom arm performs analysis of white blood cells with erythrocytes lysed while the upper arm performs counting of red blood cells and platelets.

Figure 8b shows the impedance scatter plot of cell membrane opacity (the ratio of impedance measured at 1.7 to 0.5 MHz) *vs.* the electrical cell volume (impedance magnitude at 0.5 MHz) for white blood cells. Consistent with previous studies [10,11], the three main subpopulations (lymphocytes, monocytes and neutrophils) are clearly separated while the top left region represents red blood ghost cells and other debris that are not completely eliminated by the on-chip lysis. Counting of red blood cells and platelets was performed using a single frequency of 0.5 MHz, where the cells are easily differentiated by sizes (see Figure 8c). Due to the relatively low number of platelets, platelet concordance was conducted, and the results showed an excellent linearity between the absolute platelet counts obtained from the impedance cytometry system and the hematology analyzer in hospitals.

In 2013, Bashir *et al.* proposed a microfluidic $CD4^+$ and $CD8^+$ T Lymphocyte counter for point-of-care HIV diagnostics targeting raw whole blood samples [7,127,128]. As shown in Figure 9a, the integrated microfluidic device is based on differential electrical counting and relies on five on-chip modules that, in sequence, chemically lyses erythrocytes, quenches lysis to preserve leukocytes, enumerates cells electrically, depletes the target cells (CD4 or CD8) with antibodies, and enumerates the remaining cells electrically. Target cell depletion was accomplished through shear stress-based immunocapture, and antibody-coated microposts were used to increase the contact surface areas and enhance the depletion efficiency. Based on the differential electrical counting method, which relies on two-frequency impedance data to classify lymphocytes, monocytes and neutrophils, $CD4^+$ and $CD8^+$ cell difference before and after the target cell depletion region was quantified (see Figure 9b).

Figure 9c,d show $CD4^+$ and $CD8^+$ T cell count results between chip and flow cytometry control with a close match using healthy (n = 18) and HIV-infected patient (n = 32) blood samples, respectively. By providing accurate cell counts in less than 20 min, this approach can be potentially implemented as a handheld, battery-powered instrument that would deliver simple HIV diagnostics to patients anywhere in the world, regardless of geography or socioeconomic status.

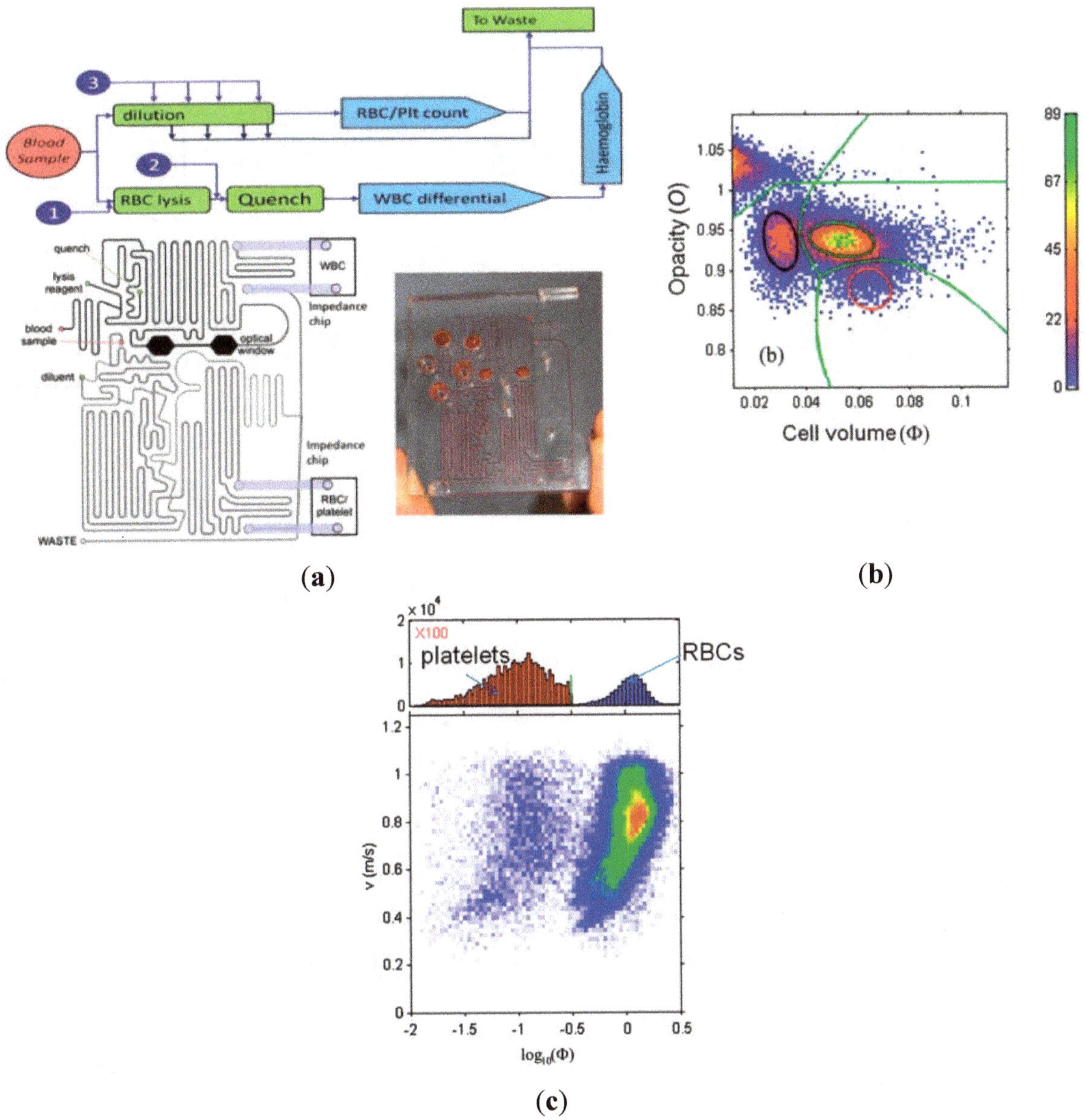

Figure 8. (**a**) The integrated point of care system based on microfluidic impedance flow cytometry enabling whole blood cell counting. The integrated system consists of two parts, an impedance measuring chip and a microfluidic sample preparation block. The bottom arm performs analysis of white blood cells with erythrocytes lysed while the upper arm performs counting of red blood cells and platelets; (**b**) The impedance scatter plot of cell membrane opacity *vs.* the electrical cell volume for classification of three main subpopulations of white blood cells (lymphocytes, monocytes and neutrophils); (**c**) Counting of red blood cells and platelets was performed based on single-frequency impedance data, where the cells are easily differentiated by sizes [88].

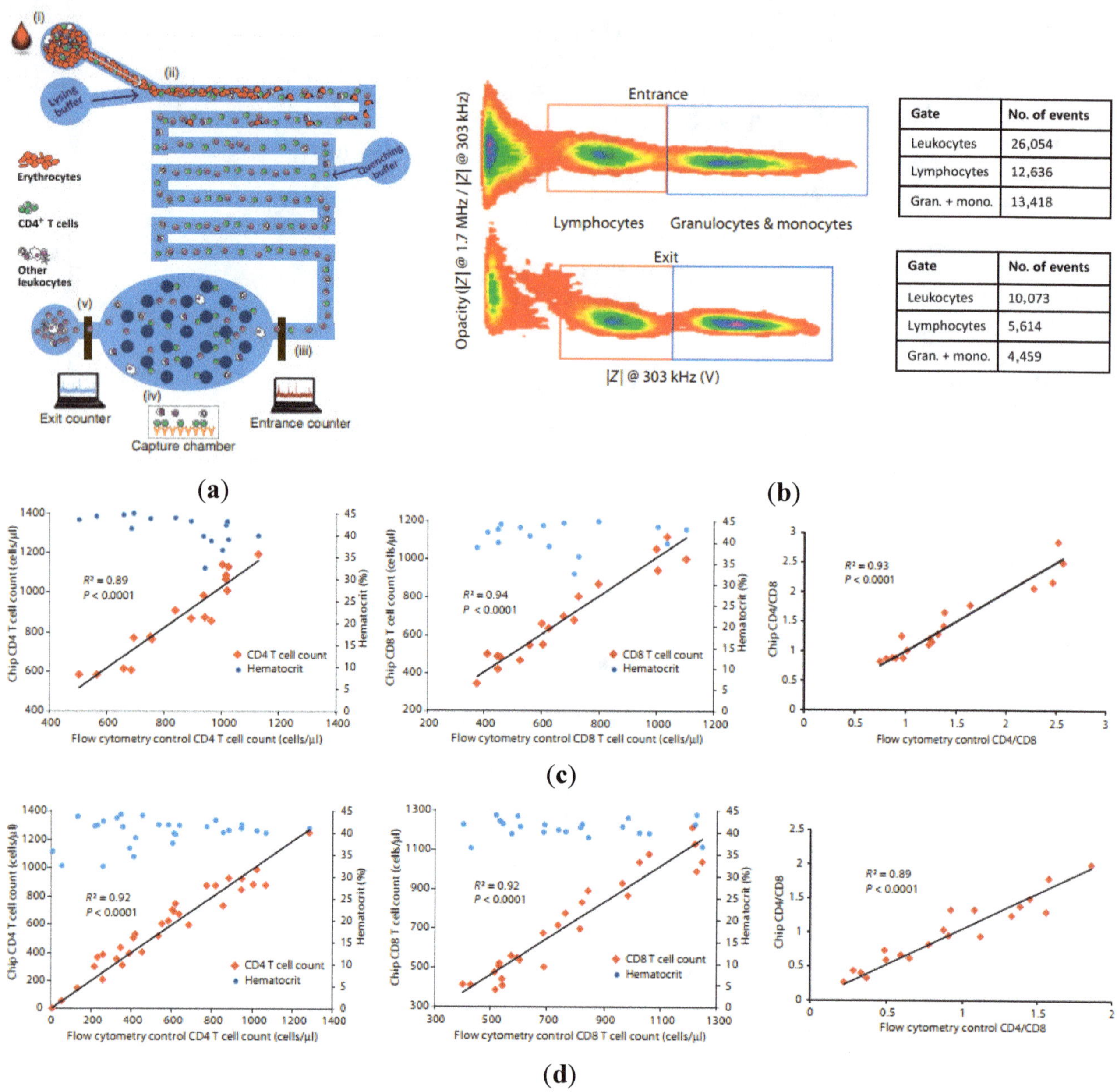

Figure 9. (**a**) The integrated point of care system based on microfluidic impedance flow cytometry enabling $CD4^+$ and $CD8^+$ T Lymphocyte counting. The integrated microfluidic device relies on five on-chip modules that are, in sequence, chemically lyses erythrocytes, quenches lysis to preserve leukocytes, enumerates cells electrically, depletes the target cells (CD4 or CD8) with antibodies, and enumerates the remaining cells electrically. Target cell depletion was accomplished through shear stress-based immunocapture; (**b**) Scatter plots of opacity *vs.* the low-frequency impedance amplitude for white blood cells before and after CD4 and CD8 depletion; $CD4^+$ and $CD8^+$ T cell count results between chip and flow cytometry control with a close match using healthy (n = 18) (**c**) and HIV-infected patient (n = 32) (**d**) blood samples, respectively [7].

6. Conclusions and Outlook

In this review, recent developments in the field of microfluidic impedance flow cytometry have been discussed from four perspectives: (1) early developments of microfluidic impedance flow cytometry for single-cell electrical property characterization; (2) microfluidic impedance flow cytometry with enhanced sensitivity; (3) microfluidic impedance and optical flow cytometry for single-cell analysis and (4) integrated point of care system based on microfluidic impedance flow cytometry.

From the aspect of technical development, microfluidic impedance flow cytometry enabling high-throughput characterization of size-independent intrinsic cellular electrical properties (e.g., specific membrane capacitance, a throughput of ~1000 cells per second) should be under intensive research. The majority of reported microfluidic impedance flow cytometry can collect cellular electrical properties in a high-throughput manner, which, however, are only capable of reporting size-dependent electrical properties (e.g., impedance values at several specific frequencies). Although these parameters can indicate membrane capacitance and cytoplasm resistance, they are dependent on cell sizes and specific experimental conditions (e.g., channel geometries and electrode dimensions). Since these parameters do not directly reflect intrinsic cellular electrical properties, it would be difficult to evaluate cellular status and classify cell types based on these parameters.

Recently, impedance spectroscopy and the constriction channel design were combined, enabling the quantification of $C_{\text{specific membrane}}$ and $\sigma_{\text{cytoplasm}}$ from hundreds of cells [85,103]. In addition, a microfluidic platform was developed where the cross-sectional area of the constriction channel is under regulation, effectively addressing the issue of constriction channel blockage [104]. However, the throughput of such microfluidic devices is roughly one cell per second, which is still low as compared to the conventional flow cytometry (~1000 cells per second). Thus, further technical development should focus on microfluidic impedance flow cytometry enabling high-throughput size-independent intrinsic electrical property characterization of single cells.

From the perspectives of clinical applications, microfluidic impedance flow cytometry can be used to classify tumor cells, stem cells, and blood cells in a label-free manner. In the field of tumor cell classification [5,64,83], paired high- and low-metastatic carcinoma strains and tumor cells as well as their counterparts with single oncogenes under regulation were successfully classified based on cellular electrical properties [5]. Further studies should be conducted to characterize electrical properties of human tumor samples and evaluate the feasibility of tumor classification based on cellular electrical properties.

As to the stem cell classification, undifferentiated and differentiated mouse embryonic carcinoma cells (P19) based on impedance data at 50 kHz, 250 kHz, 500 kHz and 1 MHz were differentiated [6] where it was speculated that the capacitance of stem cells can vary as they experience various stages of differentiation. These results provide some preliminary data along this direction but more data are needed for a decisive conclusion. For example, during stem cell differentiation, impedance data should be collected at multiple time points. This can help sketch the trend in how electrical properties of stem cells evolve as they differentiate into adult cells. Furthermore, electrical properties of human rather than mouse stem cells should be characterized to further evaluate the possibility of stem cell classification based on cellular electrical properties.

In the field of red blood cell classification based on cellular electrical properties [9,84], in 2013, Chandrakasan *et al.* developed a microfluidic impedance flow cytometry capable of differentiating

P. falciparum infected red blood cells from uninfected red blood cells based on amplitude and phase data at 2 MHz. However, multiple-frequency impedance data are suggested to further evaluate the electrical properties of various types of red blood cells. For white blood cell differentiation, since it has a close relationship with point of care applications, intensive research efforts have been devoted [7,8,10,11,88] (e.g., $CD4^{+}$ T lymphocyte counting [7]). Further studies should compare these approaches with other point of care methods and test a large number of human samples with statistical significance.

Acknowledgments

The authors would like to acknowledge financial support from National Natural Science Foundation of China (Grant No. 61201077), National Basic Research Program of China (973 Program, Grant No. 2014CB744600), National Natural Science Foundation of China (Grant No. 81261120561 and 61431019), National High Technology Research and Development Program of China (863 Program, Grant No. 2014AA093408), Chang Gung Memorial Hospital in Taiwan (Grant No. CMRPD2C0102 and CMRPD2D0041) and Beijing NOVA Program of Science and Technology. In addition, the authors would like to thank Rong Long for beneficial discussions.

Author Contributions

Jian Chen reviewed the early developments of microfluidic impedance flow cytometry for single-cell electrical property characterization. Chengcheng Xue and Yang Zhao reviewed the microfluidic impedance flow cytometry with enhanced sensitivity. Deyong Chen and Min-Hsien Wu reviewed microfluidic impedance and optical flow cytometry for single-cell analysis and Junbo Wang reviewed integrated point of care systems based on microfluidic impedance flow cytometry.

References

1. Morgan, H.; Sun, T.; Holmes, D.; Gawad, S.; Green, N.G. Single cell dielectric spectroscopy. *J. Phys. D-Appl. Phys.* **2007**, *40*, 61–70.
2. Valero, A.; Braschler, T.; Renaud, P. A unified approach to dielectric single cell analysis: Impedance and dielectrophoretic force spectroscopy. *Lab Chip* **2010**, *10*, 2216–2225.
3. Liang, X.; Graham, K.A.; Johannessen, A.C.; Costea, D.E.; Labeed, F.H. Human oral cancer cells with increasing tumorigenic abilities exhibit higher effective membrane capacitance. *Integr. Biol.* **2014**, *6*, 545–554.
4. Coley, H.M.; Labeed, F.H.; Thomas, H.; Hughes, M.P. Biophysical characterization of mdr breast cancer cell lines reveals the cytoplasm is critical in determining drug sensitivity. *Biochim. Biophys. Acta* **2007**, *1770*, 601–608.
5. Zhao, Y.; Zhao, X.T.; Chen, D.Y.; Luo, Y.N.; Jiang, M.; Wei, C.; Long, R.; Yue, W.T.; Wang, J.B.; Chen, J. Tumor cell characterization and classification based on cellular specific membrane capacitance and cytoplasm conductivity. *Biosens. Bioelectron.* **2014**, *57*, 245–253.

6. Song, H.; Wang, Y.; Rosano, J.M.; Prabhakarpandian, B.; Garson, C.; Pant, K.; Lai, E. A microfluidic impedance flow cytometer for identification of differentiation state of stem cells. *Lab Chip* **2013**, *13*, 2300–2310.
7. Watkins, N.N.; Hassan, U.; Damhorst, G.; Ni, H.; Vaid, A.; Rodriguez, W.; Bashir, R. Microfluidic CD4+ and CD8+ t lymphocyte counters for point-of-care hiv diagnostics using whole blood. *Sci. Trans. Med.* **2013**, *5*, 214ra170.
8. Han, X.; van Berkel, C.; Gwyer, J.; Capretto, L.; Morgan, H. Microfluidic lysis of human blood for leukocyte analysis using single cell impedance cytometry. *Anal. Chem.* **2012**, *84*, 1070–1075.
9. Du, E.; Ha, S.; Diez-Silva, M.; Dao, M.; Suresh, S.; Chandrakasan, A.P. Electric impedance microflow cytometry for characterization of cell disease states. *Lab Chip* **2013**, *13*, 3903–3909.
10. Holmes, D.; Pettigrew, D.; Reccius, C.H.; Gwyer, J.D.; van Berkel, C.; Holloway, J.; Davies, D.E.; Morgan, H. Leukocyte analysis and differentiation using high speed microfluidic single cell impedance cytometry. *Lab Chip* **2009**, *9*, 2881–2889.
11. Holmes, D.; Morgan, H. Single cell impedance cytometry for identification and counting of CD4 t-cells in human blood using impedance labels. *Anal. Chem.* **2010**, *82*, 1455–1461.
12. Zheng, Y.; Nguyen, J.; Wei, Y.; Sun, Y. Recent advances in microfluidic techniques for single-cell biophysical characterization. *Lab Chip* **2013**, *13*, 2464–2483.
13. Labeed, F.H.; Coley, H.M.; Thomas, H.; Hughes, M.P. Assessment of multidrug resistance reversal using dielectrophoresis and flow cytometry. *Biophys. J.* **2003**, *85*, 2028–2034.
14. Broche, L.M.; Labeed, F.H.; Hughes, M.P. Extraction of dielectric properties of multiple populations from dielectrophoretic collection spectrum data. *Phys. Med. Biol.* **2005**, *50*, 2267–2274.
15. Duncan, L.; Shelmerdine, H.; Hughes, M.P.; Coley, H.M.; Hubner, Y.; Labeed, F.H. Dielectrophoretic analysis of changes in cytoplasmic ion levels due to ion channel blocker action reveals underlying differences between drug-sensitive and multidrug-resistant leukaemic cells. *Phys. Med. Biol.* **2008**, *53*, N1–N7.
16. Vykoukal, D.M.; Gascoyne, P.R.; Vykoukal, J. Dielectric characterization of complete mononuclear and polymorphonuclear blood cell subpopulations for label-free discrimination. *Integr. Biol.* **2009**, *1*, 477–484.
17. Wu, L.; Lanry Yung, L.Y.; Lim, K.M. Dielectrophoretic capture voltage spectrum for measurement of dielectric properties and separation of cancer cells. *Biomicrofluidics* **2012**, *6*, 14113.
18. Bebarova, M. Advances in patch clamp technique: Towards higher quality and quantity. *Gen. Physiol. Biophys.* **2012**, *31*, 131–140.
19. Kornreich, B.G. The patch clamp technique: Principles and technical considerations. *J. Vet. Cardiol.* **2007**, *9*, 25–37.
20. Dale, T.J.; Townsend, C.; Hollands, E.C.; Trezise, D.J. Population patch clamp electrophysiology: A breakthrough technology for ion channel screening. *Mol. Biosyst.* **2007**, *3*, 714–722.

21. Liem, L.K.; Simard, J.M.; Song, Y.; Tewari, K. The patch clamp technique. *Neurosurgery* **1995**, *36*, 382–392.
22. Cahalan, M.; Neher, E. Patch clamp techniques: An overview. *Methods Enzymol.* **1992**, *207*, 3–14.
23. Sakmann, B.; Neher, E. Patch clamp techniques for studying ionic channels in excitable membranes. *Annu. Rev. Physiol.* **1984**, *46*, 455–472.
24. Auerbach, A.; Sachs, F. Patch clamp studies of single ionic channels. *Annu. Rev. Biophys. Bioeng.* **1984**, *13*, 269–302.
25. Rohani, A.; Varhue, W.; Su, Y.H.; Swami, N.S. Electrical tweezer for highly parallelized electrorotation measurements over a wide frequency bandwidth. *Electrophoresis* **2014**, *35*, 1795–1802.
26. Lei, U.; Sun, P.H.; Pethig, R. Refinement of the theory for extracting cell dielectric properties from dielectrophoresis and electrorotation experiments. *Biomicrofluidics* **2011**, *5*, 044109.
27. Voyer, D.; Frenea-Robin, M.; Buret, F.; Nicolas, L. Improvements in the extraction of cell electric properties from their electrorotation spectrum. *Bioelectrochemistry* **2010**, *79*, 25–30.
28. Jones, T.B. Basic theory of dielectrophoresis and electrorotation. *IEEE Eng. Med. Biol. Mag.* **2003**, *22*, 33–42.
29. Goater, A.D.; Pethig, R. Electrorotation and dielectrophoresis. *Parasitology* **1998**, *117* (Suppl.), S177–S189.
30. Fuhr, G.; Glaser, R.; Hagedorn, R. Rotation of dielectrics in a rotating electric high-frequency field-model experiments and theoretical explanation of the rotation effect of living cells. *Biophys. J.* **1986**, *49*, 395–402.
31. Egger, M.; Donath, E.; Ziemer, S.; Glaser, R. Electrorotation--a new method for investigating membrane events during thrombocyte activation. Influence of drugs and osmotic pressure. *Biochim. Biophys. Acta* **1986**, *861*, 122–130.
32. Fuhr, G.; Hagedorn, R.; Goring, H. Separation of different cell-types by rotating electric-fields. *Plant Cell Physiol.* **1985**, *26*, 1527–1531.
33. Yobas, L. Microsystems for cell-based electrophysiology. *J. Micromech. Microeng.* **2013**, *23*, 083002.
34. Sabuncu, A.C.; Zhuang, J.; Kolb, J.F.; Beskok, A. Microfluidic impedance spectroscopy as a tool for quantitative biology and biotechnology. *Biomicrofluidics* **2012**, *6*, 34103.
35. Sun, T.; Morgan, H. Single-cell microfluidic impedance cytometry: A review. *Microfluid. Nanofluid.* **2010**, *8*, 423–443.
36. Cheung, K.C.; di Berardino, M.; Schade-Kampmann, G.; Hebeisen, M.; Pierzchalski, A.; Bocsi, J.; Mittag, A.; Tarnok, A. Microfluidic impedance-based flow cytometry. *Cytom. A* **2010**, *77*, 648–666.
37. Wootton, R.C.; Demello, A.J. Microfluidics: Exploiting elephants in the room. *Nature* **2010**, *464*, 839–840.
38. Whitesides, G.M. The origins and the future of microfluidics. *Nature* **2006**, *442*, 368–373.
39. Squires, T.M.; Quake, S.R. Microfluidics: Fluid physics at the nanoliter scale. *Rev. Mod. Phys.* **2005**, *77*, 977.

40. Sackmann, E.K.; Fulton, A.L.; Beebe, D.J. The present and future role of microfluidics in biomedical research. *Nature* **2014**, *507*, 181–189.
41. El-Ali, J.; Sorger, P.K.; Jensen, K.F. Cells on chips. *Nature* **2006**, *442*, 403–411.
42. Meyvantsson, I.; Beebe, D.J. Cell culture models in microfluidic systems. *Annu. Rev. Anal. Chem.* **2008**, *1*, 423–449.
43. Paguirigan, A.L.; Beebe, D.J. Microfluidics meet cell biology: Bridging the gap by validation and application of microscale techniques for cell biological assays. *BioEssays News Rev. Mol. Cell. Dev. Biol.* **2008**, *30*, 811–821.
44. Velve-Casquillas, G.; Le Berre, M.; Piel, M.; Tran, P.T. Microfluidic tools for cell biological research. *Nano Today* **2010**, *5*, 28–47.
45. Thompson, A.M.; Paguirigan, A.L.; Kreutz, J.E.; Radich, J.P.; Chiu, D.T. Microfluidics for single-cell genetic analysis. *Lab Chip* **2014**, *14*, 3135–3142.
46. Yin, H.; Marshall, D. Microfluidics for single cell analysis. *Curr. Opin. Biotechnol.* **2012**, *23*, 110–119.
47. Lecault, V.; White, A.K.; Singhal, A.; Hansen, C.L. Microfluidic single cell analysis: From promise to practice. *Curr. Opin. Chem. Biol.* **2012**, *16*, 381–390.
48. Ryan, D.; Ren, K.; Wu, H. Single-cell assays. *Biomicrofluidics* **2011**, *5*, 21501.
49. Zare, R.N.; Kim, S. Microfluidic platforms for single-cell analysis. *Annu. Rev. Biomed. Eng.* **2010**, *12*, 187–201.
50. Sims, C.E.; Allbritton, N.L. Analysis of single mammalian cells on-chip. *Lab Chip* **2007**, *7*, 423–440.
51. Di Carlo, D.; Lee, L.P. Dynamic single-cell analysis for quantitative biology. *Anal. Chem.* **2006**, *78*, 7918–7925.
52. Cho, Y.H.; Yamamoto, T.; Sakai, Y.; Fujii, T.; Kim, B. Development of microfluidic device for electrical/physical characterization of single cell. *J. Microelectromech. Syst.* **2006**, *15*, 287–295.
53. Jang, L.S.; Wang, M.H. Microfluidic device for cell capture and impedance measurement. *Biomed. Microdevices* **2007**, *9*, 737–743.
54. Senez, V.; Lennon, E.; Ostrovidov, S.; Yamamoto, T.; Fujita, H.; Sakai, Y.; Fujii, T. Integrated 3-d silicon electrodes for electrochemical sensing in microfluidic environments: Application to single-cell characterization. *IEEE Sens. J.* **2008**, *8*, 548–557.
55. Hua, S.Z.; Pennell, T. A microfluidic chip for real-time studies of the volume of single cells. *Lab Chip* **2009**, *9*, 251–256.
56. Wang, M.H.; Jang, L.S. A systematic investigation into the electrical properties of single hela cells via impedance measurements and comsol simulations. *Biosens. Bioelectron.* **2009**, *24*, 2830–2835.
57. Jang, L.S.; Li, H.H.; Jao, J.Y.; Chen, M.K.; Liu, C.F. Design and fabrication of microfluidic devices integrated with an open-ended mems probe for single-cell impedance measurement. *Microfluid. Nanofluid.* **2010**, *8*, 509–519.
58. Malleo, D.; Nevill, J.T.; Lee, L.P.; Morgan, H. Continuous differential impedance spectroscopy of single cells. *Microfluid. Nanofluid.* **2010**, *9*, 191–198.

59. Han, K.H.; Han, A.; Frazier, A.B. Microsystems for isolation and electrophysiological analysis of breast cancer cells from blood. *Biosens. Bioelectron.* **2006**, *21*, 1907–1914.
60. Cho, S.B.; Thielecke, H. Micro hole-based cell chip with impedance spectroscopy. *Biosens. Bioelectron.* **2007**, *22*, 1764–1768.
61. James, C.D.; Reuel, N.; Lee, E.S.; Davalos, R.V.; Mani, S.S.; Carroll-Portillo, A.; Rebeil, R.; Martino, A.; Apblett, C.A. Impedimetric and optical interrogation of single cells in a microfluidic device for real-time viability and chemical response assessment. *Biosens. Bioelectron.* **2008**, *23*, 845–851.
62. Cho, Y.; Kim, H.S.; Frazier, A.B.; Chen, Z.G.; Shin, D.M.; Han, A. Whole-cell impedance analysis for highly and poorly metastatic cancer cells. *J. Microelectromech. Syst.* **2009**, *18*, 808–817.
63. Chen, J.; Zheng, Y.; Tan, Q.; Zhang, Y.L.; Li, J.; Geddie, W.R.; Jewett, M.A.; Sun, Y. A microfluidic device for simultaneous electrical and mechanical measurements on single cells. *Biomicrofluidics* **2011**, *5*, 14113.
64. Kang, G.; Kim, Y.-J.; Moon, H.-S.; Lee, J.-W.; Yoo, T.-K.; Park, K.; Lee, J.-H. Discrimination between the human prostate normal cell and cancer cell by using a novel electrical impedance spectroscopy controlling the cross-sectional area of a microfluidic channel. *Biomicrofluidics* **2013**, *7*, 044126.
65. Tan, Q.; Ferrier, G.A.; Chen, B.K.; Wang, C.; Sun, Y. Quantification of the specific membrane capacitance of single cells using a microfluidic device and impedance spectroscopy measurement. *Biomicrofluidics* **2012**, *6*, 034112.
66. Jang, L.S.; Huang, P.H.; Lan, K.C. Single-cell trapping utilizing negative dielectrophoretic quadrupole and microwell electrodes. *Biosens. Bioelectron.* **2009**, *24*, 3637–3644.
67. Lan, K.C.; Jang, L.S. Integration of single-cell trapping and impedance measurement utilizing microwell electrodes. *Biosens. Bioelectron.* **2011**, *26*, 2025–2031.
68. Hong, J.-L.; Lan, K.-C.; Jang, L.-S. Electrical characteristics analysis of various cancer cells using a microfluidic device based on single-cell impedance measurement. *Sens. Actuators B Chem.* **2012**, *173*, 927–934.
69. Chen, N.-C.; Chen, C.-H.; Chen, M.-K.; Jang, L.-S.; Wang, M.-H. Single-cell trapping and impedance measurement utilizing dielectrophoresis in a parallel-plate microfluidic device. *Sens. Actuators B Chem.* **2014**, *190*, 570–577.
70. Thein, M.; Asphahani, F.; Cheng, A.; Buckmaster, R.; Zhang, M.Q.; Xu, J. Response characteristics of single-cell impedance sensors employed with surface-modified microelectrodes. *Biosens. Bioelectron.* **2010**, *25*, 1963–1969.
71. Asphahani, F.; Wang, K.; Thein, M.; Veiseh, O.; Yung, S.; Xu, J.A.; Zhang, M.Q. Single-cell bioelectrical impedance platform for monitoring cellular response to drug treatment. *Phys. Biol.* **2011**, *8*, 015006.
72. Asphahani, F.; Zheng, X.H.; Veiseh, O.; Thein, M.; Xu, J.; Ohuchi, F.; Zhang, M.Q. Effects of electrode surface modification with chlorotoxin on patterning single glioma cells. *Phys. Chem. Chem. Phys.* **2011**, *13*, 8953–8960.
73. Wu, Y.; Han, X.; Benson, J.D.; Almasri, M. Micromachined coulter counter for dynamic impedance study of time sensitive cells. *Biomed. Microdevices* **2012**, *14*, 739–750.

74. Wu, Y.; Benson, J.D.; Critser, J.K.; Almasri, M. Note: Microelectromechanical systems coulter counter for cell monitoring and counting. *Rev. Sci. Instrum.* **2010**, *81*, 076103.
75. Swanton, E.M.; Curby, W.A.; Lind, H.E. Experiences with the coulter counter in bacteriology. *Appl. Microbiol.* **1962**, *10*, 480–485.
76. Bryan, A.K.; Engler, A.; Gulati, A.; Manalis, S.R. Continuous and long-term volume measurements with a commercial coulter counter. *PLoS ONE* **2012**, *7*, e29866.
77. Gawad, S.; Schild, L.; Renaud, P. Micromachined impedance spectroscopy flow cytometer for cell analysis and particle sizing. *Lab Chip* **2001**, *1*, 76–82.
78. Gou, H.-L.; Zhang, X.-B.; Bao, N.; Xu, J.-J.; Xia, X.-H.; Chen, H.-Y. Label-free electrical discrimination of cells at normal, apoptotic and necrotic status with a microfluidic device. *J. Chromatogr. A* **2011**, *1218*, 5725–5729.
79. Cheung, K.; Gawad, S.; Renaud, P. Impedance spectroscopy flow cytometry: On-chip label-free cell differentiation. *Cytom. Part A* **2005**, *65A*, 124–132.
80. Haandbaek, N.; Burgel, S.C.; Heer, F.; Hierlemann, A. Characterization of subcellular morphology of single yeast cells using high frequency microfluidic impedance cytometer. *Lab Chip* **2014**, *14*, 369–377.
81. Bernabini, C.; Holmes, D.; Morgan, H. Micro-impedance cytometry for detection and analysis of micron-sized particles and bacteria. *Lab Chip* **2011**, *11*, 407–412.
82. Haandbaek, N.; With, O.; Burgel, S.C.; Heer, F.; Hierlemann, A. Resonance-enhanced microfluidic impedance cytometer for detection of single bacteria. *Lab Chip* **2014**, *14*, 3313–3324.
83. Chen, J.; Zheng, Y.; Tan, Q.; Shojaei-Baghini, E.; Zhang, Y.L.; Li, J.; Prasad, P.; You, L.; Wu, X.Y.; Sun, Y. Classification of cell types using a microfluidic device for mechanical and electrical measurement on single cells. *Lab Chip* **2011**, *11*, 3174–3181.
84. Zheng, Y.; Shojaei-Baghini, E.; Azad, A.; Wang, C.; Sun, Y. High-throughput biophysical measurement of human red blood cells. *Lab Chip* **2012**, *12*, 2560–2567.
85. Zhao, Y.; Chen, D.; Li, H.; Luo, Y.; Deng, B.; Huang, S.B.; Chiu, T.K.; Wu, M.H.; Long, R.; Hu, H.; *et al.* A microfluidic system enabling continuous characterization of specific membrane capacitance and cytoplasm conductivity of single cells in suspension. *Biosens. Bioelectron.* **2013**, *43C*, 304–307.
86. Barat, D.; Spencer, D.; Benazzi, G.; Mowlem, M.C.; Morgan, H. Simultaneous high speed optical and impedance analysis of single particles with a microfluidic cytometer. *Lab Chip* **2012**, *12*, 118–126.
87. Spencer, D.C.; Elliott, G.; Morgan, H. A sheath-less combined optical and impedance micro-cytometer. *Lab Chip* **2014**, *14*, 3064–3073.
88. Van Berkel, C.; Gwyer, J.D.; Deane, S.; Green, N.G.; Holloway, J.; Hollis, V.; Morgan, H. Integrated systems for rapid point of care (poc) blood cell analysis. *Lab Chip* **2011**, *11*, 1249–1255.
89. Gawad, S.; Cheung, K.; Seger, U.; Bertsch, A.; Renaud, P. Dielectric spectroscopy in a micromachined flow cytometer: Theoretical and practical considerations. *Lab Chip* **2004**, *4*, 241–251.
90. Mernier, G.; Piacentini, N.; Tornay, R.; Buffi, N.; Renaud, P. Cell viability assessment by flow cytometry using yeast as cell model. *Sens. Actuators B Chem.* **2011**, *154*, 160–163.

91. Meissner, R.; Joris, P.; Eker, B.; Bertsch, A.; Renaud, P. A microfluidic-based frequency-multiplexing impedance sensor (fmis). *Lab Chip* **2012**, *12*, 2712–2718.
92. Mernier, G.; Hasenkamp, W.; Piacentini, N.; Renaud, P. Multiple-frequency impedance measurements in continuous flow for automated evaluation of yeast cell lysis. *Sens. Actuators B Chem.* **2012**, *170*, 2–6.
93. Shaker, M.; Colella, L.; Caselli, F.; Bisegna, P.; Renaud, P. An impedance-based flow microcytometer for single cell morphology discrimination. *Lab Chip* **2014**, *2014*, 2548–2555.
94. Choi, H.; Jeon, C.S.; Hwang, I.; Ko, J.; Lee, S.; Choo, J.; Boo, J.-H.; Kim, H.C.; Chung, T.D. A flow cytometry-based submicron-sized bacterial detection system using a movable virtual wall. *Lab Chip* **2014**, *14*, 2327–2333.
95. Spencer, D.; Morgan, H. Positional dependence of particles in microfludic impedance cytometry. *Lab Chip* **2011**, *11*, 1234–1239.
96. Shelby, J.P.; White, J.; Ganesan, K.; Rathod, P.K.; Chiu, D.T. A microfluidic model for single-cell capillary obstruction by plasmodium falciparum infected erythrocytes. *Proc. Natl. Acad. Sci. USA* **2003**, *100*, 14618–14622.
97. Guo, Q.; Reiling, S.J.; Rohrbach, P.; Ma, H. Microfluidic biomechanical assay for red blood cells parasitized by plasmodium falciparum. *Lab Chip* **2012**, *12*, 1143–1150.
98. Kwan, J.M.; Guo, Q.; Kyluik-Price, D.L.; Ma, H.; Scott, M.D. Microfluidic analysis of cellular deformability of normal and oxidatively damaged red blood cells. *Am. J. Hematol.* **2013**, *88*, 682–689.
99. Rosenbluth, M.J.; Lam, W.A.; Fletcher, D.A. Analyzing cell mechanics in hematologic diseases with microfluidic biophysical flow cytometry. *Lab Chip* **2008**, *8*, 1062–1070.
100. Hou, H.W.; Li, Q.S.; Lee, G.Y.H.; Kumar, A.P.; Ong, C.N.; Lim, C.T. Deformability study of breast cancer cells using microfluidics. *Biomed. Microdevices* **2009**, *11*, 557–564.
101. Byun, S.; Son, S.; Amodei, D.; Cermak, N.; Shaw, J.; Kang, J.H.; Hecht, V.C.; Winslow, M.M.; Jacks, T.; Mallick, P.; *et al.* Characterizing deformability and surface friction of cancer cells. *Proc. Natl. Acad. Sci. USA* **2013**, *110*, 7580–7585.
102. Luo, Y.N.; Chen, D.Y.; Zhao, Y.; Wei, C.; Zhao, X.T.; Yue, W.T.; Long, R.; Wang, J.B.; Chen, J. A constriction channel based microfluidic system enabling continuous characterization of cellular instantaneous young's modulus. *Sens. Actuators B Chem.* **2014**, *202*, 1183–1189.
103. Zhao, Y.; Chen, D.; Luo, Y.; Li, H.; Deng, B.; Huang, S.-B.; Chiu, T.-K.; Wu, M.-H.; Long, R.; Hu, H.; *et al.* A microfluidic system for cell type classification based on cellular size-independent electrical properties. *Lab Chip* **2013**, *13*, 2272–2277.
104. Huang, S.B.; Zhao, Z.; Chen, D.Y.; Lee, H.C.; Luo, Y.N.; Chiu, T.K.; Wang, J.B.; Chen, J.; Wu, M.H. A clogging-free microfluidic platform with an incorporated pneumatically-driven membrane-based active valve enabling specific membrane capacitance and cytoplasm conductivity characterization of single cells. *Sens. Actuators B Chem.* **2014**, *190*, 928–936.
105. Morgan, H.; Holmes, D.; Green, N.G. High speed simultaneous single particle impedance and fluorescence analysis on a chip. *Curr. Appl. Phys.* **2006**, *6*, 367–370.
106. Benazzi, G.; Holmes, D.; Sun, T.; Mowlem, M.C.; Morgan, H. Discrimination and analysis of phytoplankton using a microfluidic cytometer. *Iet Nanobiotechnol.* **2007**, *1*, 94–101.

107. Gawad, S.; Sun, T.; Green, N.G.; Morgan, H. Impedance spectroscopy using maximum length sequences: Application to single cell analysis. *Rev. Sci. Instrum.* **2007**, *78*, 054301.
108. Holmes, D.; She, J.K.; Roach, P.L.; Morgan, H. Bead-based immunoassays using a micro-chip flow cytometer. *Lab Chip* **2007**, *7*, 1048–1056.
109. Sun, T.; Gawad, S.; Bernabini, C.; Green, N.G.; Morgan, H. Broadband single cell impedance spectroscopy using maximum length sequences: Theoretical analysis and practical considerations. *Meas. Sci. Technol.* **2007**, *18*, 2859–2868.
110. Sun, T.; Green, N.G.; Gawad, S.; Morgan, H. Analytical electric field and sensitivity analysis for two microfluidic impedance cytometer designs. *Iet Nanobiotechnol.* **2007**, *1*, 69–79.
111. Sun, T.; Holmes, D.; Gawad, S.; Green, N.G.; Morgan, H. High speed multi-frequency impedance analysis of single particles in a microfluidic cytometer using maximum length sequences. *Lab Chip* **2007**, *7*, 1034–1040.
112. Sun, T.; van Berkel, C.; Green, N.G.; Morgan, H. Digital signal processing methods for impedance microfluidic cytometry. *Microfluid. Nanofluid.* **2009**, *6*, 179–187.
113. Barat, D.; Benazzi, G.; Mowlem, M.C.; Ruano, J.M.; Morgan, H. Design, simulation and characterisation of integrated optics for a microfabricated flow cytometer. *Opt. Commun.* **2010**, *283*, 1987–1992.
114. Gawad, S.; Holmes, D.; Benazzi, G.; Renaud, P.; Morgan, H. Impedance spectroscopy and optical analysis of single biological cells and organisms in microsystems. *Methods Mol. Biol.* **2010**, *583*, 149–182.
115. Sun, T.; Bernabini, C.; Morgan, H. Single-colloidal particle impedance spectroscopy: Complete equivalent circuit analysis of polyelectrolyte microcapsules. *Langmuir* **2010**, *26*, 3821–3828.
116. Sun, T.; Swindle, E.J.; Collins, J.E.; Holloway, J.A.; Davies, D.E.; Morgan, H. On-chip epithelial barrier function assays using electrical impedance spectroscopy. *Lab Chip* **2010**, *10*, 1611–1617.
117. Sun, T.; Tsuda, S.; Zauner, K.P.; Morgan, H. On-chip electrical impedance tomography for imaging biological cells. *Biosens. Bioelectron.* **2010**, *25*, 1109–1115.
118. Kumar, S.; Kumar, S.; Ali, M.A.; Anand, P.; Agrawal, V.V.; John, R.; Maji, S.; Malhotra, B.D. Microfluidic-integrated biosensors: Prospects for point-of-care diagnostics. *Biotechnol. J.* **2013**, *8*, 1267–1279.
119. Chin, C.D.; Linder, V.; Sia, S.K. Commercialization of microfluidic point-of-care diagnostic devices. *Lab Chip* **2012**, *12*, 2118–2134.
120. Kiechle, F.L.; Holland, C.A. Point-of-care testing and molecular diagnostics: Miniaturization required. *Clin. Lab. Med.* **2009**, *29*, 555–560.
121. Sorger, P.K. Microfluidics closes in on point-of-care assays. *Nat. Biotechnol.* **2008**, *26*, 1345–1346.
122. Sia, S.K.; Kricka, L.J. Microfluidics and point-of-care testing. *Lab Chip* **2008**, *8*, 1982–1983.
123. Linder, V. Microfluidics at the crossroad with point-of-care diagnostics. *Analyst* **2007**, *132*, 1186–1192.
124. Yu, Z.T.; Aw Yong, K.M.; Fu, J. Microfluidic blood cell sorting: Now and beyond. *Small* **2014**, *10*, 1687–1703.
125. van der Meer, A.D.; Poot, A.A.; Duits, M.H.; Feijen, J.; Vermes, I. Microfluidic technology in vascular research. *J. Biomed. Biotechnol.* **2009**, *2009*, 823148.

126. Toner, M.; Irimia, D. Blood-on-a-chip. *Annu. Rev. Biomed. Eng.* **2005**, *7*, 77–103.
127. Hassan, U.; Bashir, R. Electrical cell counting process characterization in a microfluidic impedance cytometer. *Biomed. Microdevices* **2014**, *16*, 697–704.
128. Hassan, U.; Watkins, N.N.; Edwards, C.; Bashir, R. Flow metering characterization within an electrical cell counting microfluidic device. *Lab Chip* **2014**, *14*, 1469–1476.

12

Evidence for P-Glycoprotein Involvement in Cell Volume Regulation using Coulter Sizing in Flow Cytometry

Jennifer Pasquier [1,2,*], Damien Rioult [3], Nadine Abu-Kaoud [1], Jessica Hoarau-Véchot [1], Matthieu Marin [4] and Frank Le Foll [5]

1 Stem Cell and Microenvironment Laboratory, Weill Cornell Medical College in Qatar, Education City, Qatar Foundation, Doha 24144, Qatar; E-Mails: naa2037@qatar-med.cornell.edu (N.A.-K.); jeh2036@qatar-med.cornell.edu (J.H.-V.)

2 Department of Genetic Medicine, Weill Cornell Medical College, New York, NY 10022, USA

3 UMR_I 02 INERIS-URCA-ULH SEBIO Unité Stress Environnementaux et BIOsurveillance des milieux aquatiques, Université de Reims Champagne-Ardenne, BP-1039-Reims Cedex 2, 51687 Reims, France; E-Mail: damien.rioult@univ-reims.fr

4 UGSF, UMR CNRS 8576, équipe Régulation des Signaux de Division, Université de Lille, Sciences et Technologies, 59009 Villeneuve d'Ascq, France; E-Mail: matthieu.marin@univ-lille1.fr

5 Laboratory of Ecotoxicology UPRES EA 3222, IFRMP 23, University of Le Havre, 76058 Le Havre, France; E-Mail: frank.lefoll@univ-lehavre.fr

* Author to whom correspondence should be addressed; E-Mail: jep2026@qatar-med.cornell.edu

Academic Editor: Tuhin Subhra Santra

Abstract: The regulation of cell volume is an essential function that is coupled to a variety of physiological processes such as receptor recycling, excitability and contraction, cell proliferation, migration, and programmed cell death. Under stress, cells undergo emergency swelling and respond to such a phenomenon with a regulatory volume decrease (RVD) where they release cellular ions, and other osmolytes as well as a concomitant loss of water. The link between P-glycoprotein, a transmembrane transporter, and cell volume regulation is controversial, and changes in cells volume are measured using microscopy or electrophysiology. For instance, by using the patch-clamp method, our team demonstrated that chloride currents activated in the RVD were more intense and rapid in a breast cancer cell line overexpressing the P-glycoprotein (P-gp). The Cell Lab Quanta SC is a flow cytometry system that simultaneously measures electronic volume, side scatter and three fluorescent colors; altogether this provides unsurpassed population resolution and accurate cell counting. Therefore, here we propose a novel method to follow cellular volume. By using the Coulter-type channel of the cytometer Cell Lab Quanta SC MPL (multi-platform loading), we demonstrated a role for the P-gp during different osmotic treatments, but also a differential activity of the P-gp through the cell cycle. Altogether, our data strongly suggests a role of P-gp in cell volume regulation.

Keywords: multi-drug resistance; P-glycoprotein; cell volume; flow cytometer; MCF7

---

1. Introduction

Each living cell type seems to have a form and volume that is well defined, and determined by the size of their cytoplasmic membrane as well as their cytosolic content. However, variations of osmotic pressures lead to cell volume regulation [1,2]. In the hypotonic or hypertonic context, the ability of cells to regulate their volume in a short period of time (around a minute) in order to avoid swelling or shrinkage is a fundamental mechanism [3]. Under hypo-osmotic conditions, cells are able to escape a burst after swelling by activation of a mechanism known as regulatory volume decrease (RVD) [4]. Interestingly, even under a regular osmotic pressure, the cell volume is fluctuating in response to cell events such as cell proliferation, migration, glycolysis or cell death [5–8].

P-glycoprotein (P-gp, *ABCB1*), a member of the ATP-binding cassette (ABC) superfamily of transporters, is a 170-kD transmembrane glycoprotein involved in the multi-drug resistance (MDR) phenotype [9,10]. The link between P-gp expression and cell volume regulation has been widely debated over the past decades [1,11,12]. Some studies reported that the P-gp is able to increase the magnitude of cell volume activated chloride currents, and so modulate the RVD [13–16]. On the contrary, other studies did not find any functional correlation between P-gp and volume regulation [17–20]. In 2005, our team demonstrated the role of P-gp in the regulation of volume activated chloride currents using wild-type human breast cancer MCF7 cells, and a doxorubicin-selected MDR variants [21]. Recently, it has been shown that apoptotic resistance in MDR P-gp overexpressing ascite tumour cells, involved impairment of the apoptotic volume decrease (AVD), an apoptotic event similar to the RVD [22]. Finally, in 2012, a team working on hepatocytes of the fish, *Sparus aurata*, suggested the involvement of the P-gp in the RVD [23].

The difference between all these studies could originate from the models and the methods used. Therefore, technological challenges still exist and the technique's accuracy to measure both cell volume and P-gp activity/expression seems essential.

In this paper, we develop an original method based on flow cytometry analysis to study the role of the P-gp in the RVD in response to hypo-osmotic shocks. By using the Coulter-type channel of the cytometer Cell Lab Quanta SC MPL, we were able to follow accurately the electronic volume (EV) of the cells during either osmotic and/or pharmacologic treatment types.

2. Results

2.1. Resistant and Sensitive Variants Display Different Cell Volumes, Shapes and Membrane Capacitance

Our group previously studied P-gp overexpression and activity in the MCF7/Doxo variant [24–26]; cell volumes of MCF7 and MCF7/Doxo were measured using different methods (Figure 1). In classical microscopy, morphological differences between the two variants could be verified (Figure 1A). The MCF7 appeared more birefringent and round, whereas the MCF7/Doxo were more flat and spread. Using Hoffman modulation contrast imaging on a freshly plated 50:50 co-culture, we revealed a clear morphological difference between MCF7 and MCF7/Doxo (Figure 1B top panel). After four days of co-culture, it was possible to observe a stable and unique spatial organization with the formation of MCF7 islets surrounded by MCF7/Doxo (Figure 1B bottom panel).

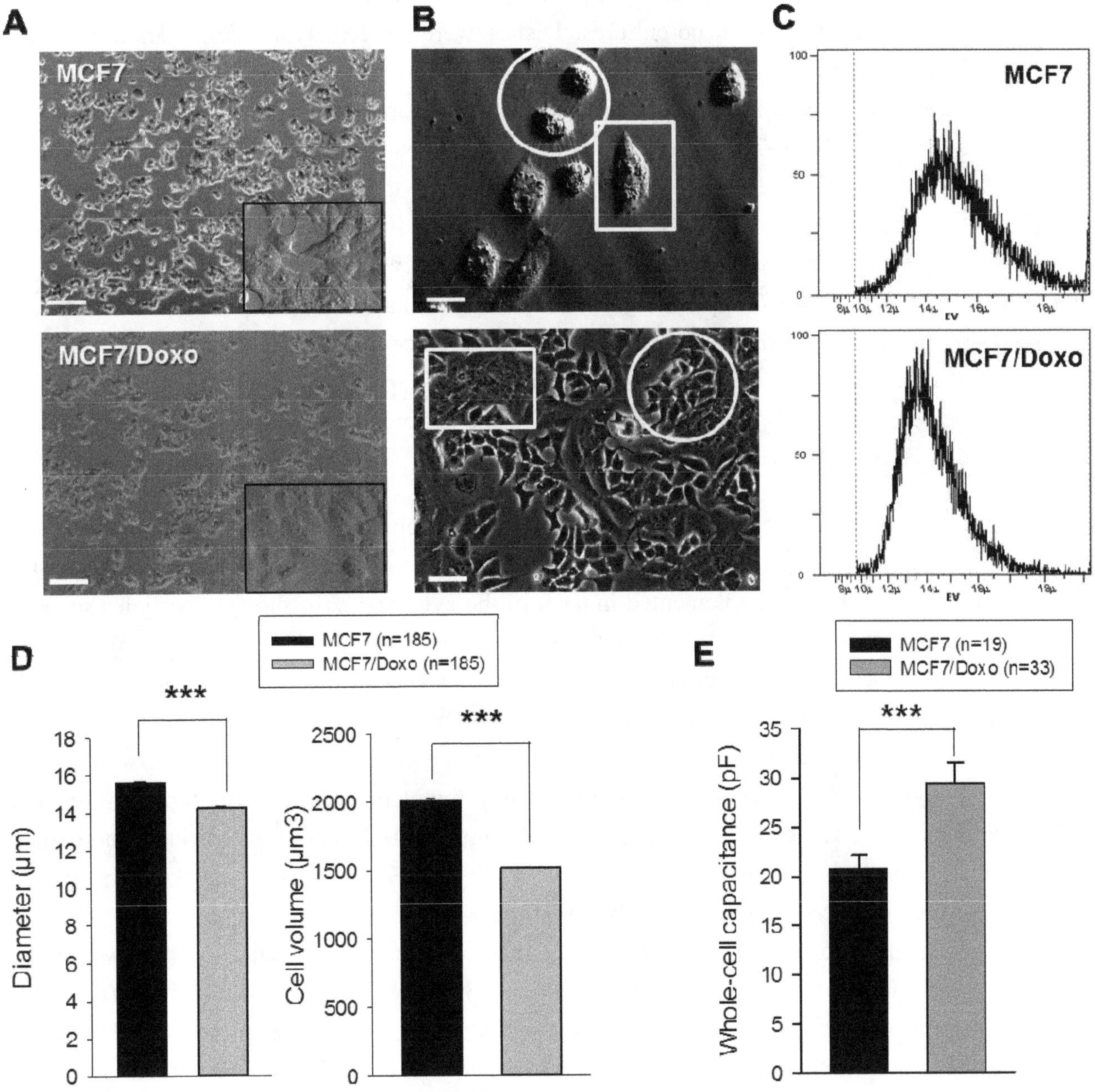

Figure 1. *Cont.*

F

	Patch Clamp		Flow cytometry	
	MCF7	MCF7/Doxo	MCF7	MCF7/Doxo
Capacitance (pF)	**20.7**	**29.4**	------	------
Surface (cm^2)	2.07×10^{-5}	2.94×10^{-5}	1.89×10^{-6}	1.54×10^{-6}
Radius (µm)	25.67	30.59	7.80	7.11
Volume (μm^3)	70 847	119 918	**2004**	**1513**

Figure 1. MCF7 and MCF7/Doxo display different morphological features. (**A**) Spatial organization of MCF7 and MCF7/Doxo in phase contrast microscopy. Scale bar: 50 µm; (**B**) Hoffman modulation contrast (**top**) and phase contrast micrographs (**bottom**) of MCF-7 and MCF-7/Doxo in co-cultures. Dishes were seeded with a 50:50 mixture of MCF-7:MCF7/Doxo at day zero. Morphological differences permit an immediate identification of each cell subpopulation islets. MCF7 appeared birefringent and round (rounds) whereas MCF-7/Doxo are more flat and spread (square). Scale bar: 20 µm; (**C**) Cell volume of MCF7 and MCF7/Doxo in flow cytometry. The electronic volume (EV) was determined by the flow cytometer according to the Coulter Principle; (**D**) Histograms giving the mean cell diameter (**left**) and volume (**right**) for 185 repeated experiments. The MCF7 appear significantly bigger than the MCF7/Doxo. Data are presented as mean ± SEM; (**E**) Whole-cell capacitance measurements of MCF7 and MCF7/Doxo. MCF7/Doxo display a higher whole-cell capacitance than the MCF7. Data are presented as mean ± SEM; (**F**) Summary of the different size measured or calculated for the MCF7 and MCF7/Doxo. *** $p < 0.001$.

These differences have been confirmed using the Cell Lab Quanta MPL flow cytometer. This system exploits the Coulter principle for an accurate volume determination instead of the low-angle laser light scattering technique implemented in most of the cytometers. In short, as particles suspended in a saline solution are drawn through the small aperture of an insulated electrical sensor, they displace an equal volume of electrolyte solution that creates a resistance and leads to a voltage pulse. The voltage pulse intensity is proportional to the particle volume, thus called Electronic Volume (EV). Figure 1C displays the diameter (in µm) of the cells in suspension in a 300 mOsmol/kg H_2O. A statistical analysis performed on more than 300 measurements revealed that the MCF7 were smaller in diameter and volume compared to the MCF7/Doxo (Figure 1D). It is also interesting to note that the MCF7 seemed to be less homogeneous in size than the MCF7/Doxo as revealed by the variation coefficient for the diameters (40.03% ± 0.05% and 30.85% ± 0.85%, respectively).

Finally, using electrophysiology, we measured the whole-cell capacitance of the two variants, 20.7 ± 1.4 pF (n = 19) for MCF7 and 29.4 ± 2.1 pF (n = 33) for MCF7/Doxo, respectively (Figure 1E). Considering a constant specific capacitance of $C_S = 1$ µF/cm^2 [27] for the plasma membrane, these results indicate that the membrane electric surface is higher in the MCF7/Doxo compared to wild-type. This observation seems to contradict the one obtained by the volume Coulter (Figure 1F).

2.2. Cell Volume Monitoring during Hypo-Osmotic Shocks

Flow cytometry coupled to Coulter EV measurements represents a valuable approach to monitor cell size variations in real-time. Thus, we have used this possibility to carry out analysis of volume change time-course of cells undergoing osmotic challenges in suspension at a low flow rate (25 µL/min). With these settings, the cell volume distributions can be determined over 20 min. This approach is better than the traditional volume coulter method that allows only a static measurement of the cell volumes. As shown in Figure 2A, cell volumes of the two variants were stable over the 20 min period. However, during 50% hypo-osmotic shocks (150 mOsmol/kg H_2O), a significant swelling of both variants was detected two minutes after the substitution of the isotonic solution with the hypotonic one (Figure 2B). The temporal monitoring of the volume compensation, RVD, revealed important differences between the two variants (Figure 2C). While the MCF7/Doxo cells were able to compensate the swelling drove by hypotonicity, the MCF7 cells could not. For the MCF7/Doxo cells, cell volume normalization appeared after less than 10 min, whilst no RVD mechanism was noticed after 20 min for the MCF7 cells.

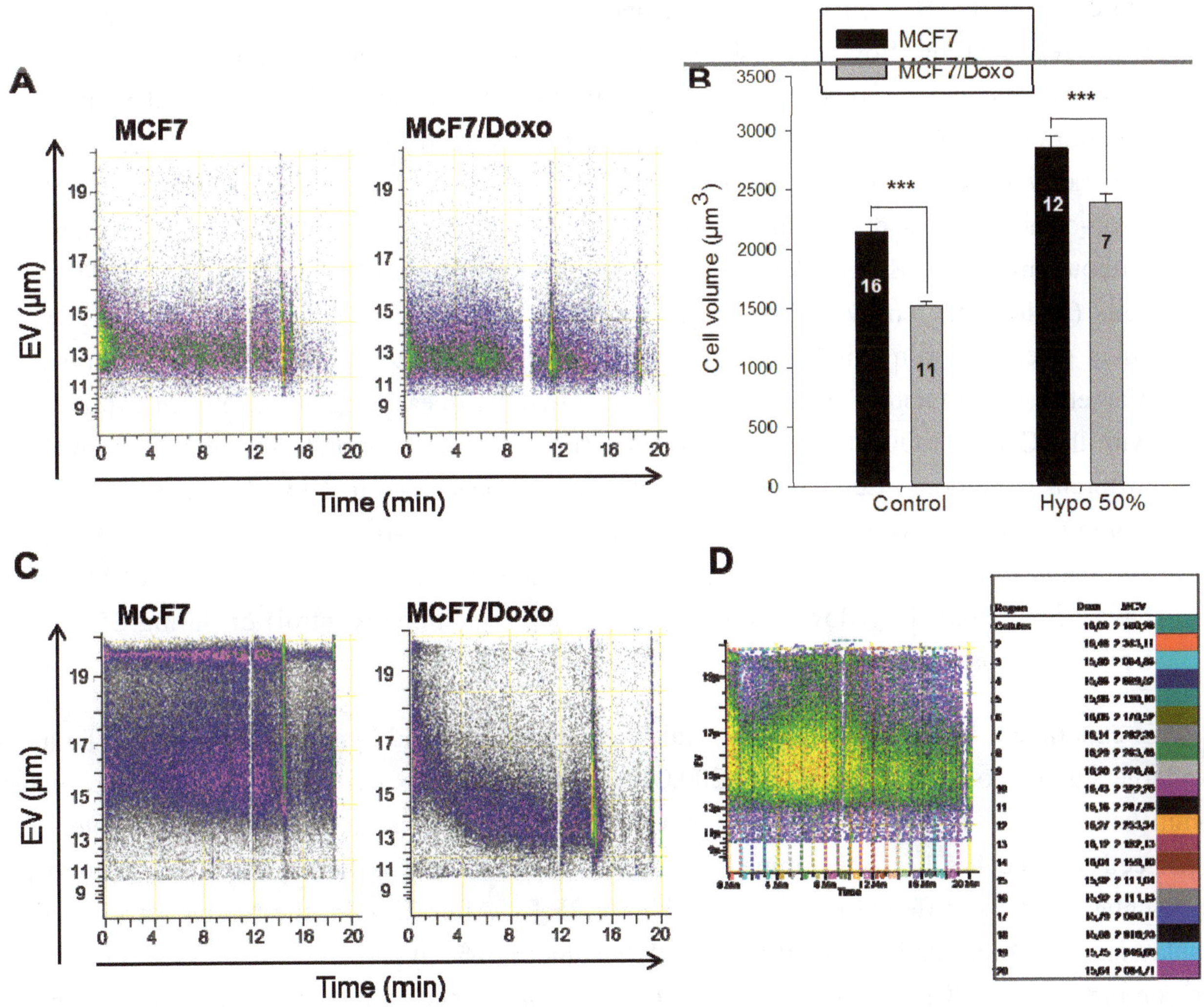

Figure 2. *Cont.*

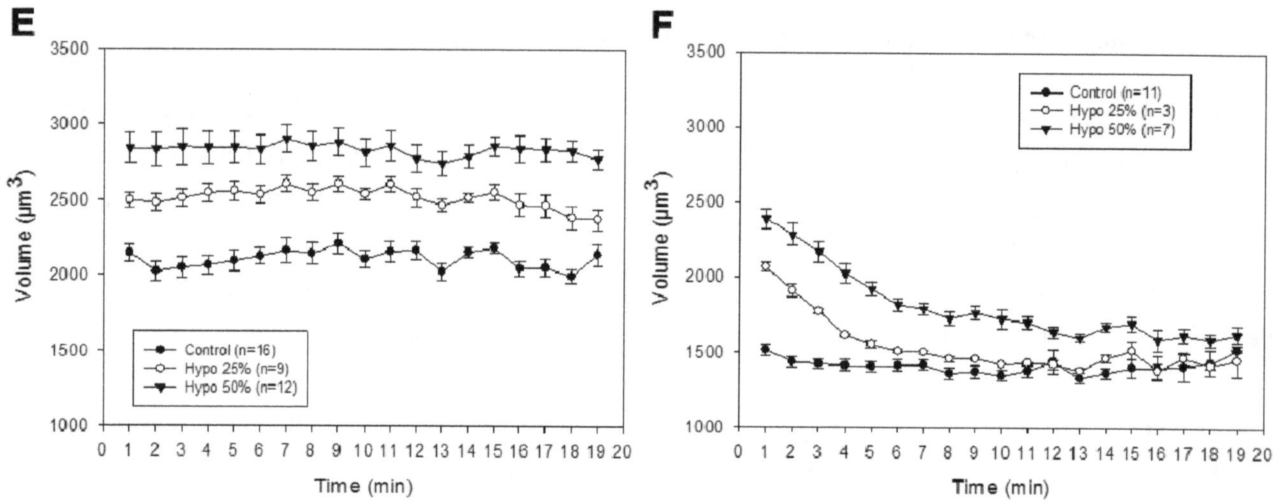

Figure 2. Cell volume monitoring during hypo-osmotic shocks. (**A**) Cell volume monitoring of MCF7 and MCF7/Doxo cells in normotonic conditions. The two graphs represent flow cytometry plots of the electronic volume (EV) *vs.* time, during the 20 min of the experiment; (**B**) Cell volumes after one minute of hypo-osmotic stress. Cell volumes were recorded by flow cytrometry before (control) or after one minute of a 50% hypo-osmotic shock (Hypo 50%). Data are presented as mean ± SEM; (**C**) Cell volumes monitoring in MCF7 and MCF7/Doxo cells during a hypo-osmotic shocks. The two flow cytometry plots represent the electronic volume (EV) of the cells subjected to a 50% hypo-osmotic stress during the 20 min. MCF7 cells (**left** plot) increased their volume without any compensation phenomenon. After a significant volume increase, MCF7/Doxo cells (**right** plot) retrieved their original volume; (**D**) Data extraction method. Nineteen equal gates of 1 min length have been created and applied to all samples. The mean cell diameter (Diam.) and of the mean cell volume (MCV) in each region have been extracted with the Cell Lab Quanta analysis software; (**E**) Cell volume monitoring of MCF7 under different conditions. Cell volumes of MCF7 have been recorded by flow cytometry in isotonic conditions (control) or during 25% or a 50% hypo-osmotic challenges; (**F**) Cell volume monitoring of MCF7/Doxo cells under different conditions. Cell volumes of MCF7 have been recorded by flow cytometry in isotonic conditions (control) or during 25% or a 50% hypo-osmotic challenges. *** $p < 0.001$.

This experiment has been repeated several times in isotonic and hypotonic conditions. To analyze the large number of points generated (200,000 cells analyzed/experiment), 20 successive gates of 1-min cell volume continuous recording have been set (Figure 2D). For each 1-min interval, the mean cell volume has been determined. In these conditions, it has been possible to draw a graph representative of the different experiments (Figure 2E,F). After 20 min, the MCF7 were not able to display any RVD mechanism, neither under 25 nor under 50% hypo-osmotic shocks (Figure 2E). On the contrary, in the MCF7/Doxo cells, a RVD was set up immediately and re-established normotonic volume values after 7 min in both 25% and 50% hypo-osmotic shocks (Figure 2F).

2.3. RVD in MCF7/Doxo Cells Is Dependent on P-gp Activity

The role of efflux activity of P-gp in the regulation of the RVD has been investigated using different ligands of P-gp chosen for their different mechanisms of action. Thus, a non-competitive inhibitor, zosuquidar (20 μM), a conformational monoclonal antibody, UIC2, and a P-gp substrate used in chemotherapy treatment of breast cancer, doxorubicin (20 μM), have been used. First we quantified the effect of each compound on the P-gp efflux activity with the calcein-AM as a fluorescent allocrite probe (Figure 3A). Cells expressing high levels of P-gp rapidly extrude nonfluorescent calcein-AM from the plasma membrane. As a result, fluorescence intensity is inversely related to P-gp [26]. The conformational antibody UIC2 and zosuquidar completely abolished the P-gp activity in MCF7/Doxo cells, while doxorubicin did not alter the ability of MCF7/Doxo to expel calcein. In isotonic conditions, the cell volumes of MCF7/Doxo were not modified by any of the three ligands (data not shown). On the contrary, after 50% hypo-osmotic shocks, RVD was inhibited by the 3 P-gp modulators (Figure 3B). Analysis at 1 min shows clearly that the untreated MCF7/Doxo cells were already engaged in a RVD process while, in the presence of zosuquidar or doxorubicin, they did not even reach their maximal volume (Figure 3C). After 7 min, the three ligands were able to significantly decrease RVD kinetics in MCF7/Doxo. Comparisons of the RVD rates revealed that doxorubicin slowed down most of the RVD process (Figure 3D). The experimental points of five different experiments, under doxorubicin treatment, were fitted with straight lines (Figure 3E). The small slope demonstrated the nearly total inhibition of the RVD in the presence of doxorubicin.

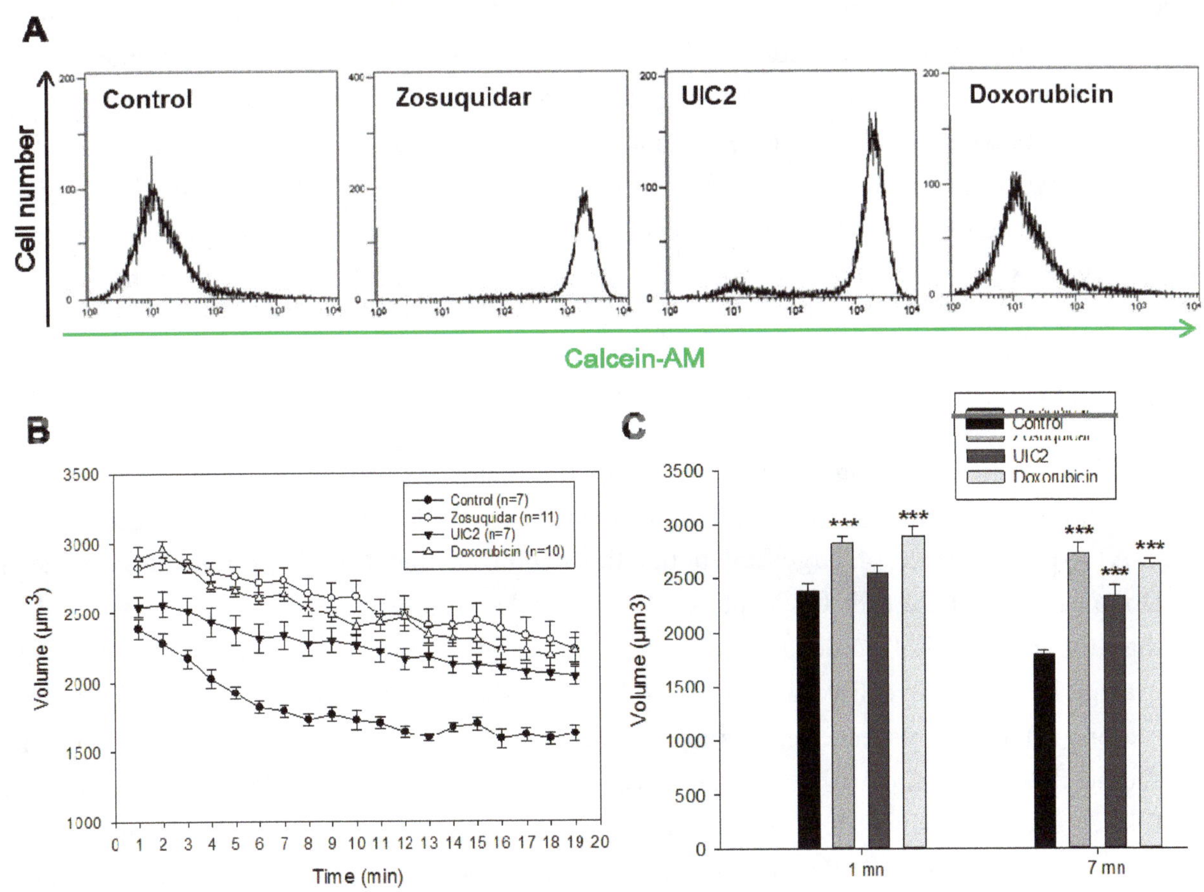

Figure 3. *Cont.*

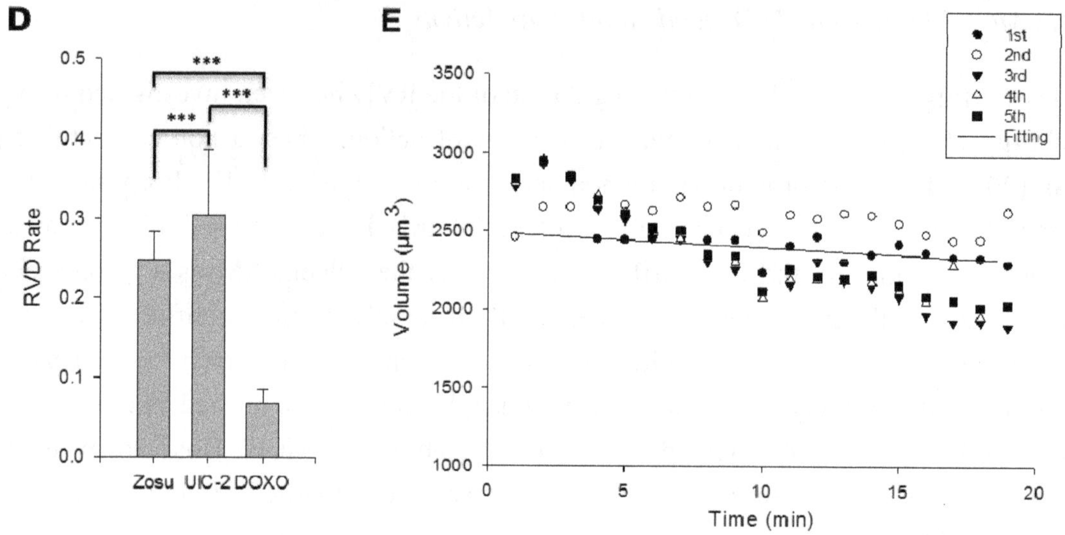

Figure 3. RVD in MCF7/Doxo is dependent on the P-gp activity. (**A**) P-gp activity. P-gp activity was followed with calcein-AM as a fluorescent probe. In each flow cytometry measurement, a sample of 10,000 cells was analyzed. Superimposed all-events histograms of calcein fluorescence distribution (log scale) in control MCF-7/Doxo cells (**left** plot) and MCF-7/Doxo pre-incubated with the P-gp antagonist zosuquidar and UIC2 (10 μM, **middle** plots) or with doxorubicin (**right** histogram); (**B**) Cell volume monitoring of MCF7/Doxo in presence of P-gp modulators; (**C**) The histogram represents the mean of the cell volume after 1 or 7 min of 50% hypo-osmotic stress for the experiments presented in **B**. While under the control conditions (black) the cells retrieved their original volume after 7 min, the cells in presence of the P-gp modulators were not able to compensate their volume increase. Data are presented as mean ± SEM, *** $p < 0.001$.; (**D**) RVD rate of MCF7/Doxo cells in 50% hypo-osmotic shock (*** $p < 0.001$); and (**E**) Graphs show MCF7/Doxo cell volumes from five independent samples analyzed by flow cytometry after 50% hypo-osmotic conditions in the presence of doxorubicin. The data were fitted with Sigma Plot.

2.4. P-gp Activity and Cell Cycle

In isotonic conditions, cell proliferation is one of the major events leading to cell size regulation. Thus, in the present work, cell volume changes were analyzed for the two variants during the cell cycle (Figure 4A). Measurements of large cell populations indicate that MCF7 exhibits larger size than MCF7/Doxo cells. In phase S or G2/M phases however, the volume of MCF7/Doxo was found to be higher than that of MCF7. In order to study the P-gp activity during the cell cycle, we developed a gating strategy based on the cell volumes defined by the cell cycle analysis (Figure 4B). However, cell size variations affect fluorescence signals and induce alteration in cell calcein content assessments. To overcome these potential variations, we have used the FL1-FC parameter of the cytometer, in which fluorescence intensity (FL1) was normalized as a ratio of fluorescence concentration to an accurate cell sizing determined by Coulter-type electronic volume (EV) [26]. Results indicate that the basal fluorescence of the calcein was more important in gate 1 (corresponding to the cells in G0/G1), which correlates with a lower P-gp activity (Figure 4C). The best P-gp activity was detected in cells in

G2/M which are the cells displaying the larger size. After this, we used a non-competitive P-gp inhibitor, the PSC833, and tested its efficacy on the calcein accumulation at the different stages of the cell cycle (Figure 4D). Dose response curves represented in Figure 4E allowed us to determine EC_{50} (half maximal effective concentration) and E_{max} (maximal effect) values of PSC833 for each gate. PSC833 maximal effect was increasing with the progression of the cells in the cell cycle, without any modifications of potency (EC_{50}).

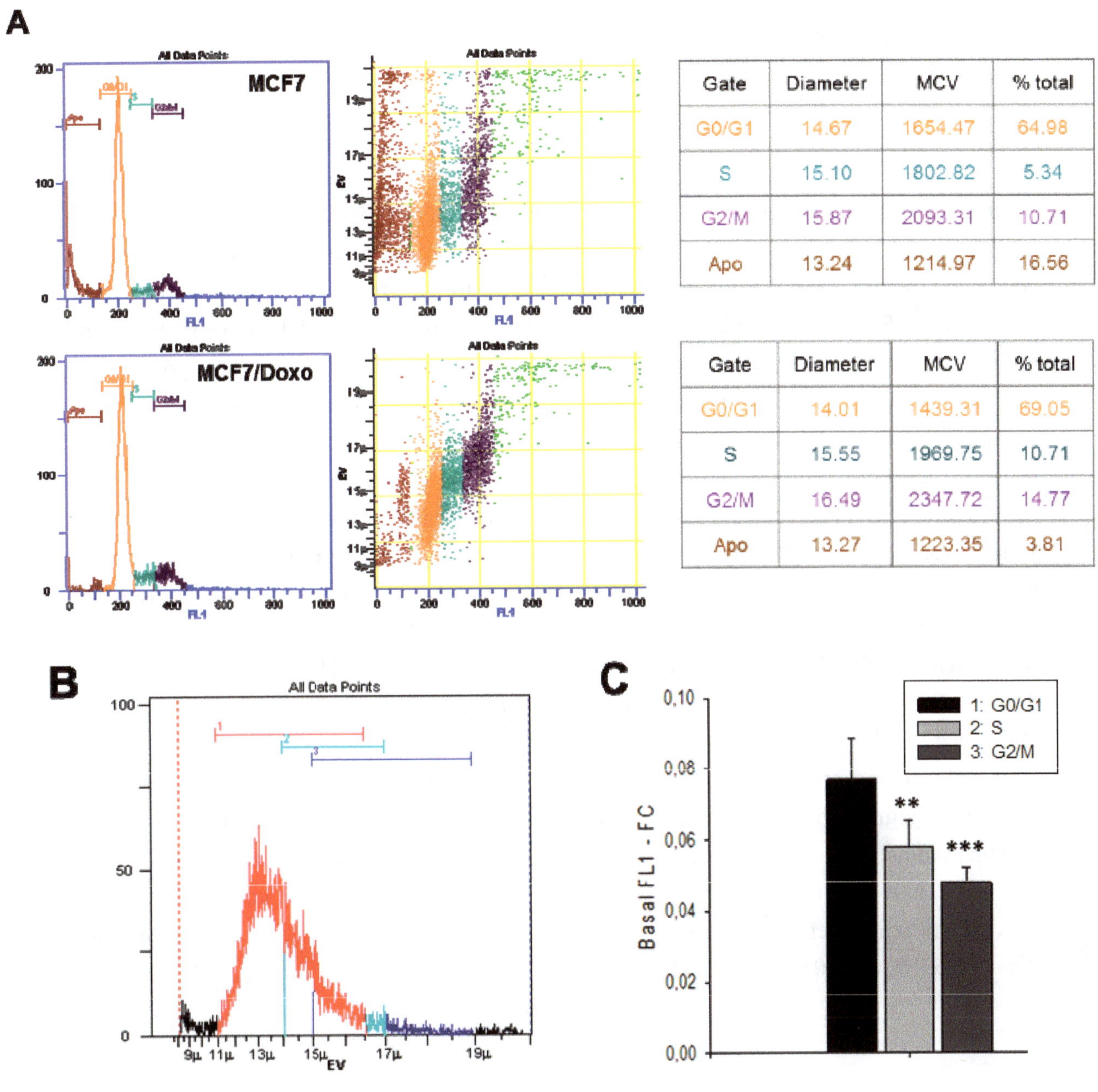

Gate	Diameter	MCV	% total
G0/G1	14.67	1654.47	64.98
S	15.10	1802.82	5.34
G2/M	15.87	2093.31	10.71
Apo	13.24	1214.97	16.56

Gate	Diameter	MCV	% total
G0/G1	14.01	1439.31	69.05
S	15.55	1969.75	10.71
G2/M	16.49	2347.72	14.77
Apo	13.27	1223.35	3.81

Figure 4. *Cont.*

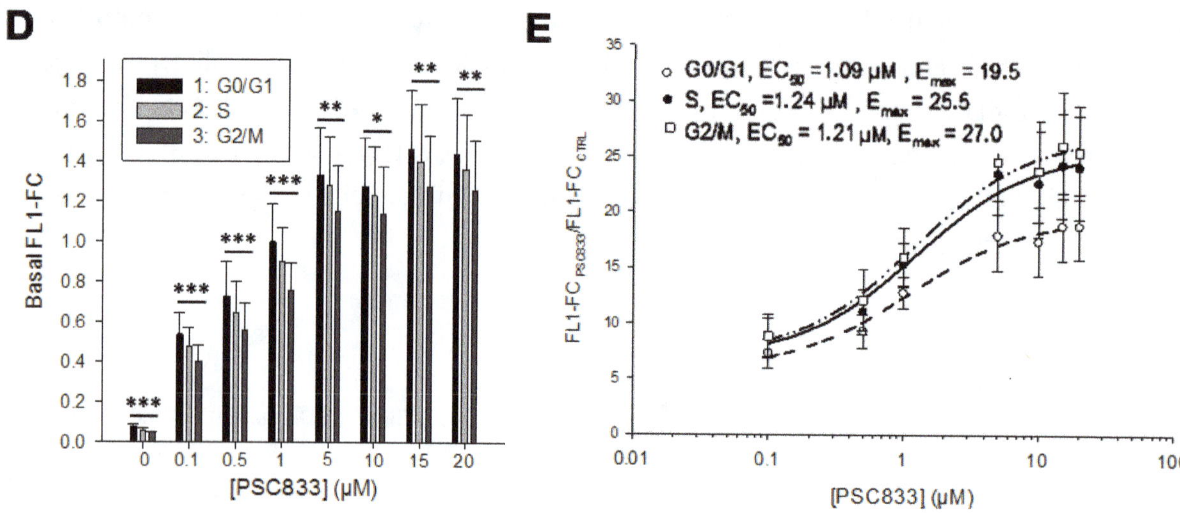

Figure 4. MCF-7/Doxo cells exhibit a cell cycle-dependent P-gp activity. (**A**) Cell cycle of MCF7 and MCF7/Doxo was determined by DNA content measurements with the nucleic acid dye Hoechst 33342. Four gates corresponding to hypoploid particles (Apo), cells in G0/G1, cells S or cells G2/M phases were defined based on the FL1 level (**left** plots) and applied on the cell volume EV (**right** plots). The mean cell volume of the MCF7 and MCF7/Doxo were extracted from each gate (tables); (**B**) The gates defined in A on the FL1 parameter were switched on the EV channel to be applied to the following experiments; (**C**) P-gp activity within the cell cycle. The P-gp activity was assayed with calcein-AM and the amount of fluorescence per cell was expressed as FL1-FC (fluorescent light in channel 1-fluorescence concentration) which is the fluorescent light (FL) divided by the electronic volume (EV) determined by the flow cytometer according to the Coulter Principle; (**D**) P-gp inhibition with PSC833 within the cell cycle. Different concentrations of PSC833 were used and the P-gp activity was assayed with calcein-AM. The histogram represents the FL1-FC mean in each gate. The stars represent the difference between the three groups after an ANOVA test; and (**E**) PSC833 dose response curves expressed as mean FL1-FC. Each point represents mean ± SEM (10 independent experiments) of FL1-FC, expressed as the ratio of signals in the presence of a blocker to signal in control conditions in each gate. Four-parameter logistic dose-response curves were fitted to the data to obtain blocker potencies (half-maximal effective concentration, EC_{50}) and efficacies (maximum response, E_{max}). Data are presented as mean ± SEM with $n = 10$ independent assays per data point. * $p < 0.05$, ** $p < 0.01$, *** $p < 0.001$.

3. Discussion

The role of P-gp in cell volume regulation has been discussed widely over the past 20 years. In the present study, we used an original, non-intrusive and label-free method based on flow cytometry coupled to Coulter volume determination to compare RVD in wild-type and doxorubicin-resistant MCF7, during hypo-osmotic challenges. This technique allowed us to follow not only the volume variation of cells in a live scenario, but also to concurrently study cell volumes and P-gp activity.

First, we demonstrated that MCF7 cells do not respond as MCF7/Doxo cells to hypo-osmotic shocks. In fact, while MCF7 cells underwent a persisting volume increase throughout the 20 min of the experiment, MCF7/Doxo cells were able to offset osmosis and to cancel swelling in less than ten minutes, suggesting strongly that the P-gp is promoting RVD. Historically, because of sequence similarities with an ABC transporter functioning as a chloride channel (the cystic fibrosis transmembrane regulator, CFTR), a role for P-gp in volume regulation has been investigated and, as a matter of fact, confirmed in P-gp overexpressing cells [13,28]. Thus, in 1992, Gill *et al.* stated that P-gp could form a channel itself, or a component of such a channel [28]. Two years later, Sadini and collaborators demonstrated in *mdr1/ABCB1* transfected NIH-3T3 cells that short-term hypotonic conditions caused an inhibition of P-gp activity [29]. In 1998, using a drug-sensitive cell line (MCF-7) and a P-gp-expressing derivative (BC19/3), the authors demonstrated an increase in the magnitude of cell volume activated chloride currents in BC19/3 cells, but ruled out the possibility of P-gp being itself the channel responsible for the volume-activated currents [14]. Conversely, other studies concluded that RVD and osmoregulatory chloride currents were not related to P-gp expression in resistant cell lines or in cells transfected with *mdr1/ABCB1* transcripts [17,19,20,30] or injected with the protein [31]. To add to the confusion, in 2002, Chen *et al.* abolished volume-activated chloride currents in bovine pigmented ciliary epithelial cells by using *mdr1/ABCB1* antisense oligonucleotides; suggesting their dependence on endogenous P-gp expression [32]. In 2005, using the same cell lines as in the present study, our team demonstrated that during hypotonic challenges, swelling-activated chloride currents were significantly activated faster and with larger densities in MCF7/Doxo cells than in MCF7 cells [21]. We also demonstrated the inhibition of this current by P-gp ligands, including conventional substrates such as doxorubicin, as well as antibodies. These results are in accordance with the total inhibition of RVD obtained in the presence of doxorubicin herein. In addition, our study is the first that combines non-invasive measurements of the cell volume and a cell model overexpressing the P-gp only by selection and not by transfection or injection.

More recently, in 2014, a relation between P-gp, AQP5 and drug resistance has been established [33]. By using AQP5-siRNA to silence AQP5 in the colon-derived cell line HT-29, the authors obtained a decrease of P-gp expression as well as of other actors of drug resistance, such as GST-π, and TOPO II. These findings are of prime importance since aquaporins are known to be involved in the early phase of apoptosis, characterized by a cell shrinkage, named apoptotic volume decrease (AVD) [34]. For instance, the under-expression of aquaporin AQP8 and 9, in hepatocellular carcinoma, is responsible for the resistance to starvation or TGFβ-induced apoptosis [35]. Resistance to apoptosis is one of the principal features of tumor cells [22,36,37], hence we could hypothesize a role of P-gp in AVD-impairment as a mechanism associated with death evasion. In the presence of cytotoxins such as doxorubicin, and while pumping the substrate out of the cells, the P-gp could be involved in pathways counteracting cell shrinkage and may thus contribute to avoid apoptosis. Alteration of AVD could therefore be added to the classical mechanisms of resistance that the cancer cells use against chemotherapeutic agents.

In parallel to the mechanisms of cell volume regulation in hypotonic conditions, we demonstrated obvious differences in term of morphology and cell volume between the two variants, in regular osmotic pressures. MCF7 cells in suspension in an isotonic solution have an average volume of about 2000 μm^3, while MCF7/Doxo cells have a much smaller volume of 1500 μm^3. This is in accordance with the results of Yang and collaborators demonstrating that human ovarian cancer cell line SKOV3 that started to express P-gp after a selection with cisplatin, exhibited dramatic changes in morphology, including reduction in cell size, loss of cellular projections and clustering [38]. In ascites tumor cell lines, a high correlation between the ability to pump out daunorubicin (by the P-gp), and the decrease in cell volume detected has also been demonstrated in resistant cell lines [39].

The measurement of the membrane capacitance in the whole cell configuration of the patch clamp technique revealed that the enclosed volume of the membrane is theoretically 70,847 μm³ for MCF7 cells and 119,918 μm³ for MCF7/Doxo cells. This technique gives some information on the total area of the plasma membrane. Therefore, although the difference between the values obtained by Coulter volume and volume per surface area may seem significant, the folding structure of the membrane, which indeed contributes to the membrane capacity and not to the volume, could be an explanation. Thus, the measurements of the membrane capacitance suggest that the two variants have a different membrane conformation, whereby MCF-7/Doxo cells contain more membrane folding than MCF-7 cells (Figure 5). The volume differences between the two variants could also be explained by the difference in terms of morphology in phase contrast on adherent cells, since it was mentioned that MCF7 cells are more birefringent, so more swollen than MCF7/Doxo cells. We can hypothesize that through peculiar interactions with lipids and favorable locations in the plasma membrane, P-gp can influence the membrane structure [40–42].

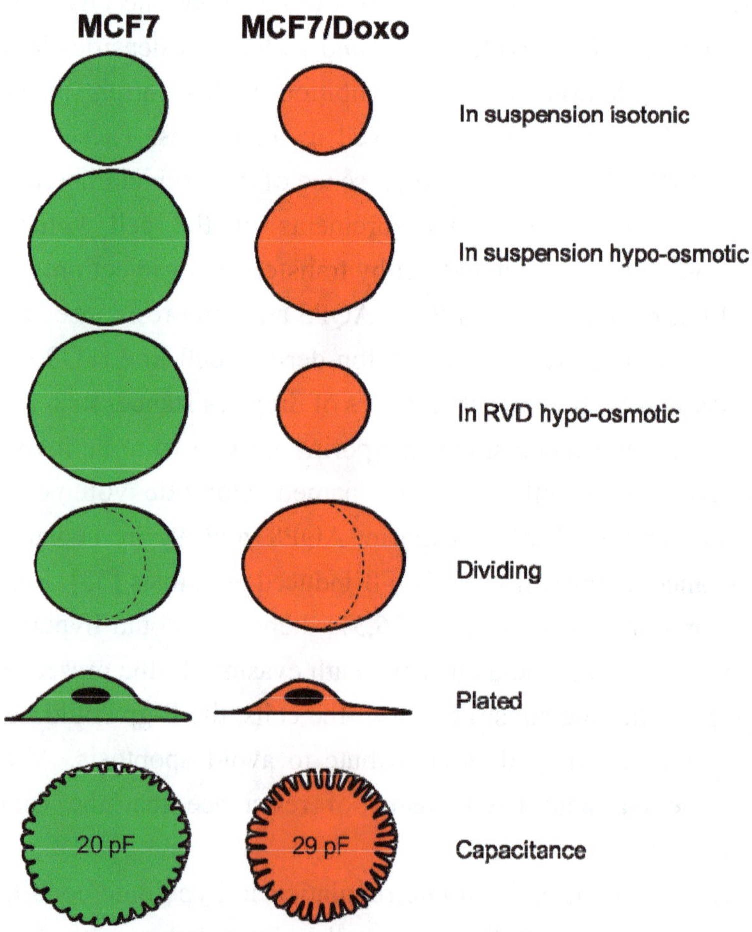

Figure 5. Schematic representation of MCF7 and MCF7/Doxo cells in each situation.

Recently, many authors raised the possibility of a G1/S volume checkpoint controlling the progression through the cell cycle [43,44]. Even if the concept of a cell size checkpoint in the cell cycle has been well established in yeast and other organisms [45], it yet remains controversial in mammalian cells [46–48]. Some authors suggest that translational checkpoints at the G1/S transition could set a lower cell size limit [49–51], while others claim the importance of an upper limit of cell volume in progression through M phase [52,53]. Even if the cell volume regulation seems mandatory for the cell cycle progression and cell division, it remains unclear if these events are triggering the cell cycle itself or if they are just actors activated through the regular cyclin dependent kinase/cyclin (Cdk/cyclin) complex [43,44]. As per our results reported here, it seems that MCF7 cells are able to undergo a complete normal mitosis. Moreover, in a previous publication, we confirmed that there is no difference between the proliferation capacity of MCF7 and MCF7/Doxo, except the resistance to chemotherapy drugs exhibited by MCF7/Doxo cells [25]. Therefore, even if during the 20 min of hypotonic stress MCF7 cells were not able to display any efficient volume compensation, it seems that they are completely able to activate volume regulatory mechanisms in the cell cycle context.

Finally, we were also able to show that the cells involved in mitosis display the highest P-gp activity. During mitosis, cells are subjected to constant volume variation that may reinforce the role of P-gp in cell volume regulation. Moreover, many drugs used in chemotherapy target cell division, and a high expression level of P-gp has been linked to resistance to these drugs [54,55]. We can thus suggest a new mechanism of resistance to this drug, independent of the drug efflux abilities of P-gp; this data therefore allows us to imagine the existence of new mechanistic consequences of P-gp overexpression.

4. Material and Methods

4.1. Cell Cultures

The study was carried out with human breast carcinoma derived cells, MCF-7 cells and its multi-drug resistant variant (MCF-7/Doxo); kindly provided by J.-P. Marie (Hôtel Dieu, Paris, France). MCF7/Doxo cells were isolated by stepwise selection with increasing concentrations of doxorubicin [56]. Cells were maintained in RPMI 1640 (Sigma, St. Louis, MO, USA) containing 5% of heat-inactivated fetal bovine serum (Sigma), 2 mM L-glutamine (Sigma, St. Louis, MO, USA), 1% antibiotic/antimycotic solution (Sigma), and incubated in a humidified atmosphere containing 5% of CO_2 at 37 °C. Doxorubicin (1 μM) was added to the culture medium for the maintenance of the multi-drug resistant phenotype of MCF7/Doxo cells.

4.2. Reagents

All reagents were of the highest grade of purity and quality available. Purified doxorubicin (DOXO), DMSO, and phosphate buffer saline (PBS buffer, pH 7.4) were purchased from Sigma. Calcein acetoxy-methylester (calcéine-AM) was supplied by Invitrogen Life Technologies (Carlsbad, CA, USA). SDZ PSC833 and zosuquidar were kindly provided by J.-P. Marie (Hôtel Dieu, Paris, France). The final concentration of DMSO and H_2O was less than 0.1%.

4.3. Flow Cytometry

For P-gp expression, activity or cell cycle, the fluorescent light (FL) was quantified using a Cell Lab Quanta SC MPL flow cytometer (Beckman Coulter, Fullerton, CA, USA) equipped with a 22 mW 488 nm excitation laser. Voltage settings of photomultipliers were not modified throughout the experiments [24,25]. Each analysis consisted of a 10,000 events record, triggered on electronic volume (EV) as primary parameter, according to a particle diameter exceeding 8 μm.

For the volume monitoring, the cytometer was operated at a flow rate of 25 μL/min. Cell sizes were accurately determined using the Coulter-type electronic volume (EV) channel of the cytometer, after calibration with 10-μm FlowCheck microspheres [57]. The cell volume distribution was measured for 20 min.

4.4. Osmostic Challenges and RVD Rate Calculation

Osmolality of solutions was checked and adjusted by using a Wescor Vapro 5520 vapor pressure osmometer (Wescor Elitech, Logan, UT, USA). RVD rate is the time constant. It has been calculated using the Software Sigma Plot 11 (Systat Software Inc., Chicago, IL, USA). Data have been fitted with the Regression wizard using the Exponential Decay function. There are three parameters and the equation is $y = y0 + a \times e(-bx)$, where "b" is the decrease constant corresponding to the time constant represented in the bar graph (Figure 3D).

4.5. P-gp Activity

The calcein-AM efflux assay was used as previously described [26]. Briefly, resuspended cells were loaded with 0.25 μM calcein acetoxy-methylester (Invitrogen) in RPMI for 15 min at 37 °C in the dark. Green FL was quantified via the FL1 channel (log scale) through a 525 nm band pass filter.

4.6. Cell Cycle

By using a cell-penetrating DNA-binding dye, Hoechst 33342, cellular DNA content could be measured without fixation and permeabilization. Briefly, cells were incubated with 20 μg/mL Hoechst 33342 in cell culture medium for 45 min at 37 °C. The stained cells were analyzed by the Quanta SC MPL using the Hg arc lamp with a 355/37BP excitation filter as excitation and detected on the FL1 with a 465/30BP filter. Data acquisition was carried out by triggering on FL1.

4.7. Electrophysiology

For patch-clamp recordings, cells were allowed to attach for 3–4 h to 20 mm diameter glass cover slips. The electrophysiological studies were performed at room temperature (23–25 °C) on the stage of an inverted microscope (Nikon TE2000, Champigny-sur-Marne, France) using the whole-cell configuration of the patch-clamp technique in the voltage-clamp mode. Patch pipettes were pulled from borosilicate glass capillaries (Harvard Apparatus, Holliston, MA, USA) with a P-97 horizontal puller (Sutter Instrument, Novato, CA, USA). The intracellular (pipette) solution contained (in mM): 134 CsCl, 1.8 $CaCl_2$, 1 $MgCl_2$, 2 ethylene glycol-*bis*(b-aminoethylether)-*N*,*N*,*N'*,*N'*,-tetraacetic acid

(EGTA), 10 4-(2-hydroxyethyl)piperazine-1-ethanesulfonic acid (HEPES) and 3 Na2ATP; adjusted to pH 7.2 with CsOH. The 20%-hypotonic bath solution (240 mOsmol/kg H_2O) contained (in mM): 2.8 tetraethyl ammonium (TEA)-Cl, 100 *N*-methyl-D-glucamine (NMDG)-Cl, 2 $MgCl_2$, 1 $CoCl_2$ and 10 HEPES, adjusted to pH 7.4 with HCl. In order to avoid any change in ECl, the osmolality was set at 300 mOsM by adding mannitol. When in the bath and filled with the intracellular solution, patch pipettes had a resistance between 2 and 4 MΩ.

Recordings were made with an Axopatch 200B patch-clamp amplifier interfaced to a 1.5 GHz computer via a Digidata 1322 and pClamp 8.0 software (Axon instruments, Foster City, CA, USA). Cells with access resistances exceeding 10 MΩ immediately after gaining the whole cell configuration or during hypotonic challenge were discarded.

4.8. Statistical Analysis

All quantitative data were expressed as mean ± standard error of the mean (SEM). Statistical analyses were performed with SigmaPlot 11 (Systat Software Inc., Chicago, IL, USA). A Shapiro-Wilk normality test, with a $p = 0.05$ rejection value, was used to test normal distribution of data prior further analysis. All pairwise multiple comparisons were performed by one way ANOVA followed by Holm-Sidak posthoc tests for data with normal distribution or by Kruskal-Wallis analysis of variance on ranks followed by Tukey posthoc tests, in case of failed normality test. Paired comparisons were performed by Student's *t*-tests or by Mann-Whitney rank sum tests in case of unequal variance or failed normality test. Statistical significance was accepted for $p < 0.05$ (*), $p < 0.01$ (**) or $p < 0.001$ (***). All experiments were performed in triplicate.

For the purpose of IC_{50} value calculation (dose-response curves of calcein-AM assay), data were fitted to a sigmoidal three parameters dose-response model [58]:

$$y = b + (a - b)/(1 + 10 (\log IC_{50} - x))$$

where (y) is response, *i.e.*, the ratio of mean FL1-FC in the presence of PSC833 to mean FL1-FC in control condition, (b) represents minimum of response, (a) represents maximum of response Emax, (x) is logarithm of PSC833 concentration and IC_{50} (or EC_{50}) is the concentration of inhibitor that corresponds to 50% of maximal effect.

5. Conclusions

In this study, we show evidence for a role of P-gp in cell volume regulation by combining microscopic observations, coulter volume, flow cytometry and electrophysiology. Taken together, our results strongly suggest that overexpression of P-gp in MCF7/Doxo cells has an impact on cell volume regulation in regular osmotic pressure and also during hypo-osmotic stresses. We highlight unexpected differences between two MCF-7 variants in regular osmotic conditions. Under osmotic challenge, P-gp overexpression influences cell volume regulation; it can then be supposed that P-gp overexpression contributes to chemoresistance not only by drug efflux, but also by volume stabilization and AVD-impairment. However, it should be emphasized that even if hypo-osmotic challenges associated with RVD monitoring remains an experimental approach to model AVD, a direct study of AVD seems necessary to validate our assumption. Overall, by multimodal cell volume monitoring, we were able to rekindle discussion on the role of P-gp in cell volume regulation.

Acknowledgments

The authors are indebted to Jean-Pierre Marie (Hôtel Dieu, Paris, France) for providing MCF7/Doxo cells, zosuquidar and PSC833. We would like to thank Guillaume Doubremelle Trébutien for his technical support.

This work was supported by European Regional Development Fund as part of the Interreg IVA Project Admin (Transchannel Advanced Microscopy network). Jennifer Pasquier and Damien Rioult were recipients for a fellowship from the Conseil Regional de Haute-Normandie.

This study was made possible by JSREP grant #4-013-3-005 from the Qatar National Research Fund (a member of Qatar Foundation). The statements made herein are solely the responsibility of the authors.

Author Contributions

Jennifer Pasquier and Frank Le Foll planned experiments; Jennifer Pasquier and Damien Rioult performed experiments; Jennifer Pasquier, Damien Rioult, Jessica Hoarau-Véchot, Nadine Abu-Kaoud, Matthieu Marin and Frank Le Foll analyzed data; Matthieu Marin contributed reagents or other essential material; and Jennifer Pasquier, Jessica Hoarau-Véchot, Nadine Abu-Kaoud, Matthieu Marin and Frank Le Foll wrote the paper.

Abbreviations

ABC: ATP-binding cassette; AVD: apoptotic volume decrease; EV: electronic volume; MDR: MultiDrug Resistance; P-gp: P-glycoprotein; RVD: regulatory volume decrease; RVI: regulatory volume increase.

References

1. Okada, Y. Volume expansion-sensing outward-rectifier Cl^- channel: Fresh start to the molecular identity and volume sensor. *Am. J. Physiol.* **1997**, *273*, C755–C789.
2. Okada, Y.; Maeno, E. Apoptosis, cell volume regulation and volume-regulatory chloride channels. *Comp. Biochem. Physiol. A Mol. Integr. Physiol.* **2001**, *130*, 377–783.
3. Agre, P.; King, L.S.; Yasui, M.; Guggino, W.B.; Ottersen, O.P.; Fujiyoshi, Y.; Engel, A.; Nielsen, S. Aquaporin water channels—From atomic structure to clinical medicine. *J. Physiol.* **2002**, *542 Pt 1*, 3–16.
4. Sardini, A.; Amey, J.S.; Weylandt, K.H.; Nobles, M.; Valverde, M.A.; Higgins, C.F. Cell volume regulation and swelling-activated chloride channels. *Biochim. Biophys. Acta* **2003**, *1618*, 153–162.

5. Lang, F.; Busch, G.L.; Volkl, H. The diversity of volume regulatory mechanisms. *Cell. Physiol. Biochem.* **1998**, *8*, 1–45.
6. Peak, M.; al-Habori, M.; Agius, L. Regulation of glycogen synthesis and glycolysis by insulin, pH and cell volume. Interactions between swelling and alkalinization in mediating the effects of insulin. *Biochem. J.* **1992**, *282*, 797–805.
7. Kroemer, G.; Galluzzi, L.; Vandenabeele, P.; Abrams, J.; Alnemri, E.S.; Baehrecke, E.H.; Blagosklonny, M.V.; El-Deiry, W.S.; Golstein, P.; Green, D.R.; *et al.* Classification of cell death: Recommendations of the Nomenclature Committee on Cell Death 2009. *Cell Death Differ.* **2009**, *16*, 3–11.
8. Okada, Y.; Maeno, E.; Shimizu, T.; Dezaki, K.; Wang, J.; Morishima, S. Receptor-mediated control of regulatory volume decrease (RVD) and apoptotic volume decrease (AVD). *J. Physiol.* **2001**, *532*, 3–16.
9. Bosch, I.; Croop, J. P-glycoprotein multidrug resistance and cancer. *Biochim. Biophys. Acta* **1996**, *1288*, F37–F54.
10. Endicott, J.A.; Ling, V. The biochemistry of P-glycoprotein-mediated multidrug resistance. *Annu. Rev. Biochem.* **1989**, *58*, 137–171.
11. Jentsch, T.J.; Gunther, W. Chloride channels: An emerging molecular picture. *Bioessays* **1997**, *19*, 117–126.
12. Nilius, B.; Eggermont, J.; Voets, T.; Droogmans, G. Volume-activated Cl^- channels. *Gen. Pharmacol.* **1996**, *27*, 1131–1140.
13. Valverde, M.A.; Diaz, M.; Sepulveda, F.V.; Gill, D.R.; Hyde, S.C.; Higgins, C.F. Volume-regulated chloride channels associated with the human multidrug-resistance P-glycoprotein. *Nature* **1992**, *355*, 830–833.
14. Horton, J.K.; Vanoye, C.G.; Reuss, L. Swelling-activated chloride currents in a drug-sensitive cell line and a P-glycoprotein-expressing derivative are underlied by channels with the same pharmacological properties. *Cell. Physiol. Biochem.* **1998**, *8*, 246–260.
15. Hardy, S.P.; Goodfellow, H.R.; Valverde, M.A.; Gill, D.R.; Sepulveda, V.; Higgins, C.F. Protein kinase C-mediated phosphorylation of the human multidrug resistance P-glycoprotein regulates cell volume-activated chloride channels. *EMBO J.* **1995**, *14*, 68–75.
16. Bond, T.D.; Valverde, M.A.; Higgins, C.F. Protein kinase C phosphorylation disengages human and mouse-1a P-glycoproteins from influencing the rate of activation of swelling-activated chloride currents. *J. Physiol.* **1998**, *508 Pt 2*, 333–340.
17. Ehring, G.R.; Osipchuk, Y.V.; Cahalan, M.D. Swelling-activated chloride channels in multidrug-sensitive and -resistant cells. *J. Gen. Physiol.* **1994**, *104*, 1129–1161.
18. Miwa, A.; Ueda, K.; Okada, Y. Protein kinase C-independent correlation between P-glycoprotein expression and volume sensitivity of Cl^- channel. *J. Membr. Biol.* **1997**, *157*, 63–69.
19. De Greef, C.; Sehrer, J.; Viana, F.; van Acker, K.; Eggermont, J.; Mertens, L.; Raeymaekers, L.; Droogmans, G.; Nilius, B. Volume-activated chloride currents are not correlated with P-glycoprotein expression. *Biochem. J.* **1995**, *307*, 713–718.
20. De Greef, C.; van der Heyden, S.; Viana, F.; Eggermont, J.; de Bruijn, E.A.; Raeymaekers, L.; Droogmans, G.; Nilius, B. Lack of correlation between MDR-1 expression and volume-activation of cloride-currents in rat colon cancer cells. *Pflugers Arch.* **1995**, *430*, 296–298.

21. Marin, M.; Poret, A.; Maillet, G.; Leboulenger, F.; le Foll, F. Regulation of volume-sensitive Cl^- channels in multi-drug resistant MCF7 cells. *Biochem. Biophys. Res. Commun.* **2005**, *334*, 1266–1278.
22. Poulsen, K.A.; Andersen, E.C.; Hansen, C.F.; Klausen, T.K.; Hougaard, C.; Lambert, I.H.; Hoffmann, E.K. Deregulation of apoptotic volume decrease and ionic movements in multidrug-resistant tumor cells: Role of chloride channels. *Am. J. Physiol. Cell Physiol.* **2010**, *298*, C14–C25.
23. Torre, A.; Trischitta, F.; Faggio, C. Purinergic receptors and regulatory volume decrease in seabream (*Sparus aurata*) hepatocytes: A videometric study. *Fish Physiol. Biochem.* **2012**, *38*, 1593–1600.
24. Pasquier, J.; Galas, L.; Boulange-Lecomte, C.; Rioult, D.; Bultelle, F.; Magal, P.; Webb, G.; le Foll, F. Different modalities of intercellular membrane exchanges mediate cell-to-cell p-glycoprotein transfers in MCF-7 breast cancer cells. *J. Biol. Chem.* **2012**, *287*, 7374–7387.
25. Pasquier, J.; Magal, P.; Boulange-Lecomte, C.; Webb, G.; Le Foll, F. Consequences of cell-to-cell P-glycoprotein transfer on acquired multidrug resistance in breast cancer: A cell population dynamics model. *Biol. Direct.* **2011**, *6*, 5.
26. Pasquier, J.; Rioult, D.; Abu-Kaoud, N.; Marie, S.; Rafii, A.; Guerrouahen, B.S.; le Foll, F. P-glycoprotein-activity measurements in multidrug resistant cell lines: Single-cell *vs.* single-well population fluorescence methods. *Biomed. Res. Int.* **2013**, *2013*, doi:10.1155/2013/676845.
27. Hille, B. *Ion Channels of Excitable Membranes*, 3rd ed.; Sinauer associates Inc.: Sunderland, MA, USA, 2001; p. 814.
28. Gill, D.R.; Hyde, S.C.; Higgins, C.F.; Valverde, M.A.; Mintenig, G.M.; Sepulveda, F.V. Separation of drug transport and chloride channel functions of the human multidrug resistance P-glycoprotein. *Cell* **1992**, *71*, 23–32.
29. Sardini, A.; Mintenig, G.M.; Valverde, M.A.; Sepulveda, F.V.; Gill, D.R.; Hyde, S.C.; Higgins, C.F.; McNaughton, P.A. Drug efflux mediated by the human multidrug resistance P-glycoprotein is inhibited by cell swelling. *J. Cell Sci.* **1994**, *107 Pt 12*, 3281–3290.
30. Weaver, J.L.; Aszalos, A.; McKinney, L. MDR1/P-glycoprotein function. II. Effect of hypotonicity and inhibitors on Cl^- efflux and volume regulation. *Am. J. Physiol.* **1996**, *270*, C1453–C1460.
31. Aleu, J.; Ivorra, I.; Lejarreta, M.; Gonzalez-Ros, J.M.; Morales, A.; Ferragut, J.A. Functional incorporation of P-glycoprotein into Xenopus oocyte plasma membrane fails to elicit a swelling-evoked conductance. *Biochem. Biophys. Res. Commun.* **1997**, *237*, 407–412.
32. Chen, L.X.; Wang, L.W.; Jacob, T. The role of *MDR1* gene in volume-activated chloride currents in pigmented ciliary epithelial cells. *Sheng Li Xue Bao* **2002**, *54*, 1–6.
33. Shi, X.; Wu, S.; Yang, Y.; Tang, L.; Wang, Y.; Dong, J.; Lu, B.; Jiang, G.; Zhao, W. AQP5 silencing suppresses p38 MAPK signaling and improves drug resistance in colon cancer cells. *Tumour Biol.* **2014**, *35*, 7035–7045.
34. Jablonski, E.; Webb, A.; Hughes, F.M., Jr. Water movement during apoptosis: A role for aquaporins in the apoptotic volume decrease (AVD). *Adv. Exp. Med. Biol.* **2004**, *559*, 179–188.

35. Jablonski, E.M.; Mattocks, M.A.; Sokolov, E.; Koniaris, L.G.; Hughes, F.M., Jr.; Fausto, N.; Pierce, R.H.; McKillop, I.H. Decreased aquaporin expression leads to increased resistance to apoptosis in hepatocellular carcinoma. *Cancer Lett.* **2007**, *250*, 36–46.
36. Hanahan, D.; Weinberg, R.A. The hallmarks of cancer. *Cell* **2000**, *100*, 57–70.
37. Johnstone, R.W.; Ruefli, A.A.; Tainton, K.M.; Smyth, M.J. A role for P-glycoprotein in regulating cell death. *Leuk. Lymphoma* **2000**, *38*, 1–11.
38. Yang, X.; Page, M. P-glycoprotein expression in ovarian cancer cell line following treatment with cisplatin. *Oncol. Res.* **1995**, *7*, 619–624.
39. Litman, T.; Nielsen, D.; Skovsgaard, T.; Zeuthen, T.; Stein, W.D. ATPase activity of P-glycoprotein related to emergence of drug resistance in Ehrlich ascites tumor cell lines. *Biochim. Biophys. Acta* **1997**, *1361*, 147–158.
40. Barakat, S.; Demeule, M.; Pilorget, A.; Regina, A.; Gingras, D.; Baggetto, L.G.; Beliveau, R. Modulation of P-glycoprotein function by caveolin-1 phosphorylation. *J. Neurochem.* **2007**, *101*, 1–8.
41. Barakat, S.; Turcotte, S.; Demeule, M.; Lachambre, M.P.; Regina, A.; Baggetto, L.G.; Beliveau, R. Regulation of brain endothelial cells migration and angiogenesis by P-glycoprotein/caveolin-1 interaction. *Biochem. Biophys. Res. Commun.* **2008**, *372*, 440–446.
42. Demeule, M.; Jodoin, J.; Gingras, D.; Beliveau, R. P-glycoprotein is localized in caveolae in resistant cells and in brain capillaries. *FEBS Lett.* **2000**, *466*, 219–224.
43. Pedersen, S.F.; Hoffmann, E.K.; Novak, I. Cell volume regulation in epithelial physiology and cancer. *Front. Physiol.* **2013**, *4*, 233.
44. Miyazaki, T.; Arai, S. Two distinct controls of mitotic cdk1/cyclin B1 activity requisite for cell growth prior to cell division. *Cell Cycle* **2007**, *6*, 1419–1425.
45. Sveiczer, A.; Novak, B.; Mitchison, J.M. The size control of fission yeast revisited. *J. Cell Sci.* **1996**, *109*, 2947–2957.
46. Wells, W.A. Does size matter? *J. Cell Biol.* **2002**, *158*, 1156–1159.
47. Cooper, S. Control and maintenance of mammalian cell size. *BMC Cell Biol.* **2004**, *5*, doi:10.1186/1471-2121-5-35.
48. Jorgensen, P.; Tyers, M. How cells coordinate growth and division. *Curr. Biol.* **2004**, *14*, R1014–R1027.
49. Saucedo, L.J.; Edgar, B.A. Why size matters: Altering cell size. *Curr. Opin. Genet. Dev.* **2002**, *12*, 565–571.
50. Conlon, I.; Raff, M. Control and maintenance of mammalian cell size: Response. *BMC Cell Biol.* **2004**, *5*, doi:10.1186/1471-2121-5-36.
51. Umen, J.G. The elusive sizer. *Curr. Opin. Cell Biol.* **2005**, *17*, 435–441.
52. Ernest, N.J.; Habela, C.W.; Sontheimer, H. Cytoplasmic condensation is both necessary and sufficient to induce apoptotic cell death. *J. Cell Sci.* **2008**, *121 Pt 3*, 290–297.
53. Habela, C.W.; Sontheimer, H. Cytoplasmic volume condensation is an integral part of mitosis. *Cell Cycle* **2007**, *6*, 1613–1620.
54. Wang, H.; Vo, T.; Hajar, A.; Li, S.; Chen, X.; Parissenti, A.M.; Brindley, D.N.; Wang, Z. Multiple mechanisms underlying acquired resistance to taxanes in selected docetaxel-resistant MCF-7 breast cancer cells. *BMC Cancer* **2014**, *14*, 37.

55. Murray, S.; Briasoulis, E.; Linardou, H.; Bafaloukos, D.; Papadimitriou, C. Taxane resistance in breast cancer: Mechanisms, predictive biomarkers and circumvention strategies. *Cancer Treat. Rev.* **2012**, *38*, 890–903.
56. Maulard, C.; Marie, J.P.; Delanian, S.; Housset, M. Clinical effects on CA 15-3 level of cyclosporin A as a chemosensitiser in chemoresistant metastatic breast cancer. *Eur. J. Cancer* **1993**, *29A*, 480.
57. Le Foll, F.; Rioult, D.; Boussa, S.; Pasquier, J.; Dagher, Z.; Leboulenger, F. Characterisation of Mytilus edulis hemocyte subpopulations by single cell time-lapse motility imaging. *Fish Shellfish Immunol.* **2010**, *28*, 372–386.
58. Rioult, D.; Pasquier, J.; Boulange-Lecomte, C.; Poret, A.; Abbas, I.; Marin, M.; Minier, C.; le Foll, F. The multi-xenobiotic resistance (MXR) efflux activity in hemocytes of Mytilus edulis is mediated by an ATP binding cassette transporter of class C (ABCC) principally inducible in eosinophilic granulocytes. *Aquat. Toxicol.* **2014**, *153*, 98–109.

13

Get to Understand More from Single-Cells: Current Studies of Microfluidic-Based Techniques for Single-Cell Analysis

Shih-Jie Lo and Da-Jeng Yao *

Institute of Nanoengineering and Microsystem, National Tsing Hua University, Hsinchu 300, Taiwan; E-Mail: roxyjay0406@gmail.com

* Author to whom correspondence should be addressed; E-Mail: djyao@mx.nthu.edu.tw

Academic Editor: Fan-Gang Tseng

Abstract: This review describes the microfluidic techniques developed for the analysis of a single cell. The characteristics of microfluidic (e.g., little sample amount required, high-throughput performance) make this tool suitable to answer and to solve biological questions of interest about a single cell. This review aims to introduce microfluidic related techniques for the isolation, trapping and manipulation of a single cell. The major approaches for detection in single-cell analysis are introduced; the applications of single-cell analysis are then summarized. The review concludes with discussions of the future directions and opportunities of microfluidic systems applied in analysis of a single cell.

Keywords: microfluidic techniques; single-cell manipulation; single-cell analysis

---

1. Introduction

A single cell is the smallest functional unit within an organism and maintains the functions of tissues through mutual cooperation. Biological research has extended from cell physiology and mechanics to gene expression. Accompanied by numerous achievements of biological research over hundreds of years, biology has progressed to becoming a data-rich subject; *omic biology* thus is integral to large-scale study (*i.e.*, genome, proteome, connectome *etc.*). Apart from the rise of omic biology, in single-cell biology there have been investigations of the effect of one or a few genes in thousands of single cells over a long period of time. Single-cell biology utilizes speed-limited approaches (*i.e.*, fluorescence *in situ* hybridization; FISH) to study functions in a single cell, whereas omic biology employs high-throughput approaches (*i.e.*, whole genome sequencing) to collect much data rapidly. Note, omic biology passes over variations between cells. In recent years, molecular biotechnology and the related techniques have progressed toward quantitative analysis of a single cell. In this way, omic biology moves in the same direction as single-cell biology, namely studying many genes in many single cells [1]. No matter in which direction biological research evolves, single-cell analysis hence remains an important and fundamental topic in the biological field.

Although analysis of a single-cell plays an important role in biological search for an understanding of functions in an organism, analytical techniques face diverse challenges: (1) the analytical tool should work well with a single cell as the size of a cell is below the micrometer scale, and the target in a single cell of interest is thus on a sub-micrometer or even nanometer scale; (2) the analysis is conducted using little sample; (3) the analytical approach could be handled with high throughput so as to acquire sufficient data for significant statistics; and (4) the analytical tool is universally affordable for laboratories. Microfluidic systems thus conform to these requirements and have exhibited great performance in recent decades [2]. Microfluidic systems transport and manipulate a liquid volume ranging from microlitres to femtolitres in microfluidic channels of which the dimensions are on a micrometer scale. A living biological specimen has typically a liquid form; as a result microfluidics is favorable for single-cell analysis. Moreover, microfluidic chips are made of inexpensive materials; the chips can certainly be designed for high-throughput use. Microfluidic chips have hitherto been fabricated from silicon or glass, elastomer, thermosets, hydrogel, thermoplastics and paper [3]. Microfluidic chips are sometimes constituted as multiple layers of varied materials for a specific purpose [4,5], *i.e.*, an elastomer bound to glass for a product with increased rigidity. Polydimethylsiloxane (PDMS) is an elastomer of one kind that has become a popular material for the fabrication of microfluidic chips because of its low cost, gas permeability and biocompatibility; most microfluidic devices related to biological research are consequently made of PDMS. In this review, we hence focus on the microfluidic-based techniques that have been applied in single-cell analysis and highlight some new developments.

2. Isolation, Trapping and Manipulation of Single Cells

Before initiating a single-cell analysis, cells of interest must be selected for further treatment. One advantage of a microfluidic device is that one can generate droplets with tiny volumes (nL to fL) as a separated environment for growth or chemical reaction of a single cell. Table 1 summarizes the microfluidic techniques for single-cell analysis and Figure 1 shows an overview of various approaches for the isolation, trapping and manipulation of a single cell. The concept of droplet-based analysis is similar to that of micro-wells, significantly decreasing the possibility of cross-contamination. The generated droplets are then selected with a flow cytometer. Flow cytometry is a historic and commonly used technique in biological laboratories for high-throughput sorting of cells of interest; it can be used for analysis of the cell properties according to various parameters, *i.e.*, cell morphology, cell viability, protein expression *etc*. Accompanying the development of microfluids, flow cytometry is unrestricted to a cumbersome instrument, and can be operated in a pocket chip. The selection of cells of interest in a microfluidic-based flow cytometer can be classified into optical sorting and magnetic sorting. In optical sorting, the design of a flow cytometer is generally constituted by switching valves with a branched microfluidic channel. The switching valve can be mechanically based or bubble-based [6]. The width and height of a fluidic channel are sufficient for a single cell but not two cells concurrently passing through. The fluorescent signal of each cell is the selection criterion of optical sorting, which is excited by a microfluidic integrated laser system and emitted from fluorescein-linked antibodies or fluorescent probes; the cells of interest become guided to the reservoir via the function of a switching valve. This optical sorting is also named fluorescence-activated cell sorting (FACS) [7]. The cell selection of interest might thus be achieved with a flow cytometer via a specific antibody–antigen conjugation and labeling with a fluorescent probe. Secondly, according to the magnetic sorting approach, magnetic-affinity cell sorting (MACS) [8], also employs the specificity of an antibody–antigen conjugation, but the difference is a magnetic-bead-linked antibody rather than a fluorescein-linked one. To select a cell in magnetic sorting involves applying a magnetic field at a cell-collection reservoir, through which a fluid flows, to capture the magnetic-bead-labeled cells. A droplet-based microfluidic device with the function of flow cytometry can thus achieve single-cell analysis of various types [9]. A droplet-based microfluidic device can also be integrated with various sorting techniques, introduced in what follows, to select the target cells or compounds for analysis.

To date, varied trap techniques are commonly utilized and combined with microfluidic systems for observation or manipulation. If enduring manipulation is required, the trapped single cells also can be cultivated on a chip with nutrition supplied in a continuous flow. In a flow cytometer, the cells of interest are selected. In what follows we introduce the commonly used techniques of cell separation and manipulation for advanced analysis of single cells.

Table 1. Comparison of various microfluidic techniques for single-cell analysis.

Approaches	Main Applications	Major Advantage	Major Disadvantage
Droplet-based microfluidics	• Single cell isolation • Cell culture in droplet	High throughput screening for specific single-cells	Challenge to encapsulate single-cells in each droplet
Hydrodynamic trap	• Single cell isolation • Cell culture on chip	Multifunction in one device	Complicated fabricating process
Magnetic trap	• Specific cell trapping	Efficient trapping of labeled cells	Requires antibodies or primers for magnetic label
Acoustic trap	• Cell manipulation	Good for cell positioning	May have negative effect to cells
Dielectrophoretic trap	• Cell manipulation • Cell selection	Easily select target cells via alternating frequency of AC	Heat problem during long-term manipulation
Optical trap	• Cell manipulation • Cell mechanics	Applicable in many fields of study	Requires expensive optical system

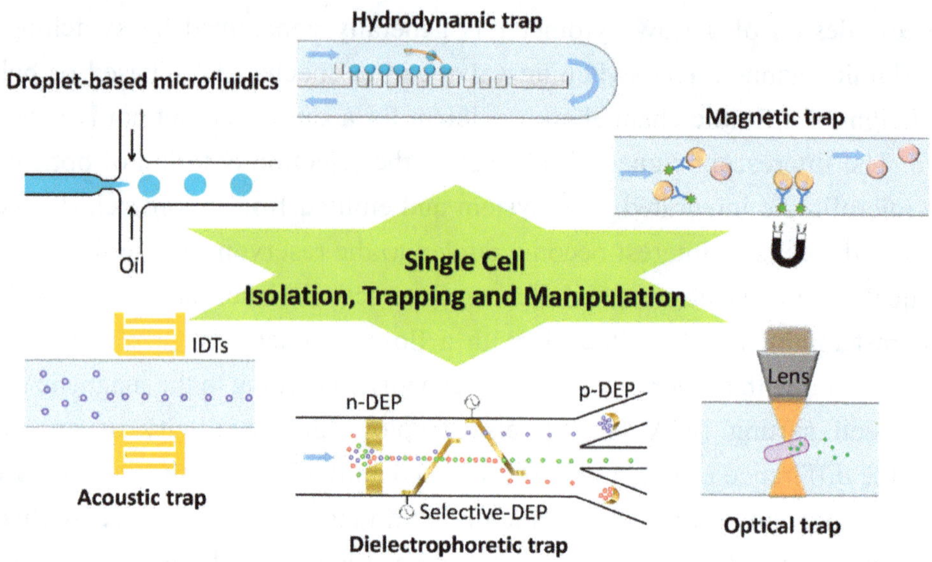

Figure 1. Schematic diagram of diverse approaches for the isolation, trapping and manipulation of single cells in a microfluidic device.

2.1. Hydrodynamic Trap

Hydrodynamic trap utilizes specific microstructures and valves in a microfluidic channel to control the fluid flow so as to collect single cells without the use of other apparatus. This method is simple, but the fabrication of a microfluidic channel might be complicated because of microstructures. The structure of a microfluidic channel acts like a filter to trap cells within the fluid flow so as to acquire many single cells, but the number of cells lost is considerable, which should be considered in applications. Long *et al.* [10] fabricated a microfluidic chip containing two large feeding channels connected with multiple trapping or growth channels on a sub-micrometer scale. This ladder-like microfluidic chip was used to study the population of *E. coli* responding to dynamic changes in their environment that was achieved on varying the composition of growth media in feeding channels. Lin *et al.* [11] demonstrated sieve-like trap arrays in a microfluidic channel to trap and to position single cells on a glass substrate for their interactive study. Various paired configurations to trap cells were efficiently investigated and discussed in this work, providing an alternative approach for cell patterning. Secondly, there is another kind of hydrodynamic trap which employs the characteristics of fluidics via alternating the flow rate, causing either laminar flows or vortex flows, so as to achieve a specific purpose such as locating targets at the desired micro-structure. Sochol *et al.* [12] demonstrated a resettable hydrodynamic arraying system for trapping and releasing the target single cells. Although the performance of target trapping is important, the efficiency of target releasing is also a major concern in device development. In their work, the loading efficiency of the device was finally 99.8% and 78% for bead-based and cell-based experiments, accordingly. Wang *et al.* [13] developed a microfluidic hydrodynamic trapping system with the capability of long-term monitoring the cellular dynamics. The microfluidic device has a special bypass structure, which alternates the hydrodynamics in flow channel, and traps single-cells at the desired locations. The microfluidic trapping array has single cell trapping efficiency of ~90% and used as a tool for evaluating the efficiency of chemotherapeutic reagents.

2.2. Optical Trap

Optical trap is also called *optical tweezers*, which is a highly precise technique for manipulating micro-scale objects. The optical trap utilizes laser beam to generate a force sub-pN in a microfluidic channel [14], which is sufficient to trap single cells. From a combination of an optical system and a microfluidic system, a new research field called optofluidics has emerged. On altering the focus of laser beams in a microfluidic channel, an object (*i.e.*, blood cells, PS beads, *etc.*) can be manipulated according to its shape, size and refractive index. The cost of using an optical trap to separate single cells is great because of the optically controlled system to locate precisely the focus of laser beams. Moreover, prolonged manipulation in an optical trap might cause a problem because the cells become damaged through the heat generated from the laser energy. Liberale *et al.* [15] developed an integrated microfluidic device containing a micro-prism structure, which was fabricated with two-photon photolithography and allowed light from an optical fiber to trap a single cell. The integrated microfluidic device is capable of on-chip manipulation, Raman and fluorescence spectra of single cells. An optical trap has been developed to alter the shape of an aperture to improve the trapping efficiency, such as a rectangle, a double nanohole (DNH) and a coaxial aperture. The DNH optical trap has been utilized to study protein–protein interaction [16] and protein–DNA interaction [17], and also to determine the size and concentration of nanoparticles in solution [18].

2.3. Magnetic Trap

The isolating technique based on magnetic force functions through an action of immunomagnetic labeling or a hybridization of a nucleic-acid probe modified with magnetic beads. The objects of interest contain antigens that can be recognized by specific antibodies; the antibodies are linked with dextran-coated magnetic particles. The magnetically labeled objects can hence be captured in a microfluidic device treated with a magnetic field. The separation can be implemented through positive selection (*i.e.*, collect objects linked with magnetic beads) or negative selection (*i.e.*, collect objects without magnetic beads linked). The only disadvantage is the specificity of a magnetic trap depending on the antibodies and the primer design. Lai *et al.* [19] developed a microfluidic chip integrated with a magnetic trap for the screening of aptamers specific to influenza A virus; the aptamer screening, also called systematic evolution of ligands and exponential enrichment (SELEX), was shortened to 60 min with this micro fluidic chip, to be compared with a conventional process that requires at least 160 min. Chen *et al.* [20] developed a mobile magnetic trap array, which was integrated with a droplet-generating microfluidic device, to encapsulate magnetically selected single cells as a powerful analytical tool for a single cell. Nawarathna *et al.* [21] developed an integrated nanoscale magnetic trap within a plastic microfluidic device; the magnetic field gradients therein were significantly increased to trap magnetic beads efficiently.

2.4. Dielectrophoretic Trap

Dielectrophoresis (DEP) is a phenomenon that involves a motion of polarizable particles under a non-uniform electric field. The types of DEP can be briefly classified into positive DEP (p-DEP) and negative DEP (n-DEP) [22,23], depending on the permittivity of the polarizable particles and the surrounding medium. When the permittivity of particles is greater than that of the medium, the particles have polarized opposite charges in the electric field; the particles move to the direction of a strong electric field, which is called p-DEP. n-DEP exhibits conversely that the particles move to the direction of a weak electric field when the permittivity of particles is less than that of a medium. The particles have polarized identical charges in the electric field because the polarized charges of the particles surface become greater than that of the particle inside. The electrical property of a particles surface and the interior is opposite as a result of the particles being induced with identical charges in the electric field. DEP is readily incorporated into a microfluidic device via a fabricated microelectrode. Cell patterning in microfluidic device is achievable via a DEP approach with an electrode array [24]. A dielectrophoretic trap has also a smaller heat problem than an optical trap during a protracted manipulation, because the heat generated is less and is removed adequately with the fluid flow. Bhattacharya *et al.* [25] employed insulator-based dielectrophoresis (iDEP) to trap single mammalian breast-cancer cells (MCF-7) from mixtures with mammalian peripheral blood mononuclear cells (PBMC). Simulation and experimental results of their tear-drop structure in a microfluidic channel indicate the weakly metastatic cancers cell (MCF-7) could also be selectively trapped from a mixture containing the highly invasive cancer cells (MDA-MB-231), so that this iDEP based microfluidic device might be applied as a diagnostic tool for breast cancer. Huang *et al.* [26] also used a DEP-integrated microfluidic device for single MCF-7 cell trapping, and investigated the effects on a trapped cell under a varied applied electric field, suggesting that an applied electric field less than 2 kV/cm was safe for cell viability. An alternative material, a carbon electrode, has been used as a function of electrode and insulator structure for DEP applications [27].

2.5. Acoustic Trap

A surface-acoustic wave can be used as a non-contact approach to sense a specific analyte or to confine single cells in a microfluidic channel. The principle of an acoustic trap is that an ultrasonic standing wave, produced with a pair of interdigital transducers (IDT) while applying electricity, traps or agglomerates cells. The IDT structure is deposited on a piezoelectric substrate in a parallel or orthogonal arrangement for varied cell patterning [28]. The negative effect of acoustic manipulation has been an issue and discussed [29,30], but little evidence verifies that exposure to ultrasound decreases the survival rate of cells. Chen *et al.* [31] developed a microfluidic device based on a standing surface-acoustic wave (SSAW) for continuous enrichment of a limited cell sample. The limited sample can be concentrated within the standing waves in a microfluidic channel and be eventually collected for further cellular study. The systems of acoustic trap and optical tweezers are currently combined as optoacoustic tweezers in a microfluidic device with the characteristics of biocompatibility and ease of fabrication for dynamically concentrating and patterning particles and cells [32].

3. Single-Cell Analysis in a Microfluidic Device

As varied techniques to trap a single are applicable in a microfluidic device, techniques of microanalysis are equally important in the analysis of a single cell. Electrophoresis is a commonly used technique to separate small molecules via varied electrophoretic mobility of charged molecules in an electric field. A major difference between electrophoresis and dielectrophoresis (DEP) depends on the objects being charged, or not. The separation principle of electrophoresis is based on the electric charge of an analyte, whereas DEP is based on an induced dipole inside an analyte. Accompanying the development of microfluidics, electrophoresis can be implemented within a narrow capillary, named capillary electrophoresis (CE), and be integrated to the microfluidic chip. Because of a small amount of single-cell analysis, CE has been applied as a treatment before an analysis and has the advantage of concentrating the analyte for the next step in the analysis.

In addition to lyse cells for CE approach, various non-invasive methods are applicable for live-cell detection. Investigation of live-cells has more restrictions than analyzing cell lysates because of the cell structure (e.g., cell membrane or nuclear membrane) that hinders the interactions between the labeling probes and targets of cells. Optical measurement is the most used non-invasive method for biological research; furthermore, fluorescence detection currently is simple and convenient for many kinds of research application because of the plentiful commercialized fluorescent probes. Fluorescent proteins, nanoparticles and quantum dots all can be utilized as the fluorescent source of probes, the specificity of which depend on the antibodies or DNA sequence (aptamers). Li *et al.* [33] developed a highly sensitive detection system for membrane protein of living single-cells by employing aptamer and enzyme assisted fluorescence amplification, which successfully carried out high throughput single-cell analysis of the low-abundance biomarker (PTK7). In what follows we introduce the major detection approaches for single-cell analysis (Figure 2).

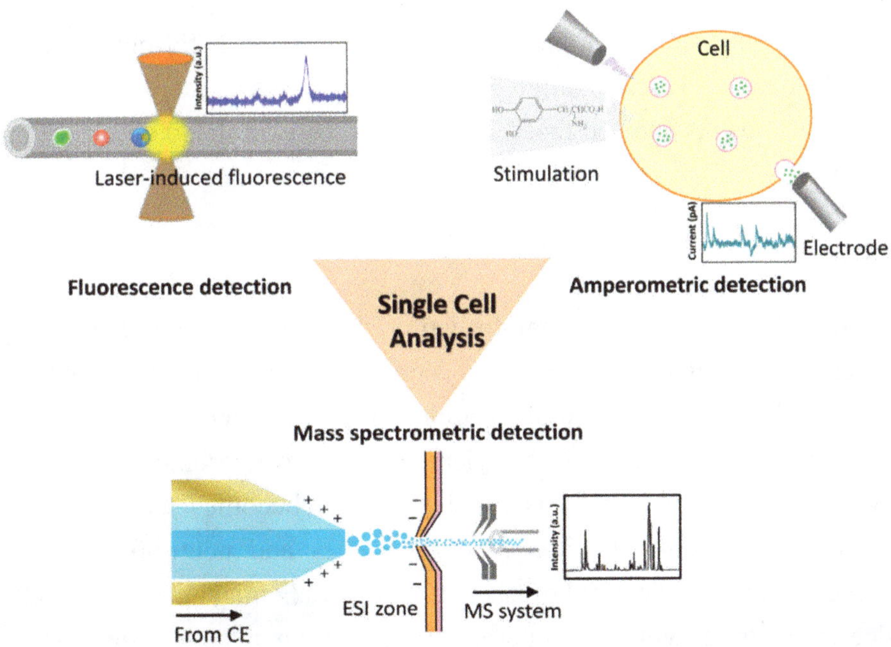

Figure 2. Schematic diagram of major approaches for the analysis of single cells with a microfluidic device.

3.1. Fluorescence Detection

Rough information about cellular contents can be acquired by lysing a trapped single cell and performing capillary electrophoresis, but such rough information to understand an organism in detail is limited. Capillary electrophoresis and laser-induced fluorescence (CE-LIF) provide scientists with more sub-cellular information about the interior of the cells. LIF employs a laser to excite fluorescence from a specific molecular moiety, and has been integrated with CE in a microfluidic device that provides a powerful technique with great sensitivity and a small detection limit for use in single-cell analysis. The fluorescein tagged substance is an analytical target of the LIF approach; the structure of the compound in the cell that signals transmission typically contains aromatic rings exhibiting self-fluorescence that is an effective candidate for LIF detection. Keithley *et al.* [34] developed a capillary electrophoresis system with a three-color laser-induced fluorescence detector to measure BODIPY fluorophores conjugated glycophingolipids. Three fluorophores (BODIPY-FL, BODIPY-TMR and BODIPY-650/665) were prepared in a chemoenzymatic synthesis and excited with diode-pumped solid-state lasers and a diode laser at wavelengths 473, 532 and 633 nm sequentially. Neuronal-like dPC12 cells were incubated with the fluorophores; their metabolic products were studied with the three-color CE-LIF system. Metto *et al.* [35] demonstrated an integrated microfluidic device to measure nitrogen oxide (NO) produced in single T-lymphocytes (Jurkat cells). The functions of cell transport, lysis, injection, electrophoretic separation and fluorescence detection were integrated in that microfluidic device. Two fluorescent probes, 4-amino-5-methylamino-2′,7′-difluorofluorescein diacetate (DAFFM DA) and 6-carboxyfluorescein diacetate (6-CFDA), were used to label NO and Jurkat cells sequentially. NO production in the cells can be stimulated on applying lipopolysaccharide (LPS) via inducible nitrogen oxide synthase (iNOS) in immune cells. The ultimate results of NO measurement from single-cell analysis compared well with the bulk cell level. Ban *et al.* employed a CE-LIF system to detect micro-RNA in cardiomyoblast cells [36] and lung-cancer cell lines [37]. Micro-RNA was captured via hybridization with DNA probes labeled with 6-FAM- or Cy5 in a separate sequence, and detected with a dual laser system. Their results showed that CE-LIF could be completed within 13 min and evaluated several endogenous miRNA.

3.2. Amperometric Detection

Although much subcellular information can be acquired via fluorescently based single-cell analysis, such as with fluorescein conjugated antibodies to capture proteins or fluorescein labeled probes to hybridize nucleic acids (DNA, mRNA), another approach that applied amperometry has been used to study specific proteins, especially in neurobiological research. Amperometry combined with capillary electrophoresis for electrochemical analysis of single cells becomes an effective tool to analyze chemical messengers (e.g., neurotransmitters, neuropeptides and neurohormones) in neurons; this method is also called capillary electrophoresis with electrochemical detection (CE-ED) [38]. The microfluidic device is integrated with an amperometric detecting system. After applying a stimulation of a trapped neuron (e.g., electric stimulation or chemical stimulation), chemical messengers in vesicles are released via exocytosis of a neuron of which the response can be measured. Moreover, the released chemical compounds can be collected concurrently in a microfluidic channel. Omiatek *et al.* [39] used this technique to measure the total vesicular content from single neuronal cells (PC12 cells), and found that a vesicle released only 40% of their transmitter load from a comparison of the single neuron measurement and a cell-free mode. They found also a phenomenon in which the release of a vesicular neurotransmitter and a hormone is not certainly all or none. In addition, the background noise decides the limit of amperometric detection, especially in a biosensor application. Larsen *et al.* [40] investigated the current noise caused by various electrode materials with varied capacitive properties, and concluded that low- capacitive materials and small electrodes have a small current noise, which benefits the design of a low-noise amperometric sensing device. Other than lowering the current noise, they investigated also the physical and electrochemical properties of poly(3,4-ethylenedioxythiophene):tosylate (PEDOT:tosylate) as a microelectrode material for neurochemical detection [41]. From the results, the capacitance of this conductive polymer is greater than of other thin-film materials, which limits the low-noise amperometric sensing, but this shortage can be overcome through sufficiently small fabrication. The neuron transmitter release of PC12 cells was measured with this chip-based device with microelectrodes made of PEDOT:tosylate that have the advantages of cheap and easy fabrication in all polymer devices.

3.3. Mass Spectrometric Detection

Mass spectrometer (MS) is a powerful tool to measure chemical components, whether used for qualitative or quantitative purpose, and is capable of providing an analysis with information about concentration, chemical structure and elemental composition. As the great sensitivity of a MS combined with CE has rapid separation and efficiency, the CE-MS analytical method becomes a suitable approach for single-cell analysis. The CE within a microfluidic device functions to selectively collect components of interest in a single cell and to concentrate the components before analyzing via a MS; the CE processed sampling approach could significantly simplify analyses entering a MS, in order to achieve a reduction of noisy signals from undesired components via the mass spectrometric analysis. With a period of CE-MS technological development, various instruments have been used for sample-ionization interfacing between a CE and a MS, such as electrospray ionization (ESI), matrix-assisted laser desorption or ionization (MALDI), laser-ablation electrospray ionization (LAESI) or inductively coupled plasma (ICP). A compatible ionizing interface is important because it affects the sensitivity of CE-MS [42]. To date, ESI and MALDI are the most popular methods for a sample–ionizing interface, but they both have advantages and disadvantages. For instance, ESI is suitable for online interfacing of a microfluidic device to a MS because of the compatible small flow rate [43]. ESI generates highly charged ions directly from a micro-fluid, which assists easy coupling between a CE and a MS, but ion suppression due to the large concentration of salts within samples becomes a challenge of analysis. In contrast, MALDI has a greater tolerance to salts, but matrix ions of MALDI limit the spectral analysis to molecular mass less than 500 Da. Aerts *et al.* [44] applied entire-cell patch-clamp recording and a CE-MS to establish the cytoplasm metabolomics of a single neuron. The physiological activity of glutamatergic thalamocortical neurons of a rat were recorded via the entire-cell patch-clamp technique, of which the metabolomics were analyzed via a CE-MS. Approximately 60 metabolites were detected and determined via a CE-MS from a tiny volume (~3 pL) of neuronal cytoplasm. This combined technique provides a new measurement tool to study the changes of cellular metabolome-related neuronal activity. Mellors *et al.* [45] developed an integrated microfluidic device that used a CE and an ESI-MS to analyze single cells automatically in real time. Human erythrocytes were lysed with this microfluidic device, of which the components were separated with the function of CE. The separated components were then ionized via an electrospray emitter and characterized in a MS. This device can verify the heme group and subunits of hemoglobin from individual erythrocytes with a throughput rate 12 cells per minute. The sensitivity of this device can ultimately be improved in optimizing the channel dimensions to decrease the possible flow rate. Smith *et al.* [46] developed an ESI-MS- integrated droplet-based microfluidic device for study of single cells with high throughput. Four populations of droplets, containing cytochrome C, α-chymotrypsinogen A, carbonic anhydrase or chicken lysozyme, were generated with a surfactant-stabilized reagent in a microfluidic device. The droplets containing sub-femtomolar quantities of an analyte were collected and then re-injected into the main channel of a microfluidic device for further characterization via an ESI-MS. This technique provides an alternative approach for analysis of single cells with high throughput.

4. Application Summary of Single-Cell Analysis

To date, microfluidic-based single-cell analysis has been applied in cellular research of diverse aspects and with multiple functions, for example, polymerase chain reaction (PCR) for gene analysis or cell differentiation for regenerative medicine. As mentioned at the beginning of this review, single-cell biology begins to investigate the functional mechanism of an organism from a single cell, the goal of which is the same as omic biology and provides more information about cellular heterogeneity. Many data of single-cell analysis can now be catalogued as single-cell genomics, single-cell transcriptomics and single-cell proteomics. Single-cell genomics can be established using single-cell qPCR [47], nanopore-based DNA sequencing [48] *etc.*, and provides researchers a complete understanding in order to verify whether a specific gene is constantly expressed in the same cell population or selectively expressed. Single-cell transcriptomics is thus generally established using a reverse transcriptase polymerase chain reaction (RT-PCR) [49,50], real-time PCR [51] or transcriptome sequencing (RNA-Seq) [52]. The study of single-cell transcriptomics provides intracellular variability of RNA profiles under varied environmental conditions. Compared with the omic biology of nucleic-acid-based methods, the construction of single-cell proteomics is much more difficult than single-cell genomics and transcriptomics because the small amount of protein in a single cell requires highly sensitive detection. A CE-integrated microfluidic device combined with MS-related techniques enables the convenient establishment of single-cell proteomics. The single-cell proteomics helps researchers in studying diverse post-translational modifications [53], translocations [54], and activity-correlating protein conformations [55]. Besides the above mentioned omic biology, epigenomics currently is a frontier in single-cell analysis [56]. Epigenomics investigates epigenetic modulations (e.g., DNA methylation or histone modification) of DNA, which is heritable and affects gene expressions in each single cell, although these single-cells have the same DNA sequence. Various bio-techniques now can be applied in establishing epigenomics [57], such as Chromatin immunoprecipitation followed by sequencing (ChIP-seq) [58] or bisulfite sequencing (BS-seq) [59], which have been used for locating histone modification.

Although the literature about microfluidic-based single-cell analysis is enormous, the contributions that matter most are in fundamental cellular research to provide new insights into the mechanisms of life. Single-cell analysis applications are now moving in the direction of diagnostic use and personalized medicine. The detection of circulating tumor cells (CTC) is a useful example of using a microfluidic device for diagnostic use. CTC are generated on metastasis of a primary tumor and are found in the peripheral blood from cancer patients. The variant number of CTC indicates the severity of the cancer and reducing it represents success of cancer therapies. The challenge of CTC detection is that CTC are rare in a heterogeneous sample (blood); high throughput and sensitive techniques are needed. A microfluidic device thus becomes an ideal tool to capture, separate, count and further analyze CTC efficiently [60–63]. Drug discovery for personalized medicine is also a prominent issue within single-cell analysis, to understand the mechanism of a drug-treated single cell, which exhibits a complicated response (*i.e.*, gene expression, metabolic alteration). The single-cell transcriptome analysis can provide a variation of RNA profiles after a drug treatment; the proteomic analysis can rapidly screen the candidates of antigen-specific antibodies for potential drug use [64].

5. Conclusions

In summary, various techniques are now available for the analysis of a single cell; with the aid of these techniques, many biological questions can be answered. A microfluidic device is perhaps now a suitable technique for single-cell analysis because a microfluidic system can be manipulated with high throughput, and the amount of sample from a single cell is limited. A microfluidic device might, however, be replaced with the further development of tools through the future efforts of physicists and engineers to answer other interesting biological questions. Although devices ready to use for diagnosis at a point of care are rare at present, most techniques have been applied in basic research. We believe that from the combination of the versatile design of microfluidic devices, flexible choice of analytical technique and increased knowledge, a portable device for personalized medical use can ultimately emerge.

Acknowledgments

The authors would like to thank the financial support from National Science Council of Taiwan under Contract No. NSC 101-2221-E-007-099-MY3.

References

1. Junker, J.P.; van Oudenaarden, A. Every cell is special: Genome-wide studies add a new dimension to single-cell biology. *Cell* **2014**, *157*, 8–11.
2. Yin, H.B.; Marshall, D. Microfluidics for single cell analysis. *Curr. Opin. Biotechnol.* **2012**, *23*, 110–119.
3. Ren, K.N.; Zhou, J.H.; Wu, H.K. Materials for microfluidic chip fabrication. *Acc. Chem. Res.* **2013**, *46*, 2396–2406.
4. Villar, G.; Graham, A.D.; Bayley, H. A tissue-like printed material. *Science* **2013**, *340*, 48–52.
5. Chen, C.C.; Chen, Y.A.; Liu, Y.J.; Yao, D.J. A multilayer concentric filter device to diminish clogging for separation of particles and microalgae based on size. *Lab Chip* **2014**, *14*, 1459–1468.
6. Mosadegh, B.; Kuo, C.H.; Tung, Y.C.; Torisawa, Y.S.; Bersano-Begey, T.; Tavana, H.; Takayama, S. Integrated elastomeric components for autonomous regulation of sequential and oscillatory flow switching in microfluidic devices. *Nat. Phys.* **2010**, *6*, 433–437.

7. Cho, S.H.; Godin, J.M.; Chen, C.H.; Qiao, W.; Lee, H.; Lo, Y.H. Review Article: Recent advancements in optofluidic flow cytometer. *Biomicrofluidics* **2010**, *4*, 43001.
8. Gao, Y.; Li, W.J.; Pappas, D. Recent advances in microfluidic cell separations. *Analyst* **2013**, *138*, 4714–4721.
9. Wang, B.L.; Ghaderi, A.; Zhou, H.; Agresti, J.; Weitz, D.A.; Fink, G.R.; Stephanopoulos, G. Microfluidic high-throughput culturing of single cells for selection based on extracellular metabolite production or consumption. *Nat. Biotechnol.* **2014**, *32*, 473–478.
10. Long, Z.; Nugent, E.; Javer, A.; Cicuta, P.; Sclavi, B.; Lagomarsino, M.C.; Dorfman, K.D. Microfluidic chemostat for measuring single cell dynamics in bacteria. *Lab Chip* **2013**, *13*, 947–954.
11. Lin, L.Y.; Chu, Y.S.; Thiery, J.P.; Lim, C.T.; Rodriguez, I. Microfluidic cell trap array for controlled positioning of single cells on adhesive micropatterns. *Lab Chip* **2013**, *13*, 714–721.
12. Sochol, R.D.; Dueck, M.E.; Li, S.; Lee, L.P.; Lin, L.W. Hydrodynamic resettability for a microfluidic particulate-based arraying system. *Lab Chip* **2012**, *12*, 5051–5056.
13. Wang, Y.; Tang, X.L.; Feng, X.J.; Liu, C.; Chen, P.; Chen, D.J.; Liu, B.F. A microfluidic digital single-cell assay for the evaluation of anticancer drugs. *Anal. Bioanal. Chem.* **2015**, *407*, 1139–1148.
14. Ramser, K.; Hanstorp, D. Optical manipulation for single-cell studies. *J. Biophotonics* **2010**, *3*, 187–206.
15. Liberale, C.; Cojoc, G.; Bragheri, F.; Minzioni, P.; Perozziello, G.; la Rocca, R.; Ferrara, L.; Rajamanickam, V.; di Fabrizio, E.; Cristiani, I. Integrated microfluidic device for single-cell trapping and spectroscopy. *Sci. Rep.* **2013**, *3*, 1258.
16. Zehtabi-Oskuie, A.; Jiang, H.; Cyr, B.R.; Rennehan, D.W.; Al-Balushi, A.A.; Gordon, R. Double nanohole optical trapping: Dynamics and protein-antibody co-trapping. *Lab Chip* **2013**, *13*, 2563–2568.
17. Kotnala, A.; Gordon, R. Double nanohole optical tweezers visualize protein p53 suppressing unzipping of single DNA-hairpins. *Biomed. Opt. Express* **2014**, *5*, 1886–1894.
18. Kotnala, A.; DePaoli, D.; Gordon, R. Sensing nanoparticles using a double nanohole optical trap. *Lab Chip* **2013**, *13*, 4142–4146.
19. Lai, H.C.; Wang, C.H.; Liou, T.M.; Lee, G.B. Influenza A virus-specific aptamers screened by using an integrated microfluidic system. *Lab Chip* **2014**, *14*, 2002–2013.
20. Chen, A.; Byvank, T.; Chang, W.J.; Bharde, A.; Vieira, G.; Miller, B.L.; Chalmers, J.J.; Bashir, R.; Sooryakumar, R. On-chip magnetic separation and encapsulation of cells in droplets. *Lab Chip* **2013**, *13*, 1172–1181.
21. Nawarathna, D.; Norouzi, N.; McLane, J.; Sharma, H.; Sharac, N.; Grant, T.; Chen, A.; Strayer, S.; Ragan, R.; Khine, M. Shrink-induced sorting using integrated nanoscale magnetic traps. *Appl. Phys. Lett.* **2013**, *102*, 63504.
22. Jubery, T.Z.; Srivastava, S.K.; Dutta, P. Dielectrophoretic separation of bioparticles in microdevices: A review. *Electrophoresis* **2014**, *35*, 691–713.
23. Gossett, D.R.; Weaver, W.M.; Mach, A.J.; Hur, S.C.; Tse, H.T.K.; Lee, W.; Amini, H.; di Carlo, D. Label-free cell separation and sorting in microfluidic systems. *Anal. Bioanal. Chem.* **2010**, *397*, 3249–3267.
24. Taff, B.M.; Voldman, J. A scalable addressable positive-dielectrophoretic cell-sorting array. *Anal. Chem.* **2005**, *77*, 7976–7983.

25. Bhattacharya, S.; Chao, T.C.; Ariyasinghe, N.; Ruiz, Y.; Lake, D.; Ros, R.; Ros, A. Selective trapping of single mammalian breast cancer cells by insulator-based dielectrophoresis. *Anal. Bioanal. Chem.* **2014**, *406*, 1855–1865.
26. Huang, C.J.; Liu, C.X.; Stakenborg, J.L.T.; Lagae, L. Single cell viability observation in cell dielectrophoretic trapping on a microchip. *Appl. Phys. Lett.* **2014**, *104*, 013703.
27. Martinez-Duarte, R.; Camacho-Alanis, F.; Renaud, P.; Ros, A. Dielectrophoresis of lambda-DNA using 3D carbon electrodes. *Electrophoresis* **2013**, *34*, 1113–1122.
28. Shi, J.J.; Ahmed, D.; Mao, X.; Lin, S.C.S.; Lawit, A.; Huang, T.J. Acoustic tweezers: Patterning cells and microparticles using standing surface acoustic waves (SSAW). *Lab Chip* **2009**, *9*, 2890–2895.
29. Evander, M.; Johansson, L.; Lilliehorn, T.; Piskur, J.; Lindvall, M.; Johansson, S.; Almqvist, M.; Laurell, T.; Nilsson, J. Noninvasive acoustic cell trapping in a microfluidic perfusion system for online bioassays. *Anal. Chem.* **2007**, *79*, 2984–2991.
30. Hultstrom, J.; Manneberg, O.; Dopf, K.; Hertz, H.M.; Brismar, H.; Wiklund, M. Proliferation and viability of adherent cells manipulated by standing-wave ultrasound in a microfluidic chip. *Ultrasound Med. Biol.* **2007**, *33*, 145–151.
31. Chen, Y.; Li, S.; Gu, Y.; Li, P.; Ding, X.; Wang, L.; McCoy, J.P.; Levine, S.J.; Huang, T.J. Continuous enrichment of low-abundance cell samples using standing surface acoustic waves (SSAW). *Lab Chip* **2014**, *14*, 924–930.
32. Xie, Y.L.; Zhao, C.L.; Zhao, Y.H.; Li, S.X.; Rufo, J.; Yang, S.K.; Guo, F.; Huang, T.J. Optoacoustic tweezers: A programmable, localized cell concentrator based on opto-thermally generated, acoustically activated, surface bubbles. *Lab Chip* **2013**, *13*, 1772–1779.
33. Li, L.; Wang, Q.; Feng, J.; Tong, L.L.; Tang, B. Highly sensitive and homogeneous detection of membrane protein on a single living cell by aptamer and nicking enzyme assisted signal amplification based on microfluidic droplets. *Anal. Chem.* **2014**, *86*, 5101–5107.
34. Keithley, R.B.; Rosenthal, A.S.; Essaka, D.C.; Tanaka, H.; Yoshimura, Y.; Palcic, M.M.; Hindsgaul, O.; Dovichi, N.J. Capillary electrophoresis with three-color fluorescence detection for the analysis of glycosphingolipid metabolism. *Analyst* **2013**, *138*, 164–170.
35. Metto, E.C.; Evans, K.; Barney, P.; Culbertson, A.H.; Gunasekara, D.B.; Caruso, G.; Huvey, M.K.; da Silva, J.A.F.; Lunte, S.M.; Culbertson, C.T. An integrated microfluidic device for monitoring changes in nitric oxide production in single T-lymphocyte (jurkat) cells. *Anal. Chem.* **2013**, *85*, 10188–10195.
36. Ban, E.; Chae, D.K.; Song, E.J. Determination of micro-RNA in cardiomyoblast cells using CE with LIF detection. *Electrophoresis* **2013**, *34*, 598–604.
37. Ban, E.; Chae, D.K.; Song, E.J. Simultaneous detection of multiple microRNAs for expression profiles of microRNAs in lung cancer cell lines by capillary electrophoresis with dual laser-induced fluorescence. *J. Chromatogr. A* **2013**, *1315*, 195–199.
38. Omiatek, D.M.; Cans, A.S.; Heien, M.L.; Ewing, A.G. Analytical approaches to investigate transmitter content and release from single secretory vesicles. *Anal. Bioanal. Chem.* **2010**, *397*, 3269–3279.
39. Omiatek, D.M.; Dong, Y.; Heien, M.L.; Ewing, A.G. Only a fraction of quantal content is released during exocytosis as revealed by electrochemical cytometry of secretory vesicles. *ACS Chem. Neurosci.* **2010**, *1*, 234–245.

40. Larsen, S.T.; Heien, M.L.; Taboryski, R. Amperometric noise at thin film band electrodes. *Anal. Chem.* **2012**, *84*, 7744–7749.
41. Larsen, S.T.; Vreeland, R.F.; Heien, M.L.; Taboryski, R. Characterization of poly(3,4-ethylenedioxythiophene):tosylate conductive polymer microelectrodes for transmitter detection. *Analyst* **2012**, *137*, 1831–1836.
42. Kleparnik, K. Recent advances in the combination of capillary electrophoresis with mass spectrometry: From element to single-cell analysis. *Electrophoresis* **2013**, *34*, 70–85.
43. Gao, D.; Liu, H.X.; Jiang, Y.Y.; Lin, J.M. Recent advances in microfluidics combined with mass spectrometry: Technologies and applications. *Lab Chip* **2013**, *13*, 3309–3322.
44. Aerts, J.T.; Louis, K.R.; Crandall, S.R.; Govindaiah, G.; Cox, C.L.; Sweedler, J.V. Patch clamp electrophysiology and capillary electrophoresis-mass spectrometry metabolomics for single cell characterization. *Anal. Chem.* **2014**, *86*, 3203–3208.
45. Mellors, J.S.; Jorabchi, K.; Smith, L.M.; Ramsey, J.M. Integrated microfluidic device for automated single cell analysis using electrophoretic separation and electrospray ionization mass spectrometry. *Anal. Chem.* **2010**, *82*, 967–973.
46. Smith, C.A.; Li, X.; Mize, T.H.; Sharpe, T.D.; Graziani, E.I.; Abell, C.; Huck, W.T.S. Sensitive, high throughput detection of proteins in individual, surfactant-stabilized picoliter droplets using nanoelectrospray ionization mass spectrometry. *Anal. Chem.* **2013**, *85*, 3812–3816.
47. Livak, K.J.; Wills, Q.F.; Tipping, A.J.; Datta, K.; Mittal, R.; Goldson, A.J.; Sexton, D.W.; Holmes, C.C. Methods for qPCR gene expression profiling applied to 1440 lymphoblastoid single cells. *Methods* **2013**, *59*, 71–79.
48. Min, S.K.; Kim, W.Y.; Cho, Y.; Kim, K.S. Fast DNA sequencing with a graphene-based nanochannel device. *Nat. Nanotechnol.* **2011**, *6*, 162–165.
49. Petriv, O.I.; Kuchenbauer, F.; Delaney, A.D.; Lecault, V.; White, A.; Kent, D.; Marmolejo, L.; Heuser, M.; Berg, T.; Copley, M.; *et al.* Comprehensive microRNA expression profiling of the hematopoietic hierarchy. *Proc. Natl. Acad. Sci. USA* **2010**, *107*, 15443–15448.
50. White, A.K.; VanInsberghe, M.; Petriv, O.I.; Hamidi, M.; Sikorski, D.; Marra, M.A.; Piret, J.; Aparicio, S.; Hansen, C.L. High-throughput microfluidic single-cell RT-qPCR. *Proc. Natl. Acad. Sci. USA* **2011**, *108*, 13999–14004.
51. Sanchez-Freire, V.; Ebert, A.D.; Kalisky, T.; Quake, S.R.; Wu, J.C. Microfluidic single-cell real-time PCR for comparative analysis of gene expression patterns. *Nat. Protoc.* **2012**, *7*, 829–838.
52. Streets, A.M.; Zhang, X.N.; Cao, C.; Pang, Y.H.; Wu, X.L.; Xiong, L.; Yang, L.; Fu, Y.S.; Zhao, L.; Tang, F.C.; *et al.* Microfluidic single-cell whole-transcriptome sequencing. *Proc. Natl. Acad. Sci. USA* **2014**, *111*, 7048–7053.
53. Faley, S.L.; Copland, M.; Reboud, J.; Cooper, J.M. Cell chip array for microfluidic proteomics enabling rapid *in situ* assessment of intracellular protein phosphorylation. *Biomicrofluidics* **2011**, *5*, 24106.
54. Zhan, Y.H.; Martin, V.A.; Geahlen, R.L.; Lu, C. One-step extraction of subcellular proteins from eukaryotic cells. *Lab Chip* **2010**, *10*, 2046–2048.
55. Bockenhauer, S.; Furstenberg, A.; Yao, X.J.; Kobilka, B.K.; Moerner, W.E. Conformational dynamics of single G protein-coupled receptors in solution. *J. Phys. Chem. B* **2011**, *115*, 13328–13338.

56. Bheda, P.; Schneider, R. Epigenetics reloaded: The single-cell revolution. *Trends Cell Biol.* **2014**, *24*, 712–723.
57. Hyun, B.R.; McElwee, J.L.; Soloway, P.D. Single molecule and single cell epigenomics. *Methods* **2015**, *72*, 41–50.
58. Giannopoulou, E.G.; Elemento, O. Inferring chromatin-bound protein complexes from genome-wide binding assays. *Genome Res.* **2013**, *23*, 1295–1306.
59. Brinkman, A.B.; Gu, H.C.; Bartels, S.J.J.; Zhang, Y.Y.; Matarese, F.; Simmer, F.; Marks, H.; Bock, C.; Gnirke, A.; Meissner, A.; *et al.* Sequential ChIP-bisulfite sequencing enables direct genome-scale investigation of chromatin and DNA methylation cross-talk. *Genome Res.* **2012**, *22*, 1128–1138.
60. Thege, F.I.; Lannin, T.B.; Saha, T.N.; Tsai, S.; Kochman, M.L.; Hollingsworth, M.A.; Rhim, A.D.; Kirby, B.J. Microfluidic immunocapture of circulating pancreatic cells using parallel EpCAM and MUC1 capture: Characterization, optimization and downstream analysis. *Lab Chip* **2014**, *14*, 1775–1784.
61. Bichsel, C.A.; Gobaa, S.; Kobel, S.; Secondini, C.; Thalmann, G.N.; Cecchini, M.G.; Lutolf, M.P. Diagnostic microchip to assay 3D colony-growth potential of captured circulating tumor cells. *Lab Chip* **2012**, *12*, 2313–2316.
62. Ozkumur, E.; Shah, A.M.; Ciciliano, J.C.; Emmink, B.L.; Miyamoto, D.T.; Brachtel, E.; Yu, M.; Chen, P.I.; Morgan, B.; Trautwein, J.; *et al.* Inertial focusing for tumor antigen-dependent and -independent sorting of rare circulating tumor cells. *Sci. Transl. Med.* **2013**, *5*, 179ra47.
63. Powell, A.A.; Talasaz, A.H.; Zhang, H.Y.; Coram, M.A.; Reddy, A.; Deng, G.; Telli, M.L.; Advani, R.H.; Carlson, R.W.; Mollick, J.A.; *et al.* Single cell profiling of circulating tumor cells: Transcriptional heterogeneity and diversity from breast cancer cell lines. *PLoS ONE* **2012**, *7*, e33788.
64. Golkowski, M.; Brigham, J.L.; Perera, B.G.K.; Romano, G.S.; Maly, D.J.; Ong, S.E. Rapid profiling of protein kinase inhibitors by quantitative proteomics. *Medchemcomm* **2014**, *5*, 363–369.

PERMISSIONS

All chapters in this book were first published by MDPI; hereby published with permission under the Creative Commons Attribution License or equivalent. Every chapter published in this book has been scrutinized by our experts. Their significance has been extensively debated. The topics covered herein carry significant findings which will fuel the growth of the discipline. They may even be implemented as practical applications or may be referred to as a beginning point for another development.

The contributors of this book come from diverse backgrounds, making this book a truly international effort. This book will bring forth new frontiers with its revolutionizing research information and detailed analysis of the nascent developments around the world.

We would like to thank all the contributing authors for lending their expertise to make the book truly unique. They have played a crucial role in the development of this book. Without their invaluable contributions this book wouldn't have been possible. They have made vital efforts to compile up to date information on the varied aspects of this subject to make this book a valuable addition to the collection of many professionals and students.

This book was conceptualized with the vision of imparting up-to-date information and advanced data in this field. To ensure the same, a matchless editorial board was set up. Every individual on the board went through rigorous rounds of assessment to prove their worth. After which they invested a large part of their time researching and compiling the most relevant data for our readers.

The editorial board has been involved in producing this book since its inception. They have spent rigorous hours researching and exploring the diverse topics which have resulted in the successful publishing of this book. They have passed on their knowledge of decades through this book. To expedite this challenging task, the publisher supported the team at every step. A small team of assistant editors was also appointed to further simplify the editing procedure and attain best results for the readers.

Apart from the editorial board, the designing team has also invested a significant amount of their time in understanding the subject and creating the most relevant covers. They scrutinized every image to scout for the most suitable representation of the subject and create an appropriate cover for the book.

The publishing team has been an ardent support to the editorial, designing and production team. Their endless efforts to recruit the best for this project, has resulted in the accomplishment of this book. They are a veteran in the field of academics and their pool of knowledge is as vast as their experience in printing. Their expertise and guidance has proved useful at every step. Their uncompromising quality standards have made this book an exceptional effort. Their encouragement from time to time has been an inspiration for everyone.

The publisher and the editorial board hope that this book will prove to be a valuable piece of knowledge for researchers, students, practitioners and scholars across the globe.

LIST OF CONTRIBUTORS

Michael Januszyk, Robert C. Rennert, Michael Sorkin, Zeshaan N. Maan, Lisa K. Wong, Alexander J. Whittam, Arnetha Whitmore, Dominik Duscher and Geoffrey C. Gurtner
Department of Surgery, Division of Plastic and Reconstructive Surgery, 257 Campus Drive West, Hagey Building GK-201, Stanford, CA 94305-5148, USA

Rajesh Kumar, Catharina de Lange Davies and Magnus B. Lilledahl
Department of Physics, Norwegian University of Science and Technology (NTNU), N-7491 Trondheim, Norway

Gajendra P. Singh
Physical Optics and Electronics Group, Massachusetts Institute of Technology (MIT), Cambridge, MA 02139, USA

Kirsten M. Grønhaug
Orthopaedic Department, Levanger Hospital, Kirkegata 2, N-7600 Levanger, Norway

Nils K. Afseth
Nofima, N-1431 Ås, Norway

Jon O. Drogset
Department of Orthopaedic Surgery, Trondheim University Hospital, N-7491 Trondheim, Norway

Yana Luo, Feng Chen and Junbo Wang
State Key Laboratory of Transducer Technology, Institute of Electronics, Chinese Academy of Sciences, Beijing 100190, China

Xiaoting Zhao, Mei Jiang and Wentao Yue
Department of Cellular and Molecular Biology, Beijing Chest Hospital, Capital Medical University, Beijing 101149, China

Rong Long
Department of Mechanical Engineering, University of Alberta, Edmonton, AB T6G 2G8, Canada

Daniel Hebenstreit
The University of Warwick, School of Life Sciences, Coventry CV4 7AL, UK

Masae Ohno, Peter Karagiannis and Yuichi Taniguchi
Laboratory for Single Cell Gene Dynamics, Quantitative Biology Center, RIKEN Address, 6-2-3 Furuedai, Suita, Osaka 565-0874, Japan

Muhammad Asraf Mansor
Faculty of Biosciences and Medical Engineering, Universiti Teknologi Malaysia, 81310-UTM Skudai, Johor, Malaysia

Chansavath Phetsouphanh
Kirby Institute, University of New South Wales, 2031 Sydney, Australia

John James Zaunders and Anthony Dominic Kelleher
Kirby Institute, University of New South Wales, 2031 Sydney, Australia
Centre for Applied Medical Research, St. Vincent's Hospital, 2010 Sydney, Australia

Andre Gross, Jonas Schoendube and Peter Koltay
Laboratory for MEMS Applications, IMTEK–Department of Microsystems Engineering, University of Freiburg, Georges-Koehler-Allee 103, Freiburg 79110, Germany
Cytena GmbH, Georges-Koehler-Allee 103, Freiburg 79110, Germany

Stefan Zimmermann and Maximilian Steeb
Laboratory for MEMS Applications, IMTEK–Department of Microsystems Engineering, University of Freiburg, Georges-Koehler-Allee 103, Freiburg 79110, Germany

Roland Zengerle
Laboratory for MEMS Applications, IMTEK–Department of Microsystems Engineering, University of Freiburg, Georges-Koehler-Allee 103, Freiburg 79110, Germany
Hahn-Schickard, Georges-Koehler-Allee 103, Freiburg 79110, Germany
BIOSS–Centre for Biological Signalling Studies, University of Freiburg, Freiburg 79110, Germany

Amelia Ahmad Khalili
Department of Control and Mechatronic Engineering, Faculty of Electrical Engineering, Universiti Teknologi Malaysia, Johor 81310, Malaysia

Mohd Ridzuan Ahmad
Department of Control and Mechatronic Engineering, Faculty of Electrical Engineering, Universiti Teknologi Malaysia, Johor 81310, Malaysia
Institute of Ibnu Sina, Universiti Teknologi Malaysia, Johor 81310, Malaysia

Jie-Long He, An-Te Chen, Jyong-Huei Lee and Shih-Kang Fan
Department of Mechanical Engineering, National Taiwan University, Taipei 10617, Taiwan

Jian Chen, Chengcheng Xue, Yang Zhao and Deyong Chen
State Key Laboratory of Transducer Technology, Institute of Electronics, Chinese Academy of Sciences, Beijing 100190, China

Min-Hsien Wu
Graduate Institute of Biochemical and Biomedical Engineering, Chang Gung University, Taoyuan 333, Taiwan

Jennifer Pasquier
Stem Cell and Microenvironment Laboratory, Weill Cornell Medical College in Qatar, Education City, Qatar Foundation, Doha 24144, Qatar
Department of Genetic Medicine, Weill Cornell Medical College, New York, NY 10022, USA

Nadine Abu-Kaoud and Jessica Hoarau-Véchot
Stem Cell and Microenvironment Laboratory, Weill Cornell Medical College in Qatar, Education City, Qatar Foundation, Doha 24144, Qatar

Damien Rioult
UMR_I 02 INERIS-URCA-ULH SEBIO Unité Stress Environnementaux et BIOsurveillance des milieux aquatiques, Université de Reims Champagne-Ardenne, BP-1039-Reims Cedex 2, 51687 Reims, France

Matthieu Marin
UGSF, UMR CNRS 8576, équipe Régulation des Signaux de Division, Université de Lille, Sciences et Technologies, 59009 Villeneuve d'Ascq, France

Frank Le Foll
Laboratory of Ecotoxicology UPRES EA 3222, IFRMP 23, University of Le Havre, 76058 Le Havre, France

Shih-Jie Lo and Da-Jeng Yao
Institute of Nanoengineering and Microsystem, National Tsing Hua University, Hsinchu 300, Taiwan

Index

Printed in the USA
CPSIA information can be obtained
at www.ICGtesting.com
LVHW080044211023
761596LV00010BA/1910

9 781647 403645